DATE DUE
D0867020

INTERNATIONAL SERIES OF MONOGRAPHS IN
NATURAL PHILOSOPHY
GENERAL EDITOR: D. TER HAAR

VOLUME 18

THEORY OF INTERMOLECULAR FORCES

OTHER TITLES IN THE SERIES
IN NATURAL PHILOSOPHY

Vol. 1. DAVYDOV—Quantum Mechanics

Vol. 2. FOKKER—Time and Space, Weight and Inertia

Vol. 3. KAPLAN—Interstellar Gas Dynamics

Vol. 4. ABRIKOSOV, GOR'KOV and DZYALOSHINSKII—Quantum Field Theoretical Methods in Statistical Physics

Vol. 5. OKUN'—Weak Interaction of Elementary Particles

Vol. 6. SHKLOVSKII—Physics of the Solar Corona

Vol. 7. AKHIEZER *et al.*—Collective Oscillations in a Plasma

Vol. 8. KIRZHNITS—Field Theoretical Methods in Many-Body Systems

Vol. 9. KLIMONTOVICH—The Statistical Theory of Non-equilibrium Processes in a Plasma

Vol. 10. KURTH—Introduction to Stellar Statistics

Vol. 11. CHALMERS—Atmospheric Electricity, 2nd edition

Vol. 12. RENNER—Current Algebras and their Applications

Vol. 13. FAIN and KHANIN—Quantum Electronics
Vol. 1.—Basic Theory

Vol. 14. FAIN and KHANIN—Quantum Electronics
Vol. 2.—Maser Amplifiers and Oscillators

Vol. 15. MARCH—Liquid Metals

Vol. 16. HORI—Spectral Properties of Disordered Chains and Lattices

Vol. 17. SAINT JAMES, THOMAS and SARMA—Type II Superconductivity

THEORY OF INTERMOLECULAR FORCES

BY

H. MARGENAU
Eugene Higgins Professor of Physics and Natural Philosophy,
Yale University

AND

N. R. KESTNER
Associate Professor of Chemistry,
Louisiana State University

THE QUEEN'S AWARD
TO INDUSTRY 1966

PERGAMON PRESS

OXFORD · LONDON · EDINBURGH · NEW YORK
TORONTO · SYDNEY · PARIS · BRAUNSCHWEIG

Pergamon Press Ltd., Headington Hill Hall, Oxford
4 & 5 Fitzroy Square, London W.1
Pergamon Press (Scotland) Ltd., 2 & 3 Teviot Place, Edinburgh 1
Pergamon Press Inc., Maxwell House Fairview Park, Elmsford, New York 10523
Pergamon of Canada Ltd., 207 Queen's Quay West, Toronto 1
Pergamon Press (Aust.) Pty. Ltd., 19a Boundary Street, Rushcutters Bay, N.S.W. 2011, Australia
Pergamon Press S.A.R.L., 24 rue des Écoles, Paris 5e
Vieweg & Sohn GmbH, Burgplatz 1, Braunschweig

Copyright © 1969
Pergamon Press Ltd.

First edition 1969

Library of Congress Catalog Card No. 68-8531

PRINTED IN HUNGARY
08 012759 2

JOINT UNIVERSITY LIBRARIES NASHVILLE, TENN.

Fixed chgs
Physics

Science
QC
173
.M3635
1969
611531

Contents

Contents

FOREWORD

ALTHOUGH the past ten years have seen many developments in the theory of intermolecular forces, the latest book which treats this topic with due thoroughness was published in 1954. It is Hirschfelder, Curtiss and Bird's *Molecular Theory of Gases and Liquids*. In spite of its wealth of material, or probably because of it, the recent edition of this book does not contain a significantly expanded section on the theory of interactions.

A notable trait of the papers published in the last decade is their increased bearing upon fundamental, quantitative understanding of molecular forces, together with a lapse of emphasis upon semi-empirical knowledge of parameters. Thus, there has been an enormous increase in the number of calculations based directly on the Schröedinger equation with little or no approximation. As a parallel development we note investigations of a more subtle and novel sort related to intermolecular forces, some of them based on techniques recently developed in many-body physics.

There has likewise been intensified activity on the experimental side. For example, the application of molecular beams to the study of intermolecular forces, at least to the relatively weak intermolecular forces, is comparatively new, and it has made possible the study of practically any type of molecule because of the availability of new general detectors. On the other hand, many fields of investigation, such as the structure of liquids and organic crystals, are waiting for more information on intermolecular forces before they can make significant progress. For all these reasons there appeared to be a growing need for a new book on the subject. We shall soon have several volumes of the *Advances in Chemical Physics*, to be edited by J. O. Hirschfelder, containing articles by many of the leaders in the field to which this treatise is devoted. However, there is at present no book available which attempts to treat the entire theory of intermolecular forces in a uniform manner, beginning from an introductory level. We hope our book will partly fill this need. Its focus is upon basic theory and it surveys other aspects, notably experimental ones, only insofar as they are relevant to basic theory.

Our plan was to introduce students of quantum mechanics and of advanced Physical Chemistry to the subject of intermolecular forces in the first few chapters of the book. Other, particularly the later, parts are designed to interest the specialist who works in neighboring areas and to bring him to the point where he understands the current literature in the field of intermolecular forces and knows where to look for information. This double aim of didactic usefulness and coverage of recent materials will, we fear, make the tenor of the book somewhat uneven. There are parts which are familiar and may seem even trivial to the expert, and others where the uninitiated will miss detail. This last reaction is bound to arise in today's presentation of subjects where, as in the theory of intermolecular forces, electronic computors are assuming an increasing role in research. Regrettably, perhaps, this commits an expositor to a judicious statement of results and a commentary on the effects included in the calculations.

As to the reader's preparation, we have assumed familiarity with elementary quantum mechanics to an extent conveyed by an advanced physical chemistry course or a good course in undergraduate atomic physics.

Two features of the book may call for comment. The first chapter presents perhaps a more extensive excursion into the history of the subject than is customary in science texts. Although this hardly adds to the practical value of the book, it may prove interesting, and we offer no apologies for including it. Secondly, we have added a feature which is likely to be useful at the end, where we have provided an extensive bibliography covering—we would like to claim nearly all, but admit fallibility—papers on the theory of intermolecular forces published between 1956 and 1966 in addition to the references cited in the text.

While it is normally desirable that the same notation be employed throughout a single book, we have come to question the wisdom of this principle while writing the present text. For there are in a subject which involves as many different techniques of calculation as ours, many places where strict adherence to the principle of notational uniformity would seem artificial. The reader will therefore occasionally find a single letter, such as E or P, serving a small variety of purposes in different places, where we hope its meaning is clear.

It would have been impossible to write this book without the interest, help and guidance of many workers in the field, most notably Professors J. O. Hirschfelder, O. Sinanoglu, R. D. Present, I. Amdur, R. Bernstein, V. McKoy, and A. McLachlan and many graduate students such as Walter Deal at Stanford, James Stamper and Antoine Royer at Yale. Various aspects of this work have also been supported by research aids. One of us (Neil R. Kestner) received two National Science Foundation grants, while the other has benefitted for years from grants by the Air Force Office of Scientific Research for studies in this and related fields. We express our gratitude to these agencies.

The book owes its existence to the initiating suggestion of Professor D. ter Haar, the General Editor of the International Series of Monographs in Natural Philosophy, to whom we are greatly indebted both for his inspiration and his forebearance.

Finally we record our thanks to several secretaries who struggled at various stages with the manuscript. We are especially grateful to Mrs. Harriet Comen at Yale, who resolved the typing problem of two authors often separated by half a continent or more. Without her effort we would have two books or none instead of one.

CHAPTER 1

History of Intermolecular Forces

THE first attempt to understand intermolecular forces in semi-quantitative terms was made by Clairault (1743) in his treatise *Théorie de la Figure de la Terre*. His book is devoted to various effects of Newtonian gravity upon plastic or fluid masses; but its author digresses to consider the shape of small portions of fluid and in particular the "strange effects of capillarity". He does this because one needs to "add but little to the theory [of gravitation] in order to make it applicable to fluids, whose parts are animated by forces the directions of which vary in the most general way."

Clairault takes issue with a theory previously proposed by Jurin, who presented the experimental facts concerning capillary attraction. He arrived at the result that the forces which suspend the liquid water column in a thin glass tube are attractive, and are exerted only along the edge at which the upper surface of the water meets the tube, where the glass exercises an attraction. Knowing that the height of rise of the water column is inversely proportional to the radius r, he reasons in a manner still found in elementary physics texts today, assuming the action of a constant force per unit length of the circle joining water and glass. If this is labeled f and the height of the liquid is h equilibrium demands that

$$\pi r^2 h = 2\pi r f,$$

so that $h \propto 1/r$, as observed. Clairault objects to this analysis on various grounds, chiefly because he thinks it incredible that the forces between particles should only exist between water and glass, and then only along the annulus in which they meet.

His own theory envisions two kinds of interparticle forces, those between water and glass and those between the liquid particles themselves. Their precise nature does not enter the account he gives of them, which is ingenious and general. He derives an expression for the height of rise in which the force laws are unspecified, and he observes that a great variety of force laws would give the correct empirical answers.

The next step in our understanding of intermolecular forces was taken by Laplace (1805) in a supplement to Book X of his treatise on celestial mechanics entitled "Capillary Action". Having discussed optical refraction, which he analyzes in Newtonian fashion by invoking forces between the constituents of the material through which light passes and the *molécules de lumière*, it occurs to him to consider the forces between the material molecules themselves. Both types of force, he shows, must have an infinitesimal range. This fact appears clearly in his theory of light refraction; for material molecules he establishes it by an observation first made by the Englishman Hawksbee, to whom he refers. Clairault, he says, (persistently omitting the second *l* in his name as do most biographers since), based one of his demonstrations upon the assumption of attractive forces between the walls of a capillary tube and the liquid filament along its axis, giving them a finite range. This, according to Laplace, is contradictory to Hawksbee's finding that the height of rise in a capillary is independent of the thickness of the tube. The range of force must therefore be insensibly small.

His considerations involve the belief that the force law is of the same form for all molecules, having a universal small range. The intensity, however, differs from substance to substance. Furthermore, he claims that these very same interactions are responsible for chemical affinity. Their strength is said to vary with temperature.

Two decades later the same problem was discussed by Gauss (1830), who dealt with it in an article entitled "Principia generalia theoriae figurae fluidorum in statu aequilibrii". Molecular forces are again treated in an incidental way, as contributors to the equilibrium of fluids in capillary vessels. Gauss attacks the problem from the starting point of the principle of virtual work, requiring equilibrium of every mass point under three types of force: (1) the force of gravity,—(2) the mutual attractive forces between the molecules,—and (3) attractive forces between the molecules of the liquid and certain fixed points, such as the molecules composing the walls of the capillary tube.

For the second of these he employs the form $dF = -mf(r)dr$, leaving the force function f largely undetermined. He introduces the mass m as a factor in this expression, thus reflecting an age-old disposition to relate these forces to gravitational attraction. Now the principle of virtual work leads to a stationary-value problem, and the quantity which plays a crucial role in the analysis is $I \equiv \int\int V(\mathbf{r}-\mathbf{r}')d\mathbf{r}d\mathbf{r}'$, where $V(\mathbf{r})$ is the intermolecular potential $-\int \mathbf{f}\cdot d\mathbf{r}$. The integral I must therefore remain finite at all distances and this requires that, if $V(\mathbf{r})$ were assumed to have the form a/r^n as $r \to \infty$, the exponent n must at least be as great as 6. This, of course, is in accord with presently known facts provided that the forces are invariably attractive, which is what Gauss assumes.

Gas phenomena were investigated from the point of view of intermolecular forces in a series of classic papers by Maxwell (1868). Their main purpose was to establish the molecular or dynamic theory of matter which was still striving for acceptance. The foundations of the molecular hypothesis had been laid by Daniel Bernoulli, Joule, Prévost, Clausius, and O. E. Meyer, all of whom omitted consideration of forces between molecules. In reviewing the history of the subject—and the review does not overlook the works of Democritus, Epicurus, and Lucretius—Maxwell notes a curious memoir by Le Sage (1862) in which the dynamical theory of gases appears as a byproduct of a theory of gravitation. Gravity, says Le Sage, is the effect of impacts of "ultra-mundane corpuscles" on terrestrial objects. "These corpuscles also set in motion the particles of light and various aetherial media, which in turn act on the molecules of gases and keep up their motions." All these men provided what Maxwell takes to be convincing evidence of the essential correctness of the dynamic-molecular theory of matter, a refutation of all continuum theories. His final verdict takes this form: "The properties of a body supposed to be a uniform *plenum* may be affirmed dogmatically, but cannot be explained mathematically."

In the pair of papers under review, Maxwell establishes the kinetic theory of gases and, incidentally, the Maxwell force law, according to which molecules repel one another with a central force proportional to the inverse fifth power of the distance. The details of his analysis have found their way into all texts on the kinetic theory and need not be reproduced here. Familiar formulas are given for the angle of deflection resulting from an encounter of two molecules which interact under a general law of the form: potential energy $V = Kr^{-n-1}$, for the diffusion coefficient, specific heats and, finally, the coefficient of viscosity. It is in connection with the last-named quantity that the value of n reveals itself.

Concerning viscosity, two important facts were known: (a) it is independent of the density of the gas,—and (b) it is proportional to the absolute temperature. Fact (a) is non-committal; it emerges for any value of n and had previously been derived on the assumption that molecules were hard, elastic spheres exerting no forces except on contact. Fact (b), however, requires n to be 5. Evidently Maxwell regards his law of interaction to be true at small distances, for he does not believe in any special feature of solidity or impenetrability of matter. "The doctrines that all matter is extended", he says, "and that no two portions of matter can coincide in the same place, being deductions from our experiments with bodies sensible to us, have no application to the theory of molecules."

We now know that inverse fifth power repulsive forces are mathematically convenient but have no basis in reality. Their advocacy by Maxwell was an instance of a logical fallacy which has proved immensely fruitful in the progress of physics, the fallacy of the inverted syllogism. One finds that the conclusions following from a premise are correct and then affirms the correctness of the premise, not realizing that the same conclusion may also flow from a host of other premises. In this case it turned out that the above-mentioned fact (b) may be derived from an infinite number of complicated force laws including Maxwell's.

The following stage of our development is therefore not surprising. Maxwell's papers are succeeded by an article of Boltzmann's which is prompted by doubt concerning the simplicity of the force law assumed by his predecessor. Noting that molecules are certainly "very complicated individuals", he thinks it useful to try out many different hypotheses, particularly some involving attractive forces, and to determine the virtues and failures of each. In this vein he repeats Maxwell's calculations of viscosity with the following three interaction models:

(a) $$V = \begin{cases} 0, & R > \delta \\ -\infty, & \varepsilon \leqslant R \leqslant \delta \\ 0, & R \leqslant \varepsilon \end{cases} \qquad \frac{\delta - \varepsilon}{\delta} \to 0$$

(b) $$V = \begin{cases} -\dfrac{c}{R^4}, & R \geqslant \varrho \\ 0, & R < \varrho \end{cases} \qquad c = \text{constant}$$

(c) $$V = \begin{cases} -\dfrac{c}{R^4}, & R \geqslant \varrho \\ \infty, & R < \varrho \end{cases} \qquad c = \text{constant}$$

The results are similar to Maxwell's except for different values of (unknown) constants, and they are equally compatible with observations. Hence Boltzmann favors attractive forces because of their added competence to account for condensation.

The next phase in the history of our understanding of molecular interactions is dominated by a series of articles by Sutherland, (1886, 1887, 1893a, 1893b) all addressed to the problem of determining the specific law of *attraction* between the molecules of a gas. For he makes clear the need for attractive forces in convincing qualitative fashion by presenting at the outset the consequences of the experiments by Thomson and Joule. The gas expanding through a porous plug is cooled, and the loss of heat is larger than can be accounted for by the external work done on the ambient medium. Hence work must have been done against the attractive forces between the molecules. This work is computed in the first paper cited above.

Empirically, the intermolecular cooling effect is proportional to $(p-p')/Ts$, $p-p'$ denoting change in pressure, T absolute temperature, and s specific heat. Sutherland demonstrates that the result follows from the assumption

$$V = -A\,\frac{m_1 m_2}{r^3}.$$

The m's are masses, and A is a further characteristic constant concerning which Sutherland initially entertains the hope that it may be a universal one. But this hope is later relinquished.

When one follows his reasoning, one sees first of all why he needs the masses as they occur in V. They produce a formula for the cooling effect in which the difference between densities ϱ (before and after expansion) appear. He is thus led to a result which contains, beside a collection of constants, the factor $(\varrho-\varrho')$; this passes into $M/k \times (p-p')/T$. In sum, then, we observe that agreement with the empirical formula is achieved, so far as its major feature $(p-p')$ is concerned, by Sutherland's assumption of a mass-proportionality of the intermolecular forces.

His confidence in the r^{-3} dependence of V has another, even more adventitious origin. One notes that in the integration $\int Vr^2dr$, which is required to compute the potential energy of a mass of gas, an inverse third-power law occasions the least difficulty. The limits are taken to be r_1, an "insensibly small" distance, and R_1, the size of the gas container. As will be evident, if $V = -c/r^n$, the value $n = 3$ produces a logarithmic dependence of the integral upon R_1/r_1, and this is inoffensive. For if, following Sutherland, we let $R_1 = nL$, "where n is a large number so that L is a small but sensible length", one may be disposed to neglect $\log n$ against $\log L/r_1$ and retain the latter as a sort of constant. On the other hand, if n is chosen to be anything but 3, the integral will either be extremely large or depend on the size of the container. It is this kind of reasoning which induced belief in the Sutherland force law.

The second paper cited above surveys a great number of experimental data on vapor pressures, specific volumes of CO_2 and Joule–Thomson coefficients in an endeavor to bring them into line with the use of the force law. On achieving a measure of success Sutherland voices the conjecture that there exists a "molic force law" including gravitation of the form

$$F = -\frac{G}{r^2} - \frac{M}{r^4}$$

which is valid "through the whole range of distances from molecular up to astronomical". And he is further "tempted to speculate whether the law of the terms representing atomic or chemic force may not be expressed by one or more higher powers of $1/r^2$, representing a force insensible at molecular distances as the molecular term of molic force is insensible at astronomical distance, but sensible at atomic distances, with the associated idea that atomic distances are exceedingly small compared to molecular". That was said in 1887.

Sutherland's third paper attempts to find the value of the constant A in the force law by fitting a great number of observed effects. Near the end of it one finds an insight which is almost modern, inasmuch as it relates A to the polarizability. The mass dependence is recognized as somewhat fortuitous and unconvincing, but for the quantity Am^2, which characterizes molecules of mass m, Sutherland feels the need of assuming that it depends on the size of the molecules. Indeed, he postulates that it is a simple function of the molecular volume, which, as we now know, is roughly proportional to the polarizability. The discovery of "dispersion" forces, which are linked to the dynamic polarizability, was therefore vaguely foreshadowed.

The last of the four papers listed is of the most enduring value for it develops a contemporary theory of viscosity. In it the author first reviews earlier conjectures, particularly one by O. E. Meyer, who obtains the correct dependence of the coefficient of viscosity on temperature without the benefit of intermolecular forces by allowing the size of the molecules to shrink with temperature. This, Sutherland holds, is most implausible, and he proceeds to introduce forces, first of unspecified mathematical form, into the collision problem. He finds, as do contemporary texts on the kinetic theory [see, for example, Present (1958, p. 112)] that a mutual potential energy of the form $-m^2f(r)$ causes an effective enlargement of the rigid-molecule collision cross section, namely $\pi(2a)^2$, by the factor $[1 + 2mf(2a)]/v^2$, where v is the relative velocity of the colliding molecules. This factor diminishes with increasing v, hence with T, and accounts for Meyer's conclusions.

Much of the article is a valiant attempt to rescue the relation $f(r) = A/r^3$ from a maze of data which do not quite fit it. Success of the attempt is limited and is brought about by manifold corrections which are perhaps no longer interesting. Yet it should be recorded that the rather single-minded and inspired persistence of the Australian physicist, Sutherland, was primarily responsible for placing the subject of molecular forces into the context of interesting physical problems.

J. H. van der Waals (1908, Part 1, pp. 207 et seq) returns to Laplace's problem of capillarity, but with special concern for the nature and mathematical form of the forces. Ignoring previous specific formulations he invokes what is now called the Yukawa potential, $V = -A/r \cdot e^{-r/a}$, A and a being empirical constants. He arrived at it by thinking of the interaction as a Newtonian attraction exerted by a molecule, but with lines of force which are "absorbed by the medium" as the distance from the center of force increases. In the book cited above, however, he disavows this conception because it presents difficulties, presumably because of the extreme weakness of the force of gravitation. Van der Waals therefore proposes his formula only as a possible and interesting one, and he repeats with it successfully the calculations of Laplace which, as we know, were largely independent of the precise form of V.

The theories of intermolecular force which appeared in the nineteenth century reflected two distinct endeavors. The first was animated by a search for a simple, basic law like Newton's. Then, when the futility of this attempt became manifest, a certain frustration gave rise to more modest, phenomenological approaches. These, aside from shedding some albeit diffuse light on the nature of the forces, provided consistent formulas for various effects (equations of state, Joule–Thomson coefficients, heats of vaporization, etc.) so that, if certain constants in the force law are adjusted to accommodate one set of data, others can be predicted. This semi-empirical procedure has proved most useful and has continued into the present era, mainly by the school of Lennard-Jones† who chose a law of the form

$$V = \lambda R^{-n} - \mu R^{-m}.$$

Here n and m are integers, $n > m$, and the two terms account, respectively, for repulsion and attraction. Many other mathematically convenient forms of V have been proposed, but we refer the reader to later chapters. A good review of them is found in Hirschfelder, Curtiss, and Bird's book, *Molecular Theory of Gases and Liquids*‡. Such so-called phenomenological approaches, despite their usefulness, will not be treated fully in this book, whose chief goal is a physical explanation of the forces and their mathematical derivation from first principles.

† For a survey of J. E. Lennard-Jones' extensive contributions see Chapter 10 in Fowler's *Statistical Mechanics*.

‡ J. E. Hirschfelder, C. F. Curtiss, and R. B. Bird (1954).

This quest arose in the twentieth century. Reinganum (1903, 1912) made the first attempt to relate the interactions intimately to the structure of molecules. At the time of his first-named publication knowledge of atomic structure was in its infancy; molecules were thought to contain positive and negative electronic charges, but these were taken to be intermingled with or lodged in ordinary uncharged matter. The view that intermolecular forces were caused by gravitational attraction was not completely dead; even in his second paper Reinganum mentions the possibility, suggested by a reflection upon van der Waals' Yukawa potential, that the basic potential between mass points is of the form

$$V = -G\,\frac{m_1 m_2 e^{\lambda/r}}{r}$$

But he subsequently discards that hypothesis.

Instead, he equips each molecule with a "bipole", a pair of opposite electronic charges separated by a distance δ, rigidly located in the interior of an otherwise chargeless molecule of linear dimension l along the dipole axis. Each such molecule surrounds itself with a cloud of others, producing by virtue of predominantly attractive forces a higher density of gas in its immediate neighborhood. This increases the mean free path, which enters into the known theory of viscosity.

Now the change in density involves the energy of interaction of two dipoles. Reinganum computes it as if all dipoles were in parallel, attractive positions, their minimum distance of separation being l. The result depends on l and δ. The first of these is known from previous evaluations of the mean free path. Comparison with coefficients of viscosity (η), yield values of δ, and these lie in the vicinity of 10^{-9} cm for all known gases. Furthermore, the calculation reflects the dependence of η on temperature remarkably well for the five gases studied and it leaves l independent of the temperature.

To test his dipole theory of forces further he applies it to the tensile strength of solids. These, he supposes, are held together by dipole forces. Under a tensile strain the molecules align their dipoles as the distance between them increases and, finally, just before the solid breaks, they have attained complete parallelism and thus maximum attraction. When this maximum dipole attraction is equated to the force (per molecular pair) at the tensile limit there results again a value of δ of the order 10^{-9} cm†.

This must have seemed like a considerable success of the dipole theory, in spite of its obvious crudeness in assuming all dipoles in parallel positions. Further corroboration came from another direction. Rutherford and McKling (1900) had used the dipole picture in explaining their observations on the absorption of X-rays in matter. The energy absorbed per ion produced, they said, is the work done in separating the charges of the dipole; their mean value for δ was 1.1×10^{-9} cm.

The idea of electronic dipoles was elaborated in Reinganum's second paper which appeared after van der Waals (J. D. van der Waals Jr. (1909)) had treated the interaction of rotating dipoles statistically, but in a manner not wholly convincing to Reinganum. He therefore carries through the calculation of the average interaction between two dipoles, $\langle Ve^{-V/kT}\rangle$, over all orientations. Since $\langle V\rangle = 0$ and $V \propto 1/R^3$, this average is proportional to $R^{-6}T^{-1}$, a result for which some evidence is available.

Langevin (1905) published his famous theory of ionic mobilities in which he employed as the ion-molecule potential the well-known

$$V = -\frac{(D-1)e^2}{4\pi n R^4}$$

featuring D the dielectric constant and n the number of molecules per cubic centimeter. Having convinced himself of the reality of permanent dipoles carried universally by all atoms, Reinganum claims that a dipole–monopole term must be added to the latter V, and this is proportional to R^{-2}. Choosing the values of δ found earlier, he concludes that when molecules are in contact the Langevin potential is 5 times as great as the dipole–monopole potential; at larger distances, of course, the latter preponderates.

The belief that all molecules carry permanent dipoles was strengthened by the early investigations of Debye (1912), who developed his theory of the dielectric constant by assuming that an electric field in general polarizes a molecule and orients its dipole. The latter effect causes the dielectric constant to depend on T. By testing his theory only for liquids now known to be polar (five alcohols) he arrived at the conclusion that δ is finite for

† For Cu, Pt, and Ag, $\delta \times 10^9 = 0.65, 0.74,$ and 0.78 cm.

all molecules, and has in fact a value close to 10^{-9} cm, as Reinganum predicted.

During the next 8 years it became apparent, however, that the simplest molecules have no permanent dipole moments. Moreover, a vague connection between one of the constants in van der Waals' equation of state (the usual *a*) and the index of refraction, i.e. the polarizability, came into sight. Coupling these two items of knowledge Debye (1920) proposed a theory which held the mobility of charges within a molecule, their displacement in response to an electric field, to be the universal cause of intermolecular forces. And he had to find the origin of the displacing field in an electrical property of molecules whose existence could at least not be denied on the basis of the observations possible at the time. Knowing that dipoles are not universal, Debye fell back upon quadrupoles as the field-producing agents.

His model has one great advantage over the "orientation effect" described by the earlier theory: it provides an attractive interaction even at high temperatures when, because of the rapid rotation of rigid dipoles, the net force is zero, all orientations being equally likely. This, Debye points out, is in accord with van der Waals' assumptions and with the observations sustaining them. In his paper, therefore, he computes first the mean square electric field $\langle E^2 \rangle$ of a rotating quadrupole which, by way of the formula $V = -(1/2)\alpha\langle E^2 \rangle$, produces the interaction (α is the polarizability). It turns out that $\langle E^2 \rangle$ is very simply related to a certain combination of the principal quadrupole moments θ_1, θ_2, and θ_3:

$$\langle E^2 \rangle = 3\tau^2/R^8, \quad \tau^2 = \theta_1^2+\theta_2^2+\theta_3^2-(\theta_1\theta_2+\theta_1\theta_3+\theta_2\theta_3).$$

The intermolecular force is therefore proportional to R^{-9}. It arises, as the derivation indicates, from the response of an induced dipole to a permanent quadrupole which became known as the "induction effect". Debye calls τ the "mean electric moment of inertia".

He is able to calculate the values of τ by fitting the van der Waals' constants he derives to the observed values. A table lists these values for twelve gases, including four rare gases, and τ is found to vary between 2.8×10^{-26} for helium and 61×10^{-26} e.s.u. for C_5H_{12}. In an addendum he employs Sutherland's theory to calculate the coefficient of viscosity with use of his force law, and this yields an independent evaluation of τ. Amazingly, these new values agree within very comfortable limits with those derived from van der Waals' equation. The agreement is especially good for *A*, *K*, and *X*, precisely those molecules which have no quadrupole moment. In the same article we find a calculation of the surface tension, based upon Debye's force law and observational material on van der Waals' constants. Again, the agreement between theory and observation is remarkable.

It is somewhat strange that Debye, on encountering the need for postulating large quadrupole moments for all his molecules to obtain his induction forces, did not consider the effect of these quadrupoles on one another. For although the mean attraction between rotating permanent quadrupoles is zero, the Boltzmann factor favors attractive states, and at low temperatures the "orientation" or "alignment" effect must predominate. The requisite calculations were published by Keesom (1921) 8 months after Debye's paper appeared.

Keesom develops formulas for the second virial coefficient B, for both dipole and quadrupole gases under the assumption that the forces become strongly repulsive when R equals the molecular diameter (hard sphere model). The quadrupoles are taken to be axially symmetric, so that the previous τ becomes identical with θ_1. Hence the theory is applicable to diatomic molecules. The induction forces are first neglected. The resulting formulas contain dipole and quadrupole moments, and by fitting them to measured values of B these moments can be determined.

One rather amazing result turns up: for a large range of temperatures B can be fitted equally well by dipole and quadrupole forces. The theory is not very discriminating. Knowing, however, that a "non-associating" gas is not dipolar, Keesom uses his information chiefly to determine the quadrupole moment of simpler molecules (H_2, O_2, N_2). There occurs a coincidence which now appears remarkable. For H_2 he calculates $\theta = 2.03 \times 10^{-26}$ e.s.u. The model of the H_2 molecule in vogue at the time was that of Bohr and Debye, in which atomic attraction results from a displacement of the two electrons revolving in phase toward the center of the molecule. J. M. Burgers (dissertation, Leyden, 1918, quoted by Keesom) computes from this model $\theta = 2.05 \times 10^{-26}$ e.s.u. True, the model had already lost its credit among most physicists because it imparted to H_2 a magnetic moment, but the coincidence is, none the less, striking. According to present knowledge the value of θ for H_2 is 0.39×10^{-26} e.s.u. This instance, as well as almost the entire history of the subject, illustrates what might well be called the occasional invariance of numerical results against models of explanation.

The induction effect of Debye required θ to be 3.2×10^{-26} e.s.u. Since alignment forces are proportional to the fourth power of θ, it is clear that they generally dominate over quadrupole induction forces, and this is borne out in later parts of Keesom's article, where he presents a theory including both. In retrospect, we observe that Debye's and Keesom's theories, though misapplied at a time when the electronic structure of molecules was poorly understood, remain adequate today in describing the interaction of polar molecules. Both dipole and quadrupoles play important roles, for example, in producing the complicated forces between water molecules.

Explanation of the forces between non-polar molecules, especially the rare gases whose charge distribution is spherically symmetric, had to await the advent of the quantum theory. Heitler and London (1927) published their famous account of the H_2 molecule, which led to an understanding of chemical binding as well as the repulsive intermolecular forces which act at small distances. It is so well known that we need not include it here, although the basic ideas will recur throughout this book. The same year also provided an understanding of the long-range forces which now go by the name dispersion forces. Wang (1927) solved the Schrödinger equation for two hydrogen-atoms at large distances of separation with inclusion of the instantaneous dipole interaction between the stationary protons and the moving electrons, using the cumbersome perturbation method of Epstein (1926, 1927). His effect may be described, and was recognized by Wang, as the instantaneous polarization of one atom by another which results in a correlation between the two electron coordinates and therefore in attraction. For the asymptotic form of V he finds

$$V = -8.7\frac{e^2a^5}{R^6} \quad (a = \text{first Bohr radius}).$$

Eisenschitz and London (1930), who 3 years later examined the same problem with greater care and related the result to the exchange forces in a more systematic way, employed a straightforward perturbation method and arrived at a similar formula, in which the coefficient 8.7 is replaced by the more accurate value, 6.47. Approximate useful formulas for the dispersion forces between other atoms and molecules were published by London (1930a, b), who subjected the whole problem to extensive scrutiny and advanced its understanding. Fuller attention will be paid to London's work in Chapter 2. After 1930 things moved rapidly. Slater and Kirkwood (1931) employed a variational method of calculation to find the same effect; quadrupole and higher interactions were included in the potential (Margenau, 1931) and various computational refinements were achieved (Pauling and Beach, 1935). These, however, form the substance of later chapters of this book. We therefore conclude the survey of the history of our subject somewhat prematurely at this point.

CHAPTER 2

Long-range Forces

2.1. Asymptotic Expansions of the Classical Potential Energy

The problem of calculating the interaction energy between two molecules reduces in principle to a solution of the Schrödinger equation with a Hamiltonian which consists of the sum of the Hamiltonians for the isolated molecules $H_1 + H_2$ plus the Coulomb interactions between all the charges in molecule 1 and those in molecule 2. The latter function will be called V; it does not include the internal Coulomb energies within 1 and within 2. In the solution of the Schrödinger equation V is treated as a perturbation, and use is made either of ordinary Rayleigh–Schrödinger perturbation theory or of the Ritz variational method. In the present chapter attention is confined to the interaction at distances which are large compared to the size of the atoms. This allows us to approximate even the classical V in a way which involves the concept of multipoles.

Indeed, the potential energy V can be written in a variety of useful forms when the charge distributions of the interacting molecules do not overlap. Let molecule 1 consist of point charges q_i momentarily situated at the points $\boldsymbol{r}_i$ with respect to an origin within it. At an outside point, a distance R from the origin, this distribution will produce a potential

$$\varphi = \sum_i \frac{q_i}{|\boldsymbol{R}-\boldsymbol{r}_i|}.$$

Expressed in a Taylor series,

$$\varphi = \frac{1}{R}\sum_i q_i + \frac{1}{R^2}\left(\frac{X}{R}\sum_i q_i x_i + \frac{Y}{R}\sum_i q_i y_i + \frac{Z}{R}\sum_i q_i z_i\right)$$
$$+\frac{1}{R^3}\left[\frac{1}{2}\left(\frac{3X^2}{R^2}-1\right)\sum_i q_i x_i^2 + \frac{1}{2}\left(\frac{3Y^2}{R^2}-1\right)\sum_i q_i y_i^2\right.$$
$$+\frac{1}{2}\left(\frac{3Z^2}{R^2}-1\right)\sum_i q_i z_i^2 + \frac{3XY}{R^2}\sum_i q_i x_i y_i + \frac{3XZ}{R^2}\sum_i q_i x_i z_i$$
$$\left.+\frac{3YZ}{R^2}\sum_i q_i y_i z_i\right] + \frac{1}{R^4}[\ldots], \tag{1}$$

where X, Y, Z and x_i, y_i, z_i are the components of $\boldsymbol{R}$ and $\boldsymbol{r}_i$, respectively. In a more compact notation ($X = X_1$, $Y = X_2$, $Z = X_3$) which employs the *monopole*

$$q = \sum_i q_i, \tag{2}$$

the *dipole* vector

$$\boldsymbol{p} = \sum_i q_i \boldsymbol{r}_i$$

and the *quadrupole* tensor

$$\overleftrightarrow{Q} = \sum_i q_i \boldsymbol{r}_i \boldsymbol{r}_i, \qquad Q_{12} = \sum_i q_i x_i y_i \quad \text{etc.} \tag{3}$$

we can write

$$\varphi = q\frac{1}{R} + \sum_\varrho p_\varrho \frac{\partial}{\partial X_\varrho}\left(\frac{1}{R}\right) + \frac{1}{2!}\sum_{\varrho\sigma} Q_{\varrho\sigma}\frac{\partial}{\partial X_\varrho}\frac{\partial}{\partial X_\sigma}\left(\frac{1}{R}\right) + \ldots \tag{4}$$

Higher terms in this series involve higher multipoles, which are tensors of higher rank constructible by an extension of (3). We note in passing that for a spherically symmetric charge distribution the term proportional to R^{-3} in φ, the so-called quadrupole term, disappears, although by definition (3) the quadrupole *tensor* remains finite.

A well-known equivalent expansion of φ which involves the Legendre polynomials P_n is (see, for example, Margenau and Murphy (1956), vol. 1, p. 102).

$$\varphi(R) = \sum_{n=0}^{\infty}\frac{1}{R^{n+1}}\sum_i q_i r_i^n P_n(\cos\theta_i) \tag{5}$$

θ_i being the angle between $\boldsymbol{r}_i$ and $\boldsymbol{R}$.

Molecule 2 is a distribution of charges q_j, disposed at points $\boldsymbol{r}_j$. These latter coordinates are measured from a new origin situated within molecule 2, and $\boldsymbol{R}$ is the vector distance from the origin of the first coordinate system located in 1 to that of the second, located within molecule 2. If every $r_j < R$, the potential energy between 1 and 2 is

$$V = \sum_j q_j \varphi(|\boldsymbol{R}+\boldsymbol{r}_j|)$$

and this can again be developed in a Taylor series which reads:

$$\begin{aligned} V &= \sum_j q_j\varphi + \sum_j q_j\left(x_j\frac{\partial\varphi}{\partial X} + y_j\frac{\partial\varphi}{\partial Y} + z_j\frac{\partial\varphi}{\partial Z}\right) \\ &\quad + \frac{1}{2!}\sum_j q_j\left(x_j^2\frac{\partial^2\varphi}{\partial^2 X} + \ldots + 2y_jz_j\frac{\partial^2\varphi}{\partial Y\partial Z}\right) + \ldots \\ &= \left\{q' + \sum_\varrho p'_\varrho\frac{\partial}{\partial X_\varrho} + \frac{1}{2!}\sum_{\varrho\sigma} Q'_{\varrho\sigma}\frac{\partial^2}{\partial X_\varrho\partial X_\sigma} + \ldots\right\}\varphi(R). \end{aligned} \tag{6}$$

The primed quantities here are charge moments for molecule 2. A compact expression for (6), which will be given below, is obtained with the aid of irreducible tensors. Before presenting it we shall write an elementary expression which is more desirable in connection with the simpler problems encountered in the present and in some later chapters. The first few terms of (6), which are the useful ones in the theory of intermolecular forces, take on a simple form if the z-axis is taken along $\boldsymbol{R}$ for all coordinates (so that $X = Y = 0$, $Z = R$):

$$\begin{aligned} V &= \frac{qq'}{R} + \frac{q'p_z - qp'_z}{R^2} + \frac{1}{R^3}(q'w_3 + qw'_3) + \frac{1}{R^4}(q'w_4 - qw'_4) \\ &\quad + \frac{1}{R^5}(q'w_5 + qw'_5) + \frac{1}{R^3}\sum_{ij} q_iq_j(x_ix_j + y_iy_j - 2z_iz_j) \\ &\quad + \frac{3}{2R^4}\sum_{ij} q_iq_j[r_i^2z_j - z_ir_j^2 + (2x_ix_j + 2y_iy_j - 3z_iz_j)(z_i - z_j)] \\ &\quad + \frac{3}{4R^5}\sum_{ij} q_iq_j[r_i^2r_j^2 - 5z_i^2r_j^2 - 5r_i^2z_j^2 - 15z_i^2z_j^2 + 2(4z_iz_j - x_ix_j - y_iy_j)^2] \\ &\quad + \frac{1}{2R^5}\sum_{ij} q_iq_j[3(r_i^2 + r_j^2)(4z_iz_j - x_ix_j - y_iy_j) \\ &\quad + 5(z_i^2 + z_j^2)(3x_ix_j + 3y_iy_j - 4z_iz_j)] + O(R^{-6}) \quad (Z\text{-axis along } \boldsymbol{R}). \end{aligned} \tag{7}$$

In this expression† we have written

$$\sum_i q_i r_i^n P_n(\cos\theta_i) \equiv w_n$$
$$\sum_j q_j' r_j^n P_n(\cos\theta_j) \equiv w_n';$$

the first five terms of (7) represent monopole–monopole, monopole–dipole, etc., interactions. They vanish if both molecules are neutral. The remaining terms proportional to R^{-3} and R^{-4} are known as the dipole–dipole and dipole–quadrupole energies; there are two components in R^{-5}, the first representing the interaction of two quadrupoles, the second that of a dipole in one with an octupole upon the other.

There are other ways of writing V. Since

$$V = \sum_{ij} \frac{q_i q_j}{|\boldsymbol{R} - \boldsymbol{r}_i + \boldsymbol{r}_j|}$$

a Taylor expansion also leads to the form

$$V = \sum_{ij} q_i q_j \exp\left[-(\boldsymbol{r}_i - \boldsymbol{r}_j)\cdot\nabla\right]\left(\frac{1}{R}\right), \qquad (6')$$

where ∇ operates only on R.

The complete evaluation of (6) or (6′) follows the work of Rose (1958) and Fontana (1961) and leads first to the symbolic result

$$V = \sum_{i,j,a,\alpha,b,\beta} \frac{16\pi^2 q_i q_j}{(2a+1)(2b+1)!!} \mathcal{Y}_a^{\alpha*}(\boldsymbol{r}_i)\mathcal{Y}_b^{\beta*}(\boldsymbol{r}_j)\mathcal{Y}_b^{\beta}(\nabla)\curlyvee_a^{\alpha}(\boldsymbol{R}). \qquad (8)$$

As to notation, $(2b+1)!! \equiv 1\cdot3\cdot5\ldots(2b+1)$; $\mathcal{Y}$ and $\curlyvee$ represent regular and irregular solid harmonics:‡

$$\mathcal{Y}_a^{\alpha*}(\boldsymbol{r}) = r^a Y_a^{\alpha*}(\boldsymbol{r}_1) = (-1)^\alpha r^a Y_a^{-\alpha}(\boldsymbol{r}_1)$$
$$\curlyvee_a^{\alpha}(\boldsymbol{R}) = R^{-a-1} Y_a^{\alpha}(\boldsymbol{R}_1)$$

$\boldsymbol{r}_1$ and $\boldsymbol{R}_1$ are unit vectors along $\boldsymbol{r}$ and $\boldsymbol{R}$ and ∇ operates on $\boldsymbol{R}$. Now

$$\mathcal{Y}_b^{\beta}(\nabla)\curlyvee_a^{\alpha}(\boldsymbol{R}) = (-1)^b(2b-1)!!$$
$$\left[\frac{(2a+1)(2b+1)(a+b-\alpha-\beta)!\,(a+b+\alpha+\beta)!}{4\pi(2a+2b+1)(a-\alpha)!\,(a+\alpha)!(b-\beta)!\,(b+\beta)!}\right]^{1/2} \curlyvee_{a+b}^{\alpha+\beta}(\boldsymbol{R}). \qquad (8a)$$

† See Margenau (1931). (A typographical error of sign in the quadrupole–quadrupole term which occurs in that paper has crept into the textbook literature.)

‡ Y_a^α denotes the usual spherical harmonic.

If again we measure all z-components along $\boldsymbol{R}$,

$$\mathcal{Y}_{a+b}^{\alpha+\beta}(\boldsymbol{R}) = R^{-a-b-1}\left[\frac{2a+2b+1}{4\pi}\right]^{1/2}\delta_{\alpha,-\beta}. \tag{8b}$$

In this coordinate system we therefore find by substitution into (8)

$$V = \sum_{i,j,a,b,\alpha} \frac{q_i q_j(-1)^{b+\alpha}4\pi(a+b)!\,\mathcal{Y}_a^{\alpha*}(\boldsymbol{r}_i)\mathcal{Y}_b^{\alpha}(\boldsymbol{r}_j)}{R^{a+b+1}[(2a+1)(2b+1)(a+\alpha)!\,(a-\alpha)!\,(b+\alpha)!\,(b-\alpha)!]^{1/2}} \tag{9a}$$

(z-axis along $\boldsymbol{R}$)

An alternate form of this (Carlson and Rushbrooke, 1950) is

$$V=\sum_{ij} q_i q_j \sum_{ab} \frac{r_i^a r_j^b}{R^{a+b+1}}\cdot B_{ab} \sum_{m=-\alpha}^{\alpha} \frac{Y_a^m(\theta_i\varphi_i)Y_b^{-m}(\theta_j\varphi_j)}{[(a+m)!(a-m)!(b+m)!(b-m)!]^{1/2}} \tag{9b}$$

$$B_{ab} = (-1)^{\alpha}\frac{4\pi(a+b)!}{[(2a+1)!\,(2b+1)!]^{1/2}},$$

where α is the smaller of a and b.

It is to be remembered that in all these formulas the index i labels the charges in the first, j those in the second molecule. There are other useful ways in which V can be written, almost too numerous to list. One, which is of special interest in connection with complicated molecules having permanent multipoles, will be discussed at the end of Chapter 7.

For some purposes it is useful to write V in a form which can be easily derived from the parent expression (6) before any stipulation about the direction of the $\boldsymbol{R}$-axis is made, viz.:

$$\begin{aligned} V &= \frac{1}{R^3}\sum_{ij} q_i q_j\left(\boldsymbol{r}_i\cdot\boldsymbol{r}_j - 3\frac{\boldsymbol{r}_i\cdot\boldsymbol{R}\,\boldsymbol{r}_j\cdot\boldsymbol{R}}{R^2}\right) \\ &+\frac{3}{2R^4}\sum_{ij} q_i q_j\{[r_i^2\boldsymbol{r}_j - r_j^2\boldsymbol{r}_i + 2\boldsymbol{r}_i\cdot\boldsymbol{r}_j\,(\boldsymbol{r}_i-\boldsymbol{r}_j)]\cdot\boldsymbol{R}/R \\ &-5\boldsymbol{r}_i\cdot\boldsymbol{R}\,\boldsymbol{r}_j\cdot\boldsymbol{R}\,(\boldsymbol{r}_i-\boldsymbol{r}_j)\cdot\boldsymbol{R}/R^3\}+\dots. \end{aligned} \tag{9}$$

This form is needed when the molecules are asymmetric and when interactions between more than two atoms are considered.

Direct use is made of the foregoing "classical" results in two older theories of intermolecular forces which were mentioned briefly in Chapter 1: Keesom's alignment theory and the multipole-induction theory of Debye and Falkenhagen. The first sought to explain molecular attractions as a direct interaction between static multipoles within the molecules, the second by assuming that a static multipole in one induces polarity in the other. While these effects are now known to be far from universal and are usually overshadowed by quantum mechanical induction, which London called the dispersion effect, they are, nevertheless, important whenever the molecules are in fact polar.

We illustrate the alignment effect in its simplest form, as it occurs for linear molecules. The dipole moment of such a structure *along its axis* will be labeled p, the quadrupole moment, which now has only one component Q, the octupole moment by $O = \sum_i q_i r_i^3$; primed letters shall refer to the second molecule. Thus, for example, $\sum_i q_i z_i = p \cos\theta$, $\sum_i q_i x_i = p \sin\theta \cos\varphi$, $\sum_i q_i z_i^2 = Q \cos^2\theta$, etc., provided θ and φ are polar coordinates of the molecular axis. With this convention eqn. (7) takes the form

$$
\begin{aligned}
V = {} & \frac{pp'}{R^3}[\sin\theta \sin\theta' \cos(\varphi-\varphi') - 2\cos\theta\cos\theta'] \\
& + \frac{3}{2R^4}\{Qp'[\cos\theta' + 2\cos\theta\sin\theta\sin\theta'\cos(\varphi-\varphi') \\
& - 3\cos^2\theta\cos\theta'] - pQ'[\cos\theta + 2\cos\theta'\sin\theta'\sin\theta\cos(\varphi-\varphi') \\
& - 3\cos^2\theta'\cos\theta]\} + \frac{3}{4R^5}QQ'\{1 - 5\cos^2\theta - 5\cos^2\theta' \\
& - 15\cos^2\theta\cos^2\theta' + 2[4\cos\theta\cos\theta' - \sin\theta\sin\theta'\cos(\varphi-\varphi')]^2\} \\
& + \frac{1}{2R^5}\{Op'[(12 - 20\cos^2\theta)\cos\theta\cos\theta' \\
& + (15\cos^2\theta - 3)\sin\theta\sin\theta'\cos(\varphi-\varphi')] \\
& + pO'[(12 - 20\cos^2\theta')\cos\theta'\cos\theta \\
& + (15\cos^2\theta' - 3)\sin\theta'\sin\theta\cos(\varphi-\varphi')]\} \\
& + \text{terms of order } R^{-6}, \text{ etc.}
\end{aligned}
\tag{10}
$$

The mean value of V over all orientations of either molecule is zero. Hence there is no average attraction between rigid multipoles if all orientations are taken to be equally probable. The Boltzmann factor in the distribution function, however, increases the weight of the attractive orientations and thereby gives rise to a potential energy which may be an appreciable fraction of the van der Waals' interaction.

The induction effect is best calculated from (1). If $\overleftrightarrow{Q}$ is reduced to principal axes, so that $Q_{\varrho\sigma} = 0$ for $\varrho \neq \sigma$, that equation is written conveniently with the use of three artificial vectors

$$\boldsymbol{Q} \equiv \boldsymbol{i}Q_{xx} + \boldsymbol{j}Q_{yy} + \boldsymbol{k}Q_{zz}$$

and

$$\boldsymbol{M} \equiv \boldsymbol{i}(R^2 - 3x^2) + \boldsymbol{j}(R^2 - 3Y^2) + \boldsymbol{k}(R^2 - 3Z^2)$$

$$\boldsymbol{N} = \boldsymbol{i}(Q_y + Q_z - 2Q_x)X + \boldsymbol{j}(Q_z + Q_x - 2Q_y)Y + \boldsymbol{k}(Q_x + Q_y - 2Q_z)Z$$

as follows:

$$\varphi = \frac{q}{R} + \frac{1}{R^3}\boldsymbol{R}\cdot\boldsymbol{p} - \frac{1}{2R^5}\boldsymbol{M}\cdot\boldsymbol{Q}. \tag{11}$$

The field strength at $\boldsymbol{R}$ is

$$\boldsymbol{F} = -\nabla\varphi = \frac{q\boldsymbol{R}}{R^3} - \frac{\boldsymbol{p}}{R^3} + \frac{3\boldsymbol{R}}{R^5}\boldsymbol{p}\cdot\boldsymbol{R} + \frac{\boldsymbol{N}}{R^5} - \frac{5\boldsymbol{R}}{2R^7}\boldsymbol{M}\cdot\boldsymbol{Q} + \dots \tag{12}$$

The induced energy in a neighboring molecule which has polarizability P is

$$V_1 = -\tfrac{1}{2}PF^2. \tag{13}$$

Insertion of (12) into (13) gives an answer which is not very interesting because the quantities involved are functions of the angular positions of the molecules, which rotate rapidly. More meaningful is the average over all orientations. Omitting quantities which vanish in the mean we find after some elementary trigonometry†

$$\overline{F^2} = \frac{p^2}{R^6} + \frac{3}{8R^8}\overline{(\boldsymbol{R}\cdot\boldsymbol{p})^2} + \frac{\overline{N^2}}{R^{10}} + \frac{25}{4R^{12}}\overline{(\boldsymbol{M}\cdot\boldsymbol{Q})^2} - \frac{5}{R^{12}}\overline{(\boldsymbol{M}\cdot\boldsymbol{Q})(\boldsymbol{N}\cdot\boldsymbol{R})} \tag{14}$$

On defining

$$\tau \equiv Q_x^2 + Q_y^2 + Q_z^2 - Q_xQ_y - Q_xQ_z - Q_yQ_z \tag{15}$$

one obtains

$$\overline{(\boldsymbol{R}\cdot\boldsymbol{p})^2} = \frac{R^2p^2}{3}, \quad \overline{N^2} = 2R^2\tau^2, \quad \overline{(\boldsymbol{M}\cdot\boldsymbol{Q})^2} = \frac{4R^4\tau^2}{5} = \overline{(\boldsymbol{M}\cdot\boldsymbol{Q})(\boldsymbol{N}\cdot\boldsymbol{R})}.$$

† In the sequel we employ bars in place of ⟨ ⟩ to designate averages.

The easiest way to deduce these formulas is to leave $\boldsymbol{p}$ and $\boldsymbol{Q}$ constant and integrate over all orientations of $\boldsymbol{R}$. We now remember that induction is a mutual process, so that actually we must write in place of (13)

$$V = -\tfrac{1}{2}\left(P\overline{F'^2}+P'\overline{F^2}\right).$$

Use of that equation, together with (14), finally gives

$$V = -P\left(\frac{p'^2}{R^6}+\frac{3}{2}\,\frac{\tau'^2}{R^8}+\cdots\right)-P'\left(\frac{p^2}{R^6}+\frac{3}{2}\,\frac{\tau^2}{R^8}+\cdots\right) \tag{16}$$

as the classical formula for the *induced* potential energy.

2.2. Quantum Theory of Long-range Forces

The interactions discussed in the previous section require the presence of a static multipole in at least one of the molecules. The existence of attractive forces, which was clearly evident in the departures of all substances, even those containing spherical molecules (e.g. helium), from the ideal gas law, was a paradox of classical physics. Quantum mechanics allows them to be explained in the following way.

An atom in an S-state, such as hydrogen or helium, has a spherical charge distribution, to be sure. But in view of the uncertainty principle the charge can not be at rest. Although its motion is not organized—there is no mean angular or linear momentum—its random excursions engender instantaneous multipoles which are able to produce rapidly fluctuating fields quite in accord with our earlier analysis. These fields polarize the other molecules and bring about a certain measure of phase agreement between the motions of the charges of the interacting partners and this, in turn, creates attractive forces.

The mathematical account of these events is rendered by perturbation theory. If the classical perturbation energy is V, and the undisturbed state of a physical system is labeled by a set of quantum numbers which are symbolically represented by 0, the others by λ, then the energy change resulting from V is

$$\triangle E = V_{00}+\sum_{\lambda}{}' \frac{|V_{0\lambda}|^2}{E_0-E_\lambda}+\cdots \tag{17}$$

We shall apply this formula to the calculation of the long-range forces between two hydrogen atoms, first in a simple highly approximative way.

2.2.1. *Simple Theory of Long-range Forces between Hydrogen Atoms*

Let our system be a pair of hydrogen atoms, numbered 1 and 2, whose ground states are represented by ψ_0. The state of the combined system for which the perturbation is to be computed is then $\psi_0^{(1)} \psi_0^{(2)}$; V is given by (7) which in this instance reduces to

$$V = \frac{e^2}{R^3}(x_1x_2+y_1y_2-2z_1z_2)+\frac{3e^2}{2R^4}[r_1^2z_2-z_1r_2^2+(2x_1x_2 + 2y_1y_2-3z_1z_2)(z_1-z_2)]+\dots \tag{18}$$

provided we restrict ourselves to the first two non-vanishing terms of the series†. The first-order perturbation vanishes because V_{00}, which should now be written more explicitly by displaying the quantum number of *each* of the atoms, in the form $V_{00,\,00}$, is zero for the spherical function $\psi_0^{(1)}\ \psi_0^{(2)}$. The index λ which appears in (17) must be split similarly into $(\lambda_1\ \lambda_2)$, E_0 becomes $2\,E_H$, i.e. twice the energy of an H-atom in its ground state, and $E_\lambda = E_{\lambda_1}+E_{\lambda_2}$. Thus

$$V_{0\lambda} = V_{00,\,\lambda_1\lambda_2} = \int \psi_0(1)\,\psi_0(2)\,V\psi_{\lambda_1}(1)\,\psi_{\lambda_2}(2)\,d\tau_1\,d\tau_2$$

and

$$\Delta E = \sum_{\lambda_1\lambda_2}{}' \frac{|V_{00,\,\lambda_1\lambda_2}|^2}{2E_H-E_{\lambda_1}-E_{\lambda_2}}\,. \tag{19}$$

As to the E_λ, they correspond to the excited states of hydrogen, the lowest of which is $\frac{1}{4}E_{H/4}$ and negative. Some of them, those in the continuous spectrum, lie above zero; hence it might be expected that all of them cluster about zero. At any rate, since every $E_\lambda > E_H$ and $E_H < 0$, ΔE is negative. We also know that for large R, $\Delta E \propto R^{-6}$ and these two facts imply that the forces, $-\frac{\partial}{\partial R}(\Delta E)$, are attractive. Clearly, this result is true for all molecules in their lowest energy states because it depends only on the sign of the denominator in (19).

† Subscripts 1 and 2 now label the electrons in atoms 1 and 2.

An upper limit for ΔE, indeed a fairly good approximation to it, results if E_{λ_1} and E_{λ_2} in (19) are simply neglected. Since, by matrix algebra,

$$\sum_{\lambda_1 \lambda_2}' |V_{00,\lambda_1\lambda_2}|^2 = (V^2)_{00,00} - (V_{00,00})^2 = (V^2)_{00,00}$$

and† $E_H = -\dfrac{e^2}{2a_0}$, we find

$$\Delta E = -\frac{a_0}{e^2}(V^2)_{00,00}.$$

When $(V^2)_{00,00}$ is computed, the integration over angles causes all cross terms, especially those with the coefficient R^{-7}, to disappear, and the result is

$$(V^2)_{00,00} = \frac{2}{3}\frac{e^4}{R^6}\left[(r^2)_{00}^2 + \frac{3}{R^2}(r^4)_{00}(r^2)_{00} + \dots\right]$$

The omitted terms drop out because averages like $\overline{xy}$, $\overline{r^2x}$, etc., are zero.

The radial matrix elements are well known:

$$(r^2)_{00} = 3a_0^2 \qquad\qquad (r^4)_{00} = 22.5a_0^4$$

where a_0, the first Bohr radius, is 0.52917 Å.

Hence

$$\Delta E = -6\frac{e^2}{a_0}\frac{a_0^6}{R^6}\left(1 + 22.5\frac{a_0^2}{R^2} + \dots\right) \tag{20}$$

or, *in atomic units*, $\left(\dfrac{e^2}{a_0} = 1,\ a_0 = 1\right)$

$$\Delta E = -\frac{6}{R^6}\left(1 + \frac{22.5}{R^2} + \dots\right) \text{[H–H interaction]} \tag{20a}$$

Convergence is evidently poor unless $R > 5a_0$; more will be said about this below. As to orders of magnitude, at $R \approx 5a_0$, where the series can not be fully trusted, $\Delta E \approx 3\times10^{-5}$ a.u. $\approx$ 1 mv, and this is, indeed, a rough measure of observed intermolecular energies.

A better calculation (Pauling and Beach, 1935), which does not neglect the E_λ's in (19) but determines them by a variational

† a_0 is the radius of the first Bohr orbit of hydrogen ($a_0 = \hbar^2/me^2$).

procedure, yields 6.5 in place of the coefficient of R^{-6} in (20) and reduces the coefficient of R^{-8} from 135 to 124.

Calculations on helium similar to those above (Margenau, 1931†) lead to

$$\Delta E = -\frac{1.62}{R^6}\left(1+\frac{7.9}{R^2}+\dots\right) \begin{matrix}\text{atomic}\\ \text{units}\end{matrix} \quad \text{[He–He interaction]} \qquad (21)$$

2.2.2. London's Formulas

Eisenschitz and London (1930) and London (1930a, b) were the first to recognize clearly the physical origin and the significance of the interactions here under consideration. They also showed that the first term of the series, which is proportional to R^{-6}, can be expressed in terms of dispersion f-values and for this reason suggested the name "dispersion forces" for the asymptotic interaction. The higher terms are not directly related to f-values. In this section we review London's work.

As to the definition of f-values we recall that a molecular (dipole) transition between states r and s has "oscillator strengths" $f^{(x)}, f^{(y)}$, and $f^{(z)}$ defined by

$$f_{rs}^{(x)} = \frac{2}{3}\frac{m}{\hbar^2} \,|\, X_{rs}\,|^2 \; (E_s - E_r), \text{ etc.,} \qquad (22)$$

(Bates, 1961, vol. 1, p. 397) where $X \equiv \sum_i x_i$, the sum extending over all electrons (notice the change in meaning of the symbols X, Y, Z in this section. They designated previously the components of $\boldsymbol{R}$); E_s is the energy of state s and m the electron mass. The quantum numbers r and s include space quantum numbers M_r and M_s on which in the absence of external fields the energies E_r and E_s do not depend. Thus, if we decompose the sets (r) and (s) into (ϱ, M_r) and (σ, M_s) we have, instead of (22),

$$f_{\varrho M_r, \sigma M_s}^{(x)} = \frac{2}{3}\frac{m}{\hbar^2} \,|\, X_{\varrho M_r, \sigma M_s}\,|^2 \,(E_\sigma - E_\varrho) \qquad (23)$$

which can conveniently be *summed* over all degenerate final states M_s and *averaged* over the initial states M_r. One thus de-

† Use was made of Slater orbitals for the helium atoms.

fines a new f-value, independent of magnetic quantum numbers,

$$f^{(x)}_{\varrho,\sigma} \equiv \frac{1}{2L_r+1} \sum_{M_r M_s} f^{(x)}_{\varrho M_r,\,\sigma M_s}, \tag{24}$$

where L_r is the angular-momentum quantum number of state r. Now it turns out that in the summations occurring here the dependence on polarization is lost, and that

$$f^{(x)}_{\varrho,\sigma} = f^{(y)}_{\varrho\sigma} = f^{(z)}_{\varrho\sigma} \equiv \tfrac{1}{3} f_{\varrho\sigma}. \tag{25}$$

We retain only the dipole–dipole part of (7). This is, in our present notation,

$$V_{dd} = \frac{e^2}{R^3}(X_1X_2+Y_1Y_2-2Z_1Z_2)$$

the indices 1 and 2 designating the molecules. If these have no permanent electric moments, the first perturbation in (17) is zero and we find for the state in which molecules 1 and 2 have quantum numbers r_1 and r_2

$$\Delta E_{\text{disp}} = \frac{e^4}{R^6} \sum_{s_1 s_2} \frac{|X_{r_1 s_1} X_{r_2 s_2} + Y_{r_1 s_1} Y_{r_2 s_2} - 2Z_{r_1 s_1} Z_{r_2 s_2}|^2}{E(r_1)+E(r_2)-E(s_1)-E(s_2)}. \tag{26}$$

On summing over the quantum numbers $M_s^{(1)}$ and $M_s^{(2)}$ which are included in s_1 and s_2 we find again that all cross terms in the numerator of the last expression vanish [Margenau, 1939; Hirschfelder, Curtiss, and Bird, 1954, p. 960]. We remember, too, that the energies in the denominator are not functions of the M's. When this summation is carried out, ΔE_{disp} still depends on the M-values of the molecular states in which the interaction takes place, and if (26) were evaluated at this stage, the result could be expressed in f-values of the form (24), which distinguishes between directions. London obtains a simpler result by *averaging* ΔE_{disp} over all M-values of the actual molecular states in which the interaction takes place:

$$\overline{\Delta E}_{\text{disp}} = [(2L_r^{(1)}+1)(2L_r^{(2)}+1)]^{-1} \sum_{M_r^{(1)} M_r^{(2)}} \Delta E_{\text{disp}}$$

$$= \frac{3}{2}\frac{e^4}{R^6}\frac{\hbar^4}{m^2} \times$$

$$\sum_{\sigma_1 \sigma_2} \frac{f_{\varrho_1 \sigma_1} f_{\varrho_2 \sigma_2}}{[E(\varrho_1)-E(\sigma_1)]\,[E(\varrho_2)-E(\sigma_2)][E(\varrho_1)+E(\varrho_2)-E(\sigma_1)-E(\sigma_2)]}. \tag{27}$$

Use has here been made of formulas (24) and (25). In case the molecular states have spherical symmetry the bar over ΔE_{disp} can be omitted. (At this point we recover formula (20) for hydrogen if we neglect all $E(\sigma)$ and use the sum rule, $\sum_{\sigma} f_{\varrho\sigma} = 1$; note that $\frac{\hbar^2}{m} = a_0 e^2$.)

A useful approximation to London's formula, (27), is available when among all the f-values a single one predominates. If this were literally true, the sum rule would force it to be n, the number of optically active (usually valence) electrons, and the formula would read

$$\overline{\Delta E}_{\text{disp}} = -\frac{3}{2} n_1 n_2 \frac{a_0^2 e^8}{R^6 h^3} [\nu_1 \nu_2 (\nu_1 + \nu_2)]^{-1}. \tag{28}$$

The ν's here are the optically active frequencies. However, this crude result can yield at best an order of magnitude estimate.

A more accurate expression involves the static polarizabilities, given by

$$P_\varrho = \frac{e^2 \hbar^2}{m} \sum_{\sigma} \frac{f_{\varrho\sigma}}{E(\varrho) - E(\sigma)}.$$

This formula can quite often be well approximated by a single term of the summation, in which case

$$f = \frac{2\pi m P}{e^2 \hbar} \nu.$$

Suppose that *both* molecules have such "single term dispersion formulas". Then

$$\overline{\Delta E}_{\text{disp}} = \frac{3h}{2R^6} \frac{\nu_1 \nu_2}{\nu_1 + \nu_2} P_1 P_2 \tag{29}$$

Estimates of dispersion forces are often based upon this simple formula, for the polarizabilities of most interesting molecules are known. As to ν_1 and ν_2, it is found that frequencies corresponding to the ionization energies often approximate them fairly well.

2.2.3. Asymptotic Series

When the higher terms in the expansion (7) or (8) (Margenau, 1931) were first computed with the use of quantum perturbation theory a peculiar situation arose. The higher terms, when combined with the best exchange interaction then known (Slater and Kirkwood, 1931), impaired the agreement with experimental findings that had been obtained by including the dipole–dipole (dispersion) term above. It was noticed that the series did not converge in a simple way in powers of R^{-1}, and practical doubt was cast upon the whole expansion. Meanwhile, mainly through the work of Brooks (1952), Roe (1952), and Dalgarno and Lewis (1956) these doubts have been dispersed; the last-named authors in particular were able to show that the series has all the reasonable properties of asymptotic series (in Poincaré's sense) and can, indeed, be used with confidence in calculations. The original dilemma arose from inadequate treatment of the exchange forces rather than the higher terms in the asymptotic series.

We shall here first derive the series and then briefly discuss its convergence. In pursuit of the first goal we follow the exposition of Fontana (1961).

Exact answers have been obtained for two general models: the "oscillator molecule" and the hydrogen atom with arbitrary screening constant (Slater orbital) (London, 1937).

(a) *Oscillator model.* The first was used by London in his calculation of the dispersion forces, then generalized (Margenau, 1938; Hornig and Hirschfelder, 1952) to include higher multipoles. It postulates the existence of a single negative charge bound by simple harmonic forces to its nucleus. The characteristics of this model are (a) that it involves but a single frequency (ν_1 for molecule 1, ν_2 for 2) and its multiples, and (b) that the matrix elements between different states are limited in number and can be calculated in closed form. Whether or not the model is meaningful for a given pair of molecules can usually be determined by seeing whether their dispersion formulas have one dominant term. It is especially appropriate for rare-gas atoms. In a sense we have already employed it in the foregoing section.

The calculation starts with (8) and employs the perturbation formula (17), again with $V_{00} = 0$. The index 0, which designates the actual state in which the two molecules interact, has as its full complement the quantum numbers n_1, l_1, m_1 for molecule 1, n_2, l_2, m_2 for molecule 2. Thus the abbreviation 0 signifies the set $\{n_1, l_1, m_1, n_2, l_2, m_2\}$; the quantum numbers of the summation states λ are, similarly, $\{n_1', l_1', m_1', n_2', l_2', m_2'\}$.

In terms of a parameter γ, which is related to f-values, frequency and polarizability P by the formula

$$\gamma = fe^2/Ph\nu \tag{29a}$$

a normalized state of a three-dimensional harmonic oscillator in spherical polar coordinates takes the form (Morse and Feshbach, 1953, p. 1663)

$$\psi_{nlm}(r) = \left[2\gamma^{3/2}\frac{\Gamma(\frac{1}{2}n-\frac{1}{2}l-\frac{1}{2})}{[\Gamma(\frac{1}{2}n+\frac{1}{2}l+1)]^3}\right]^{1/2}(\gamma r^2)^{\frac{1}{2}l}\exp(-\tfrac{1}{2}\gamma r^2)$$
$$\times L^{l+1/2]}_{\frac{1}{2}n+\frac{1}{2}l-1/2}(\gamma r^2)Y_l^m(\theta,\varphi),$$

where the Laguerre polynomial $L_a{}^b(z)$ is defined by

$$L_a{}^b(z) = \frac{[\Gamma(a+b+1)]^2}{\Gamma(a+1)\Gamma(b+1)}\,{}_1F_1(-a;b+1;z), \tag{30}$$

${}_1F_1(-a;\ b+1;\ z)$ being the Kummer confluent hyper-geometric series

$${}_1F_1(\alpha;\beta;z) = \sum_{\lambda=0}^{\infty}\frac{\Gamma(\alpha+\lambda)\Gamma(\beta)}{\Gamma(\alpha)\Gamma(\beta+\lambda)\Gamma(1+\lambda)}z^\lambda.$$

In constructing $\psi(r)$ one compounds γ from the experimental value of P and the values of f and ν which appear in the empirical dispersion formula. The total state function is $\psi_{n_1l_1m_1}(r_1)\,\psi_{n_2l_2m_2}(r_2)$; the sum over i and j in (8) disappears, and (17) now reads

$$\Delta E = \sum_\lambda\left|\sum_{a,b}\sum_{\alpha,\beta}\frac{16f_1f_2\pi^2e^2}{(2a+1)[(2b+1)!!]}\mathcal{Y}_b^\beta(\nabla)Y_a^\alpha(R)\right.$$
$$\left.\times\langle 0|\mathcal{Y}_a^{\alpha*}(r_1)\mathcal{Y}_b^{\beta*}(r_2)|\lambda\rangle\right]^2\frac{1}{E_0-E_\lambda}$$
$$=\sum_\lambda\left[\sum_{a,b}\sum_{\alpha,\beta}\frac{16f_1f_2\pi^2e^2}{(2a+1)[(2b+1)!!]}\mathcal{Y}_b^\beta(\nabla)Y_a^\alpha(R)\right.$$
$$\times\langle n_1l_1m_1|\mathcal{Y}_a^{\alpha*}(r_1)|n_1'l_1'm_1'\rangle \tag{31}$$
$$\left.\times\langle n_2l_2m_2|\mathcal{Y}_b^{\beta*}(r_2)|n_2'l_2'm_2'\rangle\right]^2\frac{1}{E_0-E_\lambda},$$

where, for example, $(2a+1)!! = 1\cdot3\cdot5\ldots(2a+1)$.

Now

$$\langle nlm|\mathcal{Y}_a^{\alpha*}(r)|n'l'm'\rangle = (-)^\alpha\langle nl|r^a|n'l'\rangle\langle lm|Y_a^{-\alpha}(\theta,\varphi)|l'm'\rangle, \tag{32}$$

where $\langle nl|r^a|n'l'\rangle$ is the radial matrix element of the 2^a-pole interaction. The factors f_1 and f_2 are inserted to account for the fact that the oscillating electrons in the two molecules, which the model treats as single charges, are f_1 and f_2 in number. To express the remaining matrix elements one may use advantageously the Clebsch–Gordan coefficients (Rose, 1957, p. 62; or Margenau and Murphy, 1964, vol. 2, p. 406) introduced by Wigner into the theory of angular momenta:

$$\begin{aligned}
\langle lm\,|\,Y_a^{-\alpha}(\theta,\varphi)\,|\,l'm'\rangle
&= \left[\frac{(2l'+1)(2a+1)}{4\pi(2l+1)}\right]^{1/2} C(l'al;\ m',\ -\alpha,\ m)C(l'al;\ 000)\\
&= \left[\frac{(2l+1)(2a+1)}{4\pi(2l'+1)}\right]^{1/2} C(lal';\ m,\ \alpha,\ m')C(lal';\ 000)(-)^{\alpha}\\
&= (-1)^{\alpha}\langle l'm'\,|\,Y_a^{\alpha}(\theta,\varphi)\,|\,lm\rangle.
\end{aligned}$$

We recall that $C(l'al;\ m',\ -\alpha,\ m) = 0$ unless $m' = m+\alpha$ and l' has one of the values $l+a,\ l+a-1,\ \ldots\ |l-a|$.

The relevant radial matrix elements which occur in (32) are

$$\langle 10\,|\,r^a\,|\,n'l'\rangle = \left[\frac{(2a+1)!!}{(2\gamma)^a}\right]^{1/2} \delta_{n',\,a+1}\ \delta_{l',\,a}\,. \tag{33}$$

As to the energy differences, they have a very simple form for the oscillator:

$$E_0-E_\lambda=\tfrac{3}{2}h\nu_a+\tfrac{3}{2}h\nu_b-(\tfrac{3}{2}+n_a')h\nu_a-(\tfrac{3}{2}+n_b')h\nu_b=-\,n_a'h\nu_a-n_b'h\nu_b\,. \tag{34}$$

The final result for ΔE is obtained by inserting (8a), (8b), (32), (33), and (34) into (31). It is to be remembered that, in using (8b), we have assumed that the z-axes of all coordinate systems are oriented along $\boldsymbol{R}$. Also, a more compact formula can be written if the summations over a and b are taken, not from 0 to ∞ but from 1 to ∞ with appropriate changes in the summands. One then obtains

$$\Delta E = -\sum_{a,\,b=1}^{\infty}\sum_{\alpha}\frac{f_1f_2e^4(2a-1)!!(2b-1)!!(a+b)!(a+b)!}{R^{2(a+b+1)}(2\gamma_1)^a(2\gamma_2)^b(a-\alpha)!(a+\alpha)!(b-\alpha)!(b+\alpha)!}\times\frac{1}{ah\nu_1+bh\nu_2}. \tag{35}$$

The lower limit of α is $-a$ or $-b$, whichever is smaller.

When evaluated for the first few terms, we find

$$\Delta E = -\frac{(3/2)e^4 f_1 f_2}{(h\nu_1+h\nu_2)\gamma_1\gamma_2 R^6} - \frac{(45/8)e^4 f_1 f_2}{(h\nu_1+2h\nu_2)\gamma_1\gamma_2^2 R^8}$$
$$- \frac{(45/8)e^4 f_1 f_2}{(2h\nu_1+h\nu_2)\gamma_1^2\gamma_2 R^8} - \frac{(315/8)e^4 f_1 f_2}{(2h\nu_1+2h\nu_2)\gamma_1^2\gamma_2^2 R^{10}} \qquad (36a)$$
$$- \frac{(210/8)e^4 f_1 f_2}{(h\nu_1+3h\nu_2)\gamma_1\gamma_2^3 R^{10}} - \frac{(210/8)e^4 f_1 f_2}{(3h\nu_1+h\nu_2)\gamma_1^3\gamma_2 R^{10}} - \ldots .$$

Successive terms here correspond to dipole–dipole, dipole–quadrupole, quadrupole–dipole, quadrupole–quadrupole, dipole–octupole, and octupole–dipole interactions. A more useful form of this result is obtained when the γ's are translated into polarizabilities P, the $h\nu$-terms into equivalent energies and *atomic units* are introduced (e^2/a_0 for all energies, a_0 for R, a_0^3 for P). It is

$$\Delta E = -\Big[\frac{3/2}{R^6}\frac{P_1E_1P_2E_2}{(E_1+E_2)} + \frac{45/8}{R^8}\frac{P_1E_1P_2^2E_2^2}{f_2(E_1+2E_2)}$$
$$+ \frac{45/8}{R^8}\frac{P_1^2E_1^2P_2E_2}{f_1(2E_1+E_2)} + \frac{315/8}{R^{10}}\frac{P_1^2E_1^2P_2^2E_2^2}{f_2(2E_1+2E_2)} \qquad (36)$$
$$+ \frac{210/8}{R^{10}}\frac{P_1E_1P_2^3E_2^3}{f_2^2(E_1+3E_2)} + \frac{210/8}{R^{10}}\frac{P_1^3E_1^3P_2E_2}{f_1^2(3E_1+E_2)} + \ldots \Big]$$

Formula (36) is useful in cases where each interacting molecule possesses a resonance frequency for which the f-value is large in comparison with all other transition frequencies; f is then the oscillator strength for that frequency, and the frequency itself is ν. The γ's involve the f's and the polarizabilities P via formula (29a). The condition for the applicability of (36) is the existence of a one-term dispersion formula, which describes the refractive index fairly accurately with the use of a single frequency in the Kramers–Heisenberg dispersion law. The rare gases satisfy this condition reasonably well. The accuracy of the terms in (36) decreases greatly as one proceeds to higher powers of $1/R$.

For identical molecules one has

$$\Delta E = -\frac{(3/2)e^4f^2}{2h\nu\gamma^2R^6} - \frac{(45/4)e^4f^2}{3h\nu\gamma^3R^8} - \frac{(735/8)e^4f^2}{4h\nu\gamma^4R^{10}} - \ldots, \qquad (37a)$$

or, because of (29a),

$$\Delta E = -\frac{3}{4}\frac{P^2h\nu}{R^6} - \frac{15}{4}\frac{P^3(h\nu)^2}{fe^2R^8} - \frac{735}{32}\frac{P^4(h\nu)^3}{f^2e^4R^{10}} \ldots \qquad (37)$$

An extensive table (Table 2.1) of the coefficients of R^{-6}, R^{-8} and R^{-10} (denoted by C_1, C_2 and C_3 respectively) has been computed on the basis of eqn. (37) by Fontana (1961). It refers to the interaction between *identical* atoms.

TABLE 2.1

	$h\nu$ (ev)	$\alpha\times10^{24}$ (cm^3)	f	$-C_1\times10^{60}$ (erg cm^6)	$-C_2\times10^{76}$ (erg cm^8)	$-C_3\times10^{92}$ (erg cm^{10})
He	24.5	0.207[a]	1.1	1.26	2.02	3.96
Ne	25.7	0.39	2.37	4.70	6.90	12.4
A	17.5	1.63	4.58	55.9	121	320
Kr	14.7	2.46	4.90	107	274	860
Xe	12.2	4.0	5.61	236	708	2622
H_2	~14.5	0.81	~1.5	11.4	31.1	104
N_2	15.8	1.74	4.61	57.5	119	302
O_2	13.6	1.57	3.11	40.3	96.0	280
CO_2	15.5	2.86	5.70	152	411	1 361
CH_4	14.1	2.58	4.60	113	310	1 044
NH_3	11.7	2.24	2.72	71	236	967
Cl_2	12.7	4.60	6.55	323	1 000	3 795
HCl	13.4	2.63	4.25	111	321	1 131
HBr	12.1	3.58	4.71	186	595	2 329
HI	10.5	5.4	5.30	368	1 367	6 221
Li	1.85	15[b]	~0.75[c]	500	6 430	101 200
Na	2.10	18[b]	0.975[d]	817	11 000	181 500
K	1.61	29[b]	0.987[d]	1630	26 700	537 800
Rb	1.57	29[b]	0.996[d]	1590	25 200	489 800
Cs	1.41	36[b]	~0.98	2200	39 500	870 200

[a] Essen, (1953)
[b] Chamberlain and Zorn, (1960).
[c] Stephenson, (1951).
[d] Stephenson, (1951).

Probably the most accurate list of C_1 values for different rare gas atoms interacting with one another was recently prepared by Kingston (1964). He employed formula (27) and sought out the best f-values and excitation energies available. His results are reproduced in Table 2.2.

More recently Kestner and Sinanoglu (1962) have solved exactly the problem of a two-electron system in which the electron–electron interaction is Coulombic but the attraction between electrons and nuclei is represented by an oscillator potential.

TABLE 2.2. *Values of* C_1 *(erg cm*^{6}*)*[a]

b/a	He	Ne	Ar	Kr	Xe
He	1.39	2.88	9.22	12.9	17.9
Ne	2.88	6.04	18.9	26.2	36.1
Ar	9.22	18.9	62.6	88.3	125
Kr	12.9	26.2	88.3	125	178
Xe	17.9	36.1	125	178	257

[a] From Kingston (1964). A more recent list, with some estimates of errors, was given by Bell (1965).

Their work sheds light on the way in which the correlation between electron motions modifies the intermolecular forces, an effect which is neglected in the work described above. Earlier unpublished work along similar lines is found in a Technical Report (No. 55) of the University of Maryland Physics Department (1956) by Tredgold and Evans. Kestner (1966c) extended this work and treated a model system for the long range R^{-6} interaction of two two-electron atoms (in Gaussian form). This permits the effects of intra-atomic electron interaction to be studied in detail.

(b) *Hydrogen model.* For atoms whose spectra are well explained by a one-electron model, the calculation of ΔE is better performed with the use of hydrogen-atom functions containing an adjusted screening constant Z^*. These functions are

$$\Psi_{nlm}(r) = \left[g^3 \frac{\Gamma(n-l)}{2n[\Gamma(n+l+1)]^3}\right]^{1/2}$$
$$\times (gr)^l e^{-gr/2} L_{n-l-1}^{2l+1}(gr) Y_l^m(\theta, \varphi),$$
$$n = 1, 2, \ldots \quad l = n-1, n-2, \ldots, 1, 0$$
$$l \geqslant m \geqslant -l \tag{38}$$

with $$g \equiv 2m_e e^2 Z^* n\hbar^2$$

The computation is made feasible by an approximation first introduced by Unsöld, and one which we have already used in section A. We replace the summation

$$\sum_\lambda{}' \frac{|V_{0\lambda}|^2}{E_0 - E_\lambda}$$

which occurs in (17), by $-(V^2)_{00}/\bar{E}$, where $\bar{E}$ is a weighted mean of the energy differences $E_\lambda - E_0$. This is permissible when $V_{00} = 0$, as is here the case. In general, we then obtain for a state of the interacting partners which is designated by the quantum numbers $n_1, l_1, m_1; n_2, l_2, m_2$, the formula

$$(V^2)_{00} = (16\pi^2 e^2)^2 \sum_{a,b,\alpha,\beta} \sum_{c,d,\gamma,\delta} \frac{\mathcal{Y}_b^\beta(\nabla) Y_a^\alpha(\boldsymbol{R}) \mathcal{Y}_d^\delta(\nabla) Y_c^\gamma(\boldsymbol{R})}{(2a+1)[(2b+1)!!](2c+1)[(2d+1)!!]}$$
$$\times \langle n_1 l_1 | r^{a+c} | n_1 l_1 \rangle \langle n_2 l_2 | r^{b+d} | n_2 l_2 \rangle \langle l_1 m_1 | Y_a^{\alpha *} Y_c^{\gamma *} | l_1 m_1 \rangle$$
$$\times \langle l_2 m_2 | Y_b^{\beta *} Y_d^{\delta *} | l_2 m_2 \rangle \tag{39}$$

simply by squaring eqn. (8). Since our interest is in the ground states of atoms, for which l_1, m_1, l_2, m_2 are zero, we require only

$$\langle 00 | Y_a^{\alpha *} Y_c^{\gamma *} | 00 \rangle = \frac{(-1)^\alpha}{4\pi} \delta_{a,c} \delta_{\alpha,-\gamma},$$

the radial matrix element

$$\langle 10 | r^s | 10 \rangle = \frac{(S+2)!}{2(2Z^*/a_0)^s}$$

and eqn. (8a) for the evaluation of (39). The result, written again in atomic units, is

$$\Delta E = -\sum_{a,b,\alpha} \frac{2(a+b)!(2a+2)!(2b+2)!}{(2R)^{2a+2b+2}(2a+1)(a+\alpha)!(a+\alpha)!(b-\alpha)!}$$
$$\times \frac{1}{(Z_1^*)^{2a}(Z_2^*)^{2b}[Z_1^{*2}+Z_2^{*2}]}. \tag{40a}$$

The value used here for $\bar{E}$ is the sum of the ionization potentials of two atoms having nuclear charges Z_1^* and Z_2^*, namely $(Z_1^{*2} + Z_2^{*2})e^2/2a_0$.

The first few terms of (40a) are

$$\Delta E = -\frac{2}{(Z_1^{*2}+Z_2^{*2})} \left[\frac{6}{R^6 Z_1^{*2} Z_2^{*2}} + \frac{135}{2R^8 Z_1^{*2} Z_2^{*4}} + \frac{135}{2R^8 Z_1^{*4} Z_2^{*2}} \right.$$
$$\left. + \frac{2835}{2R^{10} Z_1^{*4} Z_2^{*4}} + \frac{1260}{R^{10} Z_1^{*2} Z_2^{*6}} + \frac{1260}{R^{10} Z_1^{*6} Z_2^{*2}} + \ldots \right]. \tag{40}$$

All results in this section are based on second-order perturbation theory. Higher-order theory makes no contribution to the terms up to R^{-10}, the first modification appearing with a term proportional to R^{-11} which is introduced in third order (see section 2.5).

2.3. Asymmetric Molecules

Asymmetric molecules differ from those treated thus far in having different polarizabilities in different directions. We now deal with cases for which there exists an axis of symmetry, i.e. molecules for which the polarizability along the axis has a value different from that perpendicular to it, but all "perpendicular" values are the same. We designate the relevant polarizabilities by $P_\perp$ and $P_\parallel$.

The simplest model with some claim to realism is once more the harmonic oscillator; to make it fit the present situation one has to keep it anisotropic and assign to it different binding constants, our former γ's, in different directions. Its energy eigenfunctions are products of three factors, one for each Cartesian coordinate; the one for x reads

$$N_n H_n(\gamma_x^{1/2} x) e^{-1/2\gamma_x x^2}, \qquad N_n = \left(\frac{\gamma_x^{1/2}}{\pi^{1/2} 2^n n!}\right)^{1/2}. \tag{41}$$

First, only the dipole–dipole term of V will be considered. It must be written in the form (9a), but with unspecified charges q_1 and q_2. Let us also introduce at once two Cartesian systems with unit vectors $\boldsymbol{u}_1^1$, $\boldsymbol{u}_1^2$, $\boldsymbol{u}_1^3$ extending along $x_1' \equiv x_1$, $x_1^2 \equiv y_1$ and $x_1^3 \equiv z_1$, which are the coordinates of q_1, and $\boldsymbol{u}_2^1$, $\boldsymbol{u}_2^2$, $\boldsymbol{u}_2^3$ extending along x_2^1, x_2^2 and x_2^3. We now have single charges, q_1 and q_2 and the summations in the formulas which follow extend over the three space coordinates. From (9a),

$$\begin{aligned} V &= \frac{q_1 q_2}{R^3} \sum_{i,j=1}^{3} x_1^i x_2^j \left(\boldsymbol{u}_1^i \cdot \boldsymbol{u}_2^j - \frac{3\boldsymbol{u}_1^i \cdot \boldsymbol{R}\, \boldsymbol{u}_2^j \cdot \boldsymbol{R}}{R^2}\right) \\ &= \frac{q_1 q_2}{R^3} \sum_{ij} C_{ij} x_1^i x_2^j, \end{aligned} \tag{42}$$

where

$$C_{ij} = \boldsymbol{u}_1^i \cdot \boldsymbol{u}_2^j - \frac{3\boldsymbol{u}_1^i \cdot \boldsymbol{R}\, \boldsymbol{u}_2^j \cdot \boldsymbol{R}}{R^2}. \tag{43}$$

Now the dipole moment $q_1 \boldsymbol{r}_1$ can be replaced by the sum of its three components, $q_1 x_1' \boldsymbol{u}_1$, $q_1 x_1^2 \boldsymbol{u}_1^2$ and $q_1 x_1^3 \boldsymbol{u}_1^3$. Upon such decomposition we are dealing with three charges replacing molecule 1 and three charges constituting molecule 2; eqn. (43) describes their interaction and we shall consider them bound to their centers by harmonic forces of different strength, along different axes, in accordance with formula (41).

Thought must now be given to determining convenient orientations of the coordinate systems $\boldsymbol{u}^1$ and $\boldsymbol{u}^2$. The following choices are probably the simplest, even though the arrangement is difficult to draw.

Take the z_1 axis, i.e. the vector $\boldsymbol{u}_1^3$, along the symmetry axis of molecule 1 and $\boldsymbol{u}_2^3$ along the axis of 2. Denote by θ_1 and θ_2 the angles between the z-axes and $\boldsymbol{R}$.

Let $\boldsymbol{u}_1^1$ (x_1 axis) be coplanar with $\boldsymbol{R}$ and $\boldsymbol{u}_1^3$.

Let $\boldsymbol{u}_2^1$ (x_2 axis) be coplanar with $\boldsymbol{R}$ and $\boldsymbol{u}_2^3$.

Let $\boldsymbol{u}_1^2$ complete a right-handed coordinate system with $\boldsymbol{u}_1^1$ and $\boldsymbol{u}_1^3$, and let $\boldsymbol{u}_2^2$ complete a right-handed coordinate system with $\boldsymbol{u}_2^1$ and $\boldsymbol{u}_2^3$.

We project z_1 and z_2 upon a plane perpendicular to $\boldsymbol{R}$ and call the angle between the projections φ. Then

$$\begin{aligned}
C_{11} &= \cos\theta_1 \cos\theta_2 \cos\varphi - 2\sin\theta_1 \sin\theta_2,\\
C_{12} &= \cos\theta_1 \sin\varphi,\\
C_{13} &= \cos\theta_1 \sin\theta_2 \cos\varphi + 2\sin\theta_1 \cos\theta_2,\\
C_{21} &= \cos\theta_2 \sin\varphi,\\
C_{22} &= \cos\varphi,\\
C_{23} &= \sin\theta_2 \sin\varphi,\\
C_{31} &= \sin\theta_1 \cos\theta_2 \cos\varphi + 2\cos\theta_1 \sin\theta_2,\\
C_{32} &= -\sin\theta_1 \sin\varphi,\\
C_{33} &= \sin\theta_1 \sin\theta_2 \cos\varphi - 2\cos\theta_1 \cos\theta_2.
\end{aligned} \tag{44}$$

Stiffness and frequency parameters will be labeled $\gamma_\perp$ and $\nu_\perp$ for motion along the x- and y-axes, $\gamma_\parallel$ and $\nu_\parallel$ for motion along the axis of symmetry z. Each of these will carry a further subscript to designate the molecule to which it refers.

The second-order energy based on (42) is

$$\Delta E = -\left(\frac{q_1 q_2}{R^3}\right)^2 \sum_{i,j} K_{ij}$$

provided

$$K_{ij} = \sum_{\{n_1\}\{n_2\}} \frac{|(x_1^i)_{0\{n_1\}}|^2 \, |(x_2^j)_{0\{n_2\}}|^2 C_{ij}^2}{E_1\{n_1\} + E_2\{n_2\} - E_1\{0\} - E_2\{0\}}, \tag{45}$$

and $\{n_1\}$ denotes a set of oscillator quantum numbers n_1^1, n_1^2, n_1^3 pertaining to molecule 1, etc. For a single oscillator, the only non-vanishing matrix elements are $x_{01} = (2\gamma)^{-1/2}$. One is then led at once to the following table of coefficients:

$$\begin{aligned}
K_{11} &= C_{11}^2\ (\perp, \perp),\\
K_{12} &= C_{12}^2\ (\perp, \perp),\\
K_{13} &= C_{13}^2\ (\perp, \|),\\
K_{21} &= C_{21}^2\ (\perp, \perp),\\
K_{22} &= C_{22}^2\ (\perp, \perp),\\
K_{23} &= C_{23}^2\ (\perp, \|),\\
K_{31} &= C_{31}^2\ (\|, \perp),\\
K_{32} &= C_{32}^2\ (\|, \perp),\\
K_{33} &= C_{33}^2\ (\|, \|).
\end{aligned} \tag{46}$$

The symbols written here are abbreviations signifying

$$\begin{aligned}
(\perp, \perp) &\equiv \tfrac{1}{4}\,(h\nu_{1\perp} + h\nu_{2\perp})^{-1}(\gamma_{1\perp}\gamma_{2\perp})^{-1},\\
(\|, \perp) &\equiv \tfrac{1}{4}\,(h\nu_{1\|} + h\nu_{2\perp})^{-1}(\gamma_{1\|}\gamma_{2\perp})^{-1}, \quad \text{etc.}
\end{aligned} \tag{47}$$

On insertion of (46) and (47) in (45) there results

$$\Delta E = -\left(\frac{q_1q_2}{R^3}\right)^2 \{(C_{11}^2 + C_{12}^2 + C_{21}^2 + C_{22}^2)(\perp, \perp) + (C_{13}^2 + C_{23}^2)(\perp, \|) + (C_{31}^2 + C_{32}^2)(\|, \perp) + C_{33}^2(\|, \|)\}.$$

If we now substitute from (44) and define

$$\begin{aligned}
A &= (q_1q_2)^2\,(\|, \|), \qquad & B &= (q_1q_2)^2\,(\|, \perp),\\
C &= (q_1q_2)^2\,(\perp, \|), \qquad & D &= (q_1q_2)^2\,(\perp, \perp),
\end{aligned} \tag{48}$$

we find

$$\Delta E = -\frac{1}{R^6}[(A - B - C + D)(\sin\theta_1 \sin\theta_2 \cos\varphi - 2\cos\theta_1\cos\theta_2)^2 + 3(B - D)\cos^2\theta_1 + 3(C - D)\cos^2\theta_2 + B + C + 4D]. \tag{49}$$

This result can be expressed in terms of polarizabilities perpendicular ($P_\perp$) and parallel ($P_\|$) to the molecular axis by relations analogous to (29a),

$$\gamma_\perp = \frac{q^2}{P_\perp h\nu_\perp}, \qquad \gamma_\| = \frac{q^2}{P_\| h\nu_\|}, \tag{50}$$

so that, for example,

$$A = \frac{h}{4} P_{1,\parallel} P_{2,\parallel} \frac{\nu_{1,\parallel}\nu_{2,\parallel}}{\nu_{1,\parallel} + \nu_{2,\parallel}},$$

$$B = \frac{h}{4} P_{1,\parallel} P_{2,\perp} \frac{\nu_{1,\parallel}\nu_{2,\perp}}{\nu_{1,\parallel} + \nu_{2,\perp}}, \quad \text{etc.} \tag{51}$$

If the molecules are identical, all subscripts 1 and 2 may be dropped in the definitions of A, B, C, and D, and $C = B$. Then

$$\Delta E = -\frac{1}{R^6}[(A-2B+D)(\sin\theta_1 \sin\theta_2 \cos\varphi - 2\cos\theta_1 \cos\theta_2)^2 + 3(B-D)(\cos^2\theta_1 + \cos^2\theta_2) + 2B + 4D] \tag{52}$$

(molecules identical).

If the molecules are isotropic $A = B = C = D$; the dependence on angle disappears from (52) and we recover formula (37a).

If one of the molecules is isotropic (or an atom), eqn. (49) can be reduced as follows. We notice first that, if both are isotropic, $\Delta E = -(2C+4D)R^{-6}$. From this we conclude that our former coefficient $C_1 = 2C+4D$. Now let only molecule 1 be isotropic. Then $A = C$ and $B = D$, from (48). Hence

$$\begin{aligned} \Delta E &= -\frac{1}{R^6}[3(C-D)\cos^2\theta_2 + C + 5D] \\ &= -\frac{1}{R^6}[(C-D)(3\cos^2\theta_2 - 1) + 2C + 4D] \\ &= -\frac{C_1}{R^6}\left[1 + \frac{2C-2D}{2C+4D} P_2(\cos\theta_2)\right], \end{aligned} \tag{53}$$

where P_2 is the Legendre polynomial of second degree, $3/2 \cos^2\theta - 1/2$.

Equation (49) was published (without detailed derivation) by London (1942) (see also Salem (1960). Special cases of it are considered by de Boer and Heller (1937)). The present treatment is contained in a paper by A. van der Merwe (1966), who also calculated the dipole–quadrupole part of the interaction between asymmetric molecules.

This is found by the same procedure. From (9a)

$$V \equiv V_{d-q} = \frac{3}{2}\,\frac{q_1 q_2}{R^4} \sum_{ij} \left\{ (x_1^i)^2 x_2^j\, \boldsymbol{u}_2^j \cdot \frac{\boldsymbol{R}}{R} - (x_2^i)^2 x_1^j\, \boldsymbol{u}_1^j \cdot \frac{\boldsymbol{R}}{R} \right. \tag{53a}$$

$$\left. + \left(\sum_h x_1^k\, \boldsymbol{u}_1^k \cdot \frac{\boldsymbol{R}}{R} - \sum_k x_2^k\, \boldsymbol{u}_2^k \cdot \frac{\boldsymbol{R}}{R} \right) \left(2x_1^i x_2^j\, \boldsymbol{u}_1^i \cdot \boldsymbol{u}_2^j - \frac{5}{R^2}\, x_1^i x_2^j\, \boldsymbol{u}_1^i \cdot \boldsymbol{R}\, \boldsymbol{u}_2^j \cdot \boldsymbol{R} \right) \right\}.$$

We choose axes as before. In this case, the surviving matrix elements are $x_{01} = (2\gamma)^{-1/2}$, $(x^2)_{00} = (2\gamma)^{-1}$, $(x^2)_{02} = (\sqrt{2}\beta)^{-1}$. In place of (44) we encounter a more cumbersome set of trigonometric expressions. They are:

$$\begin{aligned}
A_{11} &= -\sin\theta_2 - 2\cos\theta_1\cos\theta_2\sin\theta_1\cos\varphi + 3\sin^2\theta_1\sin\theta_2 \\
B_{11} &= \sin\theta_1 + 2\cos\theta_1\cos\theta_2\sin\theta_2\cos\varphi - 3\sin^2\theta_2\sin\theta_1 \\
C_{11} &= 2\cos^2\theta_1\cos\theta_2\cos\varphi - 3\sin\theta_1\cos\theta_1\sin\theta_2 \\
D_{11} &= -2\cos\theta_1\cos^2\theta_2\cos\varphi + 3\sin\theta_1\sin\theta_2\cos\theta_2
\end{aligned} \tag{54}$$

$$\begin{aligned}
A_{12} &= 2\cos\theta_1\sin\theta_1\sin\varphi \\
B_{12} &= -2\cos^2\theta_1\sin\varphi \\
C_{12} &= 2\cos\theta_1\cos\theta_2\sin\varphi \\
D_{12} &= -2\cos\theta_1\sin\theta_2\sin\varphi
\end{aligned}$$

$$\begin{aligned}
A_{13} &= \cos\theta_2 - 2\cos\theta_1\sin\theta_2\sin\theta_1\cos\varphi - 3\sin^2\theta_1\cos\theta_2, \\
B_{13} &= -\cos\theta_1, \\
C_{13} &= 2\cos\theta_1\sin^2\theta_2\cos\varphi + 3\sin\theta_1\cos\theta_2\sin\theta_2, \\
D_{13} &= 2\cos^2\theta_1\sin\theta_2\cos\varphi + 3\sin\theta_1\cos\theta_1\cos\theta_2, \\
E_{13} &= -2\cos\theta_1\sin\theta_2\cos\theta_2\cos\varphi - 3\sin\theta_1\cos^2\theta_2
\end{aligned}$$

$$\begin{aligned}
A_{21} &= -\sin\theta_2 \\
B_{21} &= \sin\theta_1 \\
C_{21} &= -2\cos\theta_2\sin\theta_1\sin\varphi \\
D_{21} &= 2\cos\theta_2\sin\theta_2\sin\varphi \\
E_{21} &= 2\cos\theta_1\cos\theta_2\sin\varphi \\
F_{21} &= -2\cos^2\theta_2\sin\varphi
\end{aligned}$$

$$A_{22} = -2\sin\theta_1\cos\varphi$$
$$B_{22} = 2\sin\theta_2\cos\varphi$$
$$C_{22} = 2\cos\theta_1\cos\varphi$$
$$D_{22} = -2\cos\theta_2\cos\varphi$$

$$A_{23} = \cos\theta_2$$
$$B_{23} = -\cos\theta_1$$
$$C_{23} = -2\sin\theta_2\sin\theta_1\sin\varphi$$
$$D_{23} = 2\sin^2\theta_2\sin\varphi$$
$$E_{23} = 2\sin\theta_2\cos\theta_1\sin\varphi$$
$$F_{23} = -2\sin\theta_2\cos\theta_2\sin\varphi$$

$$A_{31} = -\sin\theta_2+2\sin\theta_1\cos\theta_1\cos\theta_2\cos\varphi+3\cos^2\theta_1\sin\theta_2$$
$$B_{31} = \sin\theta_1$$
$$C_{31} = -2\sin^2\theta_1\cos\theta_2\cos\varphi-3\cos\theta_1\sin\theta_2\sin\theta_1$$
$$D_{31} = 2\sin\theta_1\cos\theta_2\sin\theta_2\cos\varphi+3\cos\theta_1\sin^2\theta_2$$
$$E_{31} = -2\sin\theta_1\cos^2\theta_2\cos\varphi-3\cos\theta_1\sin\theta_2\cos\theta_2$$

$$A_{32} = 2\sin^2\theta_1\sin\varphi$$
$$B_{32} = -2\sin\theta_1\sin\theta_2\sin\varphi$$
$$C_{32} = -2\sin\theta_1\cos\theta_1\sin\varphi$$
$$D_{32} = 2\sin\theta_1\cos\theta_2\sin\varphi$$

$$A_{33} = \cos\theta_2+2\sin\theta_1\cos\theta_1\sin\theta_2\cos\varphi-3\cos^2\theta_1\cos\theta_2$$
$$B_{33} = -\cos\theta_1-2\sin\theta_1\sin\theta_2\cos\theta_2\cos\varphi+3\cos\theta_1\cos^2\theta_2$$
$$C_{33} = -2\sin^2\theta_1\sin\theta_2\cos\varphi+3\cos\theta_1\cos\theta_2\sin\theta_1$$
$$D_{33} = 2\sin\theta_1\sin^2\theta_2\cos\varphi-3\cos\theta_1\cos\theta_2\sin\theta_2.$$

At this stage the result is

$$-\Delta E_{dq} = \left(\frac{3}{4}\frac{q_1q_2}{R^4}\right)^2\Bigg\{(2\gamma_{1,\perp}h\nu_{1,\perp})^{-1}[(B_{11}+B_{21})^2 + D_{21}^2]\left(\frac{1}{\gamma_{2,\perp}}-\frac{1}{\gamma_{2,\parallel}}\right)^2 + (2\gamma_{1,\parallel}h\nu_{1,\parallel})^{-1}B_{33}^2\left(\frac{1}{\gamma_{2,\perp}}-\frac{1}{\gamma_{2,\parallel}}\right)^2 + [2\gamma_{1,\perp}\gamma_{2,\perp}^2h(\nu_{1,\perp} + 2\nu_{2,\perp})]^{-1}(D_{12}^2+2B_{11}^2+2B_{21}^2+B_{22}^2+2D_{21}^2)$$

$$+[\gamma_{1,\perp}\gamma_{2,\parallel}^2 h(\nu_{1,\perp}+2\nu_{2,\parallel})]^{-1}[(E_{13}+B_{31})^2+F_{13}^2]$$
$$+[2\gamma_{1,\parallel}\gamma_{2,\perp}\gamma_{2,\parallel}h(\nu_{1,\perp}+\nu_{2,\perp}+\nu_{2,\parallel})]^{-1}[C_{12}^2$$
$$+(D_{11}+C_{13})^2+D_{22}^2+(F_{21}+D_{23})^2]+[2\gamma_{1,\parallel}\gamma_{2,\perp}^2 h(\nu_{1,\parallel}$$
$$+2\nu_{2,\perp})]^{-1}[B_{32}^2+2(B_{13}+D_{31})^2+2B_{23}^2]$$
$$+[\gamma_{1,\parallel}\gamma_{2,\parallel}^2 h(\nu_{1,\parallel}+2\nu_{2,\parallel})]^{-1}B_{33}^2$$
$$+[2\gamma_{1,\parallel}\gamma_{2,\perp}\gamma_{2,\parallel}h(\nu_{1,\parallel}+\nu_{2,\perp}+\nu_{2,\parallel})]^{-1}[D_{32}^2$$
$$\left.+(E_{31}+D_{33})^2]+\text{“transpose”}\right\}. \qquad (55)$$

By "transpose" we mean the entire preceding expression with the subscripts 1 and 2 transposed.

For isotropic molecules, one finds that

$$B_{11}+B_{21}+D_{13}+B_{31} = D_{21}+F_{23} = B_{13}+B_{23}+D_{31}+B_{33} = 0,$$

and one obtains the relevant parts of (36); if the molecules are identical use can be made of (29a) with the result that

$$\Delta E_{d-q} = -\frac{15}{4}\frac{P^3(h\nu)^2}{f e^2 R^8}.$$

Van der Merwe has also found that (55) takes on a somewhat more compact form if it is written in terms of double angles,

$$\Theta_1 = 2\theta_1, \qquad \Theta_2 = 2\theta_2.$$

We may then define the smaller set of expressions

$$F_0 = \tfrac{1}{8}(9-\cos\Theta_1+6\cos\Theta_2+\cos^2\Theta_2-6\cos\Theta_1\cos\Theta_2$$
$$-9\cos\Theta_1\cos^2\Theta_2)$$
$$G_0 = \tfrac{1}{4}(17-\cos\Theta_1-10\cos\Theta_2+\cos^2\Theta_2+2\cos\Theta_1\cos\Theta_2$$
$$-9\cos\Theta_1\cos^2\Theta_2)$$
$$H_0 = \tfrac{1}{2}(11-7\cos\Theta_1+2\cos\Theta_2-\cos^2\Theta_2+2\cos\Theta_1\cos\Theta_2$$
$$+9\cos\Theta_1\cos^2\Theta_2)$$
$$F_1 = \tfrac{1}{2}\sin\Theta_1\sin\Theta_2(1+3\cos\Theta_2)$$
$$G_1 = \sin\Theta_1\sin\Theta_2(1-3\cos\Theta_2)$$
$$H_1 = 6\sin\Theta_1\sin\Theta_2\cos\Theta_2$$
$$F_2 = \tfrac{1}{2}(1-\cos\Theta_1)\sin^2\Theta_2$$
$$G_2 = \cos\Theta_2(1-\cos\Theta_1)(1-\cos\Theta_2)$$
$$H_2 = (1-\cos\Theta_1)(1-\cos\Theta_2)(1+2\cos\Theta_2). \qquad (56)$$

The result now reads

$$
\begin{aligned}
-\Delta E_{d-q}
&= \left(\frac{3}{4}\frac{q_1q_2}{R^4}\right)^2\left\{\frac{1}{2\gamma_{1,\perp}h\nu_{1,\perp}}\left(\frac{1}{\gamma_{2,\perp}}-\frac{1}{\gamma_{2,\parallel}}\right)(F_0+F_1\cos\varphi-F_2\cos^2\varphi)\right.\\
&+\frac{1}{8\gamma_{1,\parallel}}\frac{1}{h\nu_{1,\parallel}}\left(\frac{1}{\gamma_{2,\perp}}-\frac{1}{\gamma_{2,\parallel}}\right)\left[\frac{1}{4}(5+6\cos\Theta_2+5\cos^2\Theta_2)\right.\\
&\qquad\left.-F_0-F_1\cos\varphi+F_2\cos^2\varphi\right]\\
&+\frac{1}{8\gamma_{1,\perp}\gamma_{2,\perp}^2}\frac{1}{h(\nu_{1,\perp}+2\nu_{2,\perp})}(G_0-G_1\cos\varphi-G_2\cos^2\varphi)\\
&+\frac{1}{8\gamma_{1,\parallel}\gamma_{2,\perp}^2}\frac{1}{h(\nu_{1,\parallel}+2\nu_{2,\perp})}\left[\frac{1}{2}(13-10\cos\Theta_2+5\cos^2\Theta_2)G_0\right.\\
&\qquad\left.+G_1\cos\varphi+G_2\cos^2\varphi\right]\\
&+\frac{1}{8\gamma_{1,\perp}\gamma_{2,\parallel}^2}\frac{1}{h(\nu_{1,\perp}+2\nu_{2,\parallel})}(2F_0+2F_1\cos\varphi-2F_2\cos^2\varphi)\\
&+\frac{1}{8\gamma_{1,\parallel}\gamma_{2,\parallel}^2}\frac{1}{h(\nu_{1,\parallel}+2\nu_{2,\parallel})}\left[\frac{1}{2}(5+6\cos\Theta_2+5\cos^2\Theta_2)\right.\\
&\qquad\left.-2F_0-2F_1\cos\varphi+2F_2\cos^2\varphi\right]\\
&+\frac{1}{8\gamma_{1,\perp}\gamma_{2,\perp}\gamma_{2,\parallel}}\frac{1}{h(\nu_{1,\perp}+\nu_{2,\perp}+\nu_{2,\parallel})}(H_0-H_1\cos\varphi+H_2\cos^2\varphi)\\
&+\frac{1}{8\gamma_{1,\perp}\gamma_{2,\perp}\gamma_{2,\parallel}}\frac{1}{h(\nu_{1,\parallel}+\nu_{2,\perp}+\nu_{2,\parallel})}\left[(11+2\cos\Theta_2-5\cos^2\Theta_2)\right.\\
&\qquad\left.\left.-H_0+H_1\cos\varphi-H_2\cos^2\varphi\right]\right\}+\text{transpose.}
\end{aligned}
\tag{57}
$$

To make these general formulas more meaningful we consider a few special orientations of the molecules. Furthermore, we take them to be similar, so that the subscripts 1 and 2 become unnecessary. In writing them we shall also eliminate the γ's and replace them by polarizabilities P via relations (50). Equation (57) then reduces as follows:

Case A

$$\theta_1 = \theta_2 = \frac{\pi}{2}, \qquad \varphi = 0$$

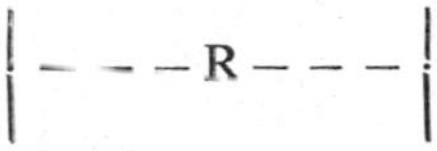

$$\Delta E_{d-d} = -\frac{h}{8R^6}\left(P_\parallel^2 \nu_\parallel + 5P_\perp^2 \nu_\perp\right) \tag{58'}$$

$$\Delta E_{d-q} = -\left(\frac{3h}{2qR^4}\right)^2 \left\{\frac{1}{4} P_\perp (P_\perp \nu_\perp - P_\parallel \nu_\parallel)^2 + \frac{7}{6} P_\perp^3 \nu_\perp^2 + \frac{3}{2}\,\frac{P_\perp \nu_\perp (P_\parallel \nu_\parallel)^2}{\nu_\perp + 2\nu_\parallel}\right\} \tag{58''}$$

Case B

$$\theta_1 = \frac{\pi}{2}, \qquad \theta_2 = 0$$

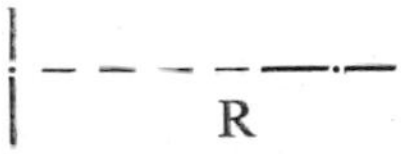

$$\Delta E_{d-d} = -\frac{h}{8R^6}\left(\frac{10 P_\parallel P_\perp \nu_\parallel \nu_\perp}{\nu_\parallel + \nu_\perp} + P_\perp^2 \nu_\perp\right) \tag{59'}$$

$$\Delta E_{d-q} = -\left(\frac{3h}{2qR^4}\right)^2 \left\{P_\perp (P_\perp \nu_\perp - P_\parallel \nu_\parallel)^2 + \frac{1}{3} P_\perp^3 \nu_\perp^2 + \frac{3 P_\perp \nu_\perp (P_\parallel \nu_\parallel)^2}{\nu_\perp + 2\nu_\parallel} + \frac{P_\parallel \nu_\parallel (P_\perp \nu_\perp)^2}{\nu_\parallel + 2\nu_\perp}\right\} \tag{59''}$$

Case C

$$\theta_1 = \theta_2 = 0$$

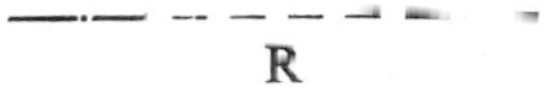

$$\Delta E_{d-d} = -\frac{h}{8R^6}\left(4P_\parallel^2 \nu_\parallel + 2P_\perp^2 \nu_\perp\right) \tag{60'}$$

$$\Delta E_{d-q} = -\left(\frac{3h}{2qR^4}\right)^2 \left\{P_\parallel (P_\perp \nu_\perp - P_\parallel \nu_\parallel)^2 + 3\,\frac{P_\parallel \nu_\parallel (P_\perp \nu_\perp)^2}{\nu_\parallel + 2\nu_\perp} + \frac{2}{3} P_\parallel^3 \nu_\parallel^3\right\} \tag{60''}$$

Case D

$$\theta_1 = \theta_2 = \frac{\pi}{2}, \quad \varphi = \frac{\pi}{2}.$$

$$\Delta E_{d-d} = -\frac{h}{8R^6}\left(\frac{4 P_\parallel P_\perp \nu_\parallel \nu_\perp}{\nu_\perp + \nu_\parallel} + 4P_\perp^2 \nu_\perp\right) \tag{61'}$$

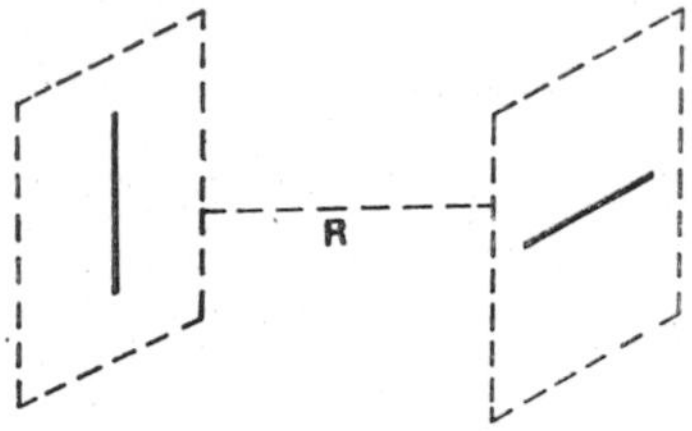

FIG. 1. Illustrating case D.

$$\Delta E_{d-q} = -\left(\frac{3h}{2qR^4}\right)^2 \left\{\frac{9}{32} P_\perp (P_\perp \nu_\perp - P_\parallel \nu_\parallel)^2 + \frac{1}{32} P_\parallel (P_\perp \nu_\perp - P_\parallel \nu_\parallel)^2 \right. \tag{61''}$$

$$\left. + \frac{17}{48} P_\perp^3 \nu_\perp^2 + \frac{31}{16} \frac{P_\perp \nu_\perp (P_\parallel \nu_\parallel)^2}{\nu_\perp + 2\nu_\parallel} + \frac{31}{16} \frac{P_\parallel \nu_\parallel (P_\perp \nu_\perp)^2}{\nu_\parallel + 2\nu_\perp} + \frac{1}{48} P_\parallel^3 \nu_\parallel^2 \right\}.$$

If all orientations are given equal weights (i.e., $\overline{\Delta E} = (4\pi)^{-1} \times \int \Delta E \sin\theta d\theta d\varphi$) we obtain

$$\overline{\Delta E}_{d-d} = -\frac{h}{12R^6}\left(P_\parallel{}^2 \nu_\parallel + 8P_\parallel P_\perp \frac{\nu_\parallel \nu_\perp}{\nu_\parallel + \nu_\perp} + 4P_\perp^2 \nu_\perp\right) \tag{62'}$$

$$\overline{\Delta E}_{d-q} = -\left(\frac{3h}{2qR^4}\right)^2 \left\{\frac{21}{40} P_\perp (P_\perp \nu_\perp - P_\parallel \nu_\parallel)^2 \right.$$

$$+ \frac{47}{240} P_\parallel (P_\perp \nu_\perp - P_\parallel \nu_\parallel)^2 + \frac{23}{60} P_\perp^3 \nu_\perp^2 + \frac{21}{40} \frac{P_\perp \nu_\perp (P_\parallel \nu_\parallel)^2}{\nu_\perp + 2\nu_\parallel}$$

$$\left. + \frac{59}{30} \frac{P_\parallel \nu_\parallel (P_\perp \nu_\perp)^2}{\nu_\parallel + 2\nu_\perp} + \frac{163}{360} P_\parallel^3 \nu_\parallel^2 \right\}. \tag{62''}$$

These formulas can be written in still different ways. For a single oscillating charge one has

$$P_\perp \nu_\perp^2 = P_\parallel \nu_\parallel^2 = \left(\frac{q}{2\pi}\right)^2 \frac{1}{m} = P_0 \nu_0^2. \tag{63}$$

With these replacements, (58′), (59′), and (60′) become identical with the corresponding expressions of de Boer and Heller (1937).

Finally, all results can be expressed in terms of f-values by virtue of the relation

$$P_0 \nu_0^2 = \left(\frac{e}{2\pi}\right)^2 \frac{f}{m}. \tag{64}$$

When this is done the formulas appear in a form similar to those previously obtained with the oscillator model (Margenau, 1938; Hornig and Hirschfelder, 1952); for example in case A,

$$\Delta E_{d-d} = -\frac{1}{8R^6}\hbar e\left(\frac{f}{m}\right)^{1/2}(P_{\parallel}^{3/2}+5P_{\perp}^{3/2}), \quad \text{etc.},$$

and the ΔE_{d-q}'s do not depend on f at all.

Some numerical data will exemplify the magnitude of the asymmetry effects. Let r_{d-d} be the ratio of ΔE_{d-d} for a pair of asymmetric molecules (in the positions indicated in Table 2.3) to ΔE_{d-d} for a symmetric molecule ($P_{\parallel} = P_{\perp}$). Similarly, r_{d-q} is the ratio of ΔE_{d-q} (asymmetric) to a ΔE_{d-q} (symmetric). These are given in Table 2.3.

TABLE 2.3

	r_{d-d}				r_{d-q}			
	Case A	B	C	D	A	B	C	D
Cl_2	0.87	1.02	1.37	0.83	0.85	1.16	1.45	1.05
H_2	1.07	0.96	0.90	1.06	1.07	0.94	0.87	0.96
HCl	0.93	1.03	1.15	0.93	0.93	1.08	1.17	1.03
HBr	0.95	1.03	1.14	0.94	0.94	1.07	1.15	1.03
HI	0.93	1.03	1.15	0.92	0.92	1.08	1.19	1.03
N_2	0.88	1.02	1.32	0.84	0.87	1.15	1.38	1.05
O_2	0.85	1.00	1.46	0.79	0.82	1.20	1.57	1.05

The values of $P_{\parallel}$ and $P_{\perp}$ from which the table was computed are those given by Denbigh (1940).

The simple molecules listed in Table 2.3 have relatively weak asymmetries, the polarizabilities rarely differing by as much as a factor 2. Large organic molecules are expected to exhibit greater differences with orientation than those here recorded. Nevertheless, the variation listed in the Table for these simple cases is not negligible, and it may be noted that the asymmetry effects are somewhat more pronounced in the dipole–quadrupole interaction. The behavior of H_2 is anomalous because for this molecule $P_{\perp} > P_{\parallel}$.

Some of the entries in Table 2.3 have permanent dipole moments. For these the dispersion energies listed are overshadowed by static interactions. Terms proportional to R^{-10} have not been calculated; a survey of known cases would indicate that they are of lesser importance. An application of the results to concrete observations like crystal energies and virial coefficients is hampered, of course, by our ignorance of the short-range repulsive forces.

2.4. Variational Theory of Long-range Forces

Variational theory is based on the Ritz theorem according to which the quantity $\bar{H} \equiv \int \varphi^* H \varphi \, d\tau / \int \varphi^* \varphi \, d\tau$ is never smaller than the eigenvalue E of the Schrödinger equation $H\psi = E\psi$ for any function φ which has the proper symmetry and satisfies the boundary conditions to be imposed on ψ. If, then, we take for φ a function with arbitrary parameters, compute $\bar{H}$ and then vary the parameters to make $\bar{H}$ a minimum, it is expected that this minimum is close to E.

In our problem, we assume that

$$H = H_0 + V, \tag{65}$$

$$H_0\psi_0 = E_0\psi_0, \tag{66}$$

and we take as variation function

$$\varphi = \psi_0(1+v)$$

leaving v, which will turn out to be a real function, for the present unspecified. However, it is supposed to be small, so that v^2 can be neglected against v. In that case, if ψ_0 is normalized to 1,

$$\int \varphi^* \varphi d\tau = 1 + 2\int \psi_0^* v \psi_0 d\tau = 1 + 2v_{00}. \tag{67}$$

We proceed to calculate $\bar{H}$.

It is convenient to split H into kinetic and potential energy, the latter including the perturbation V. Thus

$$H = -\frac{\hbar^2}{2}\sum_i \frac{\nabla_i^2}{m_i} + U = \sum_i T_i + U. \tag{68}$$

Let us first compute

$$\overline{\nabla_i^2} = \int \varphi^* \nabla_i^2 \varphi \, d\tau = \int \psi_0^*(1+v)^2 \nabla_i^2 \psi_0 \, d\tau + \int \psi_0^*(1+v)[2\nabla_i v \cdot \nabla_i \psi_0 + \psi_0 \nabla_i^2 v] \, d\tau.$$

The last integral can be modified. We shall be dealing with real functions ψ_0. A partial integration, in which the integrated part vanishes, establishes the identity

$$\int \psi_0^* \psi_0 (\nabla v)^2 \, d\tau = \int \psi_0^2 [\nabla (1+v)]^2 \, d\tau = -\int (1+v) \nabla \cdot [\psi_0^2 \nabla (1+v)] \, d\tau$$
$$= -\int \psi_0 (1+v) [2\nabla v \cdot \nabla \psi_0 + \psi_0 \nabla^2 v] \, d\tau$$

and we find

$$\overline{\nabla_i^2} = \int \psi_0 (1+v)^2 \nabla_i^2 \psi_0 \, d\tau - \int \psi_0^2 (\nabla v)^2 \, d\tau. \tag{69}$$

Thus we may write in place of

$$\int \varphi T_i \varphi d\tau - \frac{\hbar^2}{2m_i} \int \psi_0 (1+v) \nabla_i^2 \psi_0 (1+v) d\tau$$

the expression

$$-\frac{\hbar^2}{2m_i} \int [\psi_0 (1+v^2) \nabla_i^2 \psi_0 \, d\tau - \int \psi_0^2 (\nabla_i v)^2 \, d\tau].$$

In the calculation of $\overline{U}$, the position of the factor U relative to $(1+v)$ is immaterial. Hence we find

$$\int \varphi^* H \varphi \, d\tau = \int \psi_0 (1+v)^2 H \psi_0 \, d\tau + \frac{\hbar^2}{2} \int \psi_0^2 \sum_i \frac{1}{m_i} (\nabla_i v)^2 \, d\tau. \tag{70}$$

We now recall that $H = H_0 + V$. Then in view of (66) and (67), (70) leads to

$$\overline{H} = E_0 + (1+2v_{00})^{-1} \left\{ [(1+v)^2 V]_{00} + \frac{\hbar^2}{2} \sum_i \frac{1}{m_i} [(\nabla_i v)^2]_{00} \right\}$$

and

$$\Delta E = \overline{H} - E_0 = \frac{V_{00} + (2vV)_{00} + \frac{\hbar^2}{2} \sum_i \frac{1}{m_i} [(\nabla_i v)^2]_{00}}{1+2v_{00}} \tag{71}$$

if v^2 is again neglected. In our applications we shall see that $v_{00} = 0$, so that the denominator, too, may be omitted.

Slater and Kirkwood (1931), who first employed the variational method in the problem at hand, related atomic polarizabilities to interatomic forces. Hence, to follow their reasoning, we first derive a formula for the polarizability. This quantity P may be defined as follows. When an atom is placed in a constant electric field of strength F, it suffers a change of internal energy

$$\Delta E = -\tfrac{1}{2} P F^2. \tag{72}$$

If the field extends along the z-axis, and there are N electrons at points $(x_i y_i z_i)$, the potential energy of the atom is

$$V = Fe \sum_{i=1}^{N} z_i. \tag{73}$$

We consider here an atom containing only closed shells of electrons and simplify the problem further by ignoring the contribution of the inner shells to P. This will introduce unknown errors in the case of the heavier rare gases, but the errors are expected to be small because the interior electrons are tightly bound with respect to small perturbing forces.

The function ψ_0, too, will be idealized: we take it to be a product of individual electron functions a and b, each of which has spherical symmetry. As a consequence, $(x_i^2)_{00} = (y_i^2)_{00} = (z_i^2)_{00}$ for every i, and $(x_i)_{00} = (y_i)_{00} = (z_i)_{00} = 0$. For the same reason $V_{00} = 0$.

A suitable form for v is a linear function of the electron coordinates which are involved in V. Hence we take

$$v = \sum_i \lambda_i z_i \tag{74}$$

and treat the λ's as variation parameters to be adjusted for minimum $\overline{H}$. Then

$$(vV)_{00} = Fe \sum_{ij} \lambda_i (z_i z_j)_{00} = Fe \sum_i \lambda_i (z_i^2)_{00}. \tag{75}$$

To see the last result one must remember the meaning of our somewhat peculiar notation; ψ_0 contains the factors $a(i)b(j)$; and $(z_i z_j)_{00}$, when explicitly written, means $\int a(i)b(j)z_i z_j a(i)b(j) \times d\tau_i \, d\tau_j$. This clearly vanishes when $i \neq j$, for it then becomes $(z_i)_{00}(z_j)_{00}$. Notice also that $\partial v/\partial x_i = \partial v/\partial y_i = 0$, $\partial v/\partial z_i = \lambda_i$. Equation (71) therefore reads

$$\Delta E = 2Fe \sum_i \lambda_i (z_i^2)_{00} + \frac{\hbar^2}{2m} \sum_i \lambda_i^2.$$

If this is to be a minimum

$$\lambda_i = -\frac{2mFe}{\hbar^2} (z_i^2)_{00}$$

and the minimum energy itself is

$$\Delta E = -\frac{2me^2F^2}{\hbar^2} (z_i^2)_{00}^2.$$

Comparison with (48) yields

$$P = \frac{4me^2}{\hbar^2} \sum_i (z_i^2)_{00}^2. \tag{76}$$

The equivalence of this formula with (29a) will be recognized when one recalls that, for an oscillator, $(z^2)_{00} = (2\gamma)^{-1}$, and $\gamma = 4\pi^2 \nu m/h$.

Dispersion forces are easily calculated by the same method. Denoting now the interacting atoms by arguments 1 and 2, we must take $\psi_0 = \psi_0(1)\psi_0(2)$ and $V = \sum_{ij} V_{ij}$.

$$V_{ij} = \frac{e^2}{R}(x_{1i}x_{2j}+y_{1i}y_{2j}-2z_{1i}z_{2j}),$$

where i, j label the electrons, whose numbers are, respectively, N_1 and N_2. We confine ourselves in this section to the dipole–dipole interaction. Again, $V_{00} = 0$. In analogy with the previous method, we put

$$v = \sum_{ij} \lambda_{ij}(x_{1i}x_{2j}+y_{1i}y_{2j}-2z_{1i}z_{2i})$$

whence $v_{00} = 0$.

$$(vV)_{00} = \frac{6e^2}{R^3}\sum_{ij}(z_{1i}^2)_{00}(z_{2j}^2)_{00}.$$

This result emerges when use is made of the fact that every $(r_{i1})_{00}$ and $(r_{j2})_{00}$ vanishes. Further,

$$\sum_{i=1}^{N_1}(\nabla_{1i}v)^2 = \sum_{i=1}^{N_1}\sum_{j,k=1}^{N_2}\lambda_{ij}\lambda_{ik}(x_{2j}x_{2k}+y_{2j}y_{2k}+4z_{2j}z_{2k})$$

and the 00-element of this quantity reads

$$\sum_{ij}\lambda_{ij}^2[(x_{2j}^2)_{00}+(y_{2j}^2)_{00}+4(z_{2j}^2)_{00}] = 6\sum_{ij}\lambda_{ij}^2(z_{2j}^2)_{00}.$$

A similar result, with z_{2j} replaced by z_{1j}, holds for $\sum_{i=1}^{N}(\nabla_{2j}v)^2$. Hence (71) reads

$$\Delta E = 12\sum_{ij}\left\{\frac{e^2}{R^3}\lambda_{ij}(z_{1i}^2)_{00}(z_{2j}^2)_{00}+\frac{\hbar^2}{4m}\lambda_{ij}^2[(z_{1i}^2)_{00}+(z_{2j}^2)_{00}]\right\}. \quad (77)$$

If it is to be a minimum,

$$\lambda_{ij} = -\frac{2e^2M}{R^3\hbar^2}\frac{(z_{1i}^2)_{00}(z_{2j}^2)_{00}}{(z_{1i}^2)_{00}+(z_{2j}^2)_{00}},$$

so that the minimum energy is

$$\Delta E = -\frac{12M}{\hbar^2}\frac{e^4}{R^6}\sum_{ij}\frac{(z_{1i}^2)_{00}(z_{2j}^2)_{00}}{(z_{1i}^2)_{00}+(z_{2j}^2)_{00}}. \quad (78)$$

Interesting semi-empirical relations can be established by comparing the formulas for P and $\triangle E$. For electrons in closed shells the $(z_{1i}^2)_{00}$ and $(z_{2j}^2)_{00}$ are independent of i and j. Then, according to (52),

$$P = \frac{4Nme^2}{\hbar^2}(z^2)_{00}^2$$

and (78) becomes

$$\Delta E = -\frac{3}{2R^6}\frac{e\hbar}{(m)^{1/2}}\frac{P_1P_2}{\left(\frac{P_1}{N_1}\right)^{1/2}+\left(\frac{P_2}{N_2}\right)^{1/2}}. \quad (79)$$

This result was obtained by Slater and Kirkwood (1931). Like the London formula, it emphasizes the dependence of the R^{-6} force upon polarizabilities, but it does not involve dispersion f-values.

Kirkwood (1932) and Muller (1936) proposed an easily derivable modification of formula (79) which introduces the diamagnetic susceptibilities

$$\chi = N_0e^2/6mc^2 \sum_i (r_i^2)_{00}, \quad N_0 = \text{Avogadro's number.}$$

Their formula is

$$\Delta E = -\frac{6mc^2}{N_0R^6}\frac{P_1P_2}{P_1/\chi_1+P_2/\chi_2}. \quad (80)$$

To convey an idea of the measure of agreement between these different approaches we list in Table 2.4 the values of the coefficients C_1 of $1/R^6$ in ΔE as obtained by three methods for helium and argon, where quite accurate results are available.

TABLE 2.4. $C_1 \times 10^{60}$ (*erg cm*^{6})

	Oscillator formula	Slater–Kirkwood formula	Probable value
He	1.23	1.49	1.41[a]
Ne	4.70	7.94	6.04
A	55.4	69.5	58.7[b]

[a] Margenau (1939), Dalgarno and Lynn (1957), and Bell (1965).
[b] Stamper (unpublished).

The so-called probable values are based mainly on London's exact formula (27) with a sufficient number of f-values and energy differences inserted. Formula (79) appears to give values which are too high.

There are several modifications and refinements of the variational method as it applies to long-range forces. Perhaps the most important advance has come from a consideration which takes as its starting point eqn. (26), the result of second-order perturbation theory, and converts it into an integral tractable by variational methods.

We illustrate the method for spherical atoms in their lowest states ($r_1 = r_2 = 0$). There (26) permits averaging over all orientations of the atoms and leads to

$$\Delta E = -\frac{6e^4}{R^6}\sum_{s_1 s_2}\frac{|Z_{0s_1}|^2\,|Z_{0s_2}|^2}{E(s_1, 0)+E(s_2, 0)}, \tag{81}$$

where we have written $E(s)-E(0) \equiv E(s, 0)$.

Note first that (81) is equivalent to the integral

$$\Delta E = -\frac{6e^4}{\pi R^6}\sum_{s_1 s_2}\int_{-\infty}^{\infty}\frac{E(s_1, 0)\,|Z_{0s_1}|^2\,E(s_2, 0)\,|Z_{0s_2}|^2}{[E(s_1, 0)^2+u^2][E(s_2, 0)^2+u^2]}\,du. \tag{82}$$

The poles of the integrand of this expression which lie in the positive imaginary half-plane occur at $u = iE(s_1, 0)$ and $u = iE(s_2, 0)$. Hence, by the theorem of residues (the integrand vanishes at $u = i\infty$ and permits closing the contour there)

$$\Delta E = -2\pi i\frac{6e^4}{\pi R^6}\sum_{s_1 s_2}\left\{\frac{E(s_1, 0)\,|Z_{0s_1}|^2\,|Z_{0s_2}|^2}{2i[E^2(s_2, 0)-E^2(s_1, 0)]}\right.$$
$$\left.-\frac{E(s_1, 0)\,|Z_{0s_1}|^2\,|Z_{0s_2}|^2}{2i[E^2(s_2, 0)-E^2(s_1, 0)]}\right\}.$$

Cancellation here leads at once to (81).

Now (82) has an interesting interpretation. The frequency-dependent polarizability of an atom satisfies the well-known dispersion formula (Bates, 1961, vol. 1, p. 280)

$$P(\nu) = 2e^2\sum_s\frac{E(s,0)\,|Z_{0s}|^2}{E(s,0)^2-(h\nu)^2}. \tag{83}$$

On introducing an imaginary frequency $u = ih\nu$ we have

$$P(u) = 2e^2\sum_s\frac{E(s,0)\,|Z_{0s}|^2}{E^2(s,0)+u^2}$$

so that (82) takes the integral form

$$\Delta E = -\frac{3}{\pi R^6}\int_0^\infty P_1(u)P_2(u)\,du. \tag{84}$$

If we knew the polarizabilities as analytic functions of u this result would represent the dispersion part of the interaction exactly. In the absence of such knowledge a proposal by Mavroyannis and Stephen (1962) is useful. They take

$$P(u) = \frac{A}{a^2+u^2}$$

and adjust the unknown constants A and a so that the static polarizability and the polarizability for large frequencies are reproduced correctly. That is to say,

$$\frac{A}{a^2} = P(0) \equiv P \tag{85}$$

and

$$\frac{A}{u^2} = \frac{e^2h^2}{4\pi^2 m}\,\frac{\sum_s f_{0s}}{u^2},$$

the latter because of eqn. (22) et seq. But $\sum_s f_{0s} = N$, the number of atomic electrons, by virtue of the sum rule. Hence

$$A = \frac{e^2\hbar^2 N}{m} \quad \text{and} \quad a = \left(\frac{A}{P}\right)^{1/2}. \tag{86}$$

From (60) one then obtains

$$\begin{aligned}
\Delta E &= -\frac{3}{\pi R^6}\int_0^\infty \frac{A_1A_2}{(a_1^2+u^2)(a_2^2+u^2)}\,du \\
&= -\frac{3A_1A_2}{\pi R^6}\,\frac{1}{a_2^2-a_1^2}\int\left(\frac{1}{a_1^2+u^2}-\frac{1}{a_2^2+u^2}\right)du \\
&= -\frac{3A_1A_2}{\pi R^6}\,\frac{\pi/2}{a_1a_2(a_1+a_2)} \\
&= -\frac{3}{2R^6}\,\frac{e\hbar}{m^{1/2}}\,\frac{P_1P_2}{\left(\dfrac{P_1}{N_1}\right)^{1/2}+\left(\dfrac{P_2}{N_2}\right)^{1/2}}.
\end{aligned}$$

This is once more formula (79).

Comparison with the oscillator results (cf. (36) et seq.) shows that they also agree with (79). The Slater–Kirkwood formula remains identical with London's as long as the molecules have a single absorption frequency and N in the former is replaced by f. The reason why the former gives larger results than London's (and is less reliable) is simply that f is always smaller than N, which is a rather dubious quantity at best.

It is possible to make more sophisticated use of (84) and to obtain better results. Karplus and Kolker (1964) have developed a variational method for calculating $P(u)$ in *a priori* fashion without any reliance upon experimental data. The details are complicated and involve in most instances resort to computers. Hence they will not be reproduced here. The results, however, being non-empirical, are worth listing and are given in Table 2.5,

TABLE 2.5 $C_1 \times 10^{60}$ *(erg cm⁶)*

	Karplus and Kolker	Semi-empirical	Experimental estimate
H–H	6.2206	6.46[a]	
H–He	2.8826	2.70[b]	
He–He	1.5842	1.53[a] 1.40[b]	1.573[c]
Ne–He	3.359	2.94[b]	
Ne–Ne	7.241	6.35[b]	9.95[d]
A–He	13.042	9.46[b]	11.7[e] 25[f]
A–Ne	26.977	19.75[b]	60.3[f]
A–A	112.9	65.16[b]	180[f] 103[g]

The values of C_1 are often expressed in atomic units.
To convert from (erg cm⁶) to atomic units it is necessary to multiply by 1.0448.

[a] Salem (1950).
[b] Dalgarno and Kingston (1961).
[c] Mason and Rice (1954).
[d] Corner (1948) (Lennard-Jones potential fit to second virial-coefficient data).
[e] Srivastava and Madon (1953) (Lennard-Jones potential and thermal-diffusion data).
[f] Rothe, Marino, Neynaber, Roe and Trujillo (1962) (molecular beam scattering experiments).
[g] Hirschfelder, Curtiss, and Bird (1954) (Lennard-Jones potential and second virial coefficient data).

which is taken directly from Karplus and Kolker. The analytical reason for the usefulness of complex polarizabilities in this context lies evidently in the fact that $P(u)$ has no poles on the real axis, in contrast to the real P which has a large number.

Table 2.5 reflects the genuine uncertainties which still exist with respect even to the simplest part of the van der Waals' forces, especially for the heavier atoms. Comparison with experimental values is not always meaningful because the latter usually include contributions from C_2 and C_3, which tend to be neglected when experimental data are adjusted to theory; the comparison suffers further from uncertainties in the repulsive part of the potential which enters the analysis.

Further discussion of the experimental determination of long-range intermolecular forces and their accuracy is given in Appendix B of this book. At present the most accurate values of C_1 are those given by the use of formula (27).

2.5. Third-order Forces; Matter–Antimatter Interactions

The results of third-order perturbation theory are rarely of physical interest in the calculation of long-range forces between two systems. They do have significance in the interaction between more than two partners, being non-additive (see Chapter 5) and also perhaps in considerations concerning antimatter. We shall indicate here how they are calculated and give estimates of their magnitude for the H–H problem.

The third-order energy can be computed by simply extending formula (17):

$$\Delta_3 E = \sum_{\lambda\mu}{}' \frac{V_{0\lambda}V_{\lambda\mu}V_{\mu 0}}{(E_0-E_\lambda)(E_0-E_\mu)}, \tag{87}$$

and V is given by (7), (8), or (9). For present purposes, (7) is most convenient, and we write its relevant parts in the form

$$V = \frac{s}{R^3}+\frac{t}{R^4}+\frac{u}{R^5} \tag{88}$$

assuming the absence of monopoles and ignoring octupoles and higher multipoles. s, t and u can be identified from (7). The function u does not include the dipole–octupole expression (the last one written out in (7)). As will be seen, this constituent makes no contribution to (81).

The average denominator method will once more be used in evaluating (87). A prime on the summation sign means omission of terms for which the denominator vanishes, that is, λ, $\mu \neq 0$. But since $V_{00} = 0$, this represents no restriction. Thus, by the rule of matrix multiplication,

$$\Delta_3 E = \frac{1}{\langle E^2 \rangle}(V^3)_{00}, \tag{89}$$

where $\langle E^2 \rangle$ denotes some undetermined mean energy value. For atomic hydrogen and rare gas atoms, it would presumably be in the neighborhood of twice the ionization energy—for reasons given in section 2.2.

Let us then evaluate (89) for the H–H interaction. The sums over i and j now disappear from s, t, and u; i must be replaced by 1 and j by 2:

$$V^3 = \frac{s^3}{R^9} + \frac{3s^2 t}{R^{10}} + \frac{3s^2 u + 3st^2}{R^{11}} + O(R^{-12}). \tag{90}$$

To form $(V^3)_{00}$ we average over all directions of the electrons in both atoms and perform the radial integrations $(r^n)_{00} = \int R_0^2 r^n r^2 \, dr$. Denoting averages over angles by bars, we observe that

$$\overline{x^4} = \overline{y^4} = \overline{z^4} = \frac{r^4}{5}$$

$$\overline{x^2} = \overline{y^2} = \overline{z^2} = \frac{r^2}{3}$$

$$\overline{x^2 y^2} = \overline{x^2 z^2} = \ldots = \frac{r^4}{15}. \tag{91}$$

Averages over all odd powers of a coordinate are zero. With this information it becomes clear at once that the first two terms of (90) vanish in the formation of $(V^3)_{00}$. A typical term of s is $x_1 x_2$, and $\overline{s^3}$ behaves as $\overline{x_1^3 y_1^3} = 0$. A typical term of t is $r_1^2 z_2$, so that $\overline{s^2 t} = \overline{r_1^2 x_1^2 x_2^2 z_2} = 0$. However, $s^2 u$ leads to expressions like $\overline{x_1^2 x_2^2 r_1^2 r_2^2}$ which survive. Likewise, st^2 has some components of the form $\overline{x_1^2 z_1^2 x_2^2 z_2^2}$, each of which equals $r_1^4 r_2^2/225$.

The compounding of all relevant terms in $s^2 u$ and st^2 is an elementary if somewhat lengthy task. One finds, with the use of relation (91), the following results:

$$(s^2u)_{00} = \frac{24}{25} e^6 (r_1^4)_{00}(r_2^4)_{00},$$

$$(st^2)_{00} = \frac{16}{25} e^6 (r_1^4)_{00}(r_2^4)_{00}.$$

But for hydrogen functions, $(r^4)_{00} = 6!/2^5 a_0^4 = 22.5 a_0^4$, a_0 = first Bohr radius. Hence from (90)

$$(V^3)_{00} = 3 \times \frac{8}{5} \times e^6 \times (22.5)^2 a_0^8 R^{-11} = 2430 \left(\frac{e^2}{a_0}\right)^3 \left(\frac{a_0}{R}\right)^{11}.$$

For $\langle E^2 \rangle$ in (89) we choose 1.84 times the ionization energy of hydrogen, $e^2/2a_0$, squared, this being the value which yields the correct second-order energy. Therefore

$$\Delta_3 E \cong 2900 \left(\frac{a_0}{R}\right)^{11} \frac{e^2}{a_0}. \tag{92}$$

This disagrees, both in sign and (slightly) in magnitude, with values published by Dalgarno and Lewis (1956). (Their negative sign is probably a misprint; the error leaves these authors' conclusion with respect to convergence of the multipole series unaffected.)

The numerical factor in (92) is rather uncertain. A more careful treatment of $\langle E^2 \rangle$ was conducted by Morgan (1966) who concludes that

$$\Delta_3 E = 3700 \left(\frac{a_0}{R}\right)^{11} \frac{e^2}{a_0} \pm 17\%,$$

which would mean that the proper energy denominator to be used in the calculation of the third-order perturbation is not $0.92\, e^2/a_0$ but $0.81\, e^2/a_0$.

Formulas (89) and (90) can be used for complex atoms; but the summations over electrons in s and t must then be retained, and the ground state is more complicated.

It may be premature to speculate about the forces between atoms of matter and antimatter. Yet there is at least one instance in which they may be of interest. Muonium $(\mu^+ e^-)$, is known to be formed when a beam of high energy positive muons is sent into dense argon gas (Hughes, McColm, Ziock, and Prepost, 1960). There is a small but perhaps observable probability for the conversion of muonium into antimuonium ($\mu^- e^+$) (Feinberg and Weinberg, 1961), and this, aside from other factors, depends on the energy difference between the muonium and the antimuonium atom in the environment in which the conversion takes place.† That difference, however, arises from the intermolecular forces between muonium on the one hand, antimuonium on the other, and the surrounding gas atoms. Only the long-range forces matter; if the interaction in question exceeds a certain small value the present argument fails.

Let us consider, first, the van der Waals' forces between muonium and, for simplicity's sake, a hydrogen atom. Except for the smaller mass of the nucleus, muonium has the structure of hydrogen; hence its energy levels are those of hydrogen but for a small isotope shift. One is interested in distances of separation at which exchange forces vanish, and the long-range forces will be identical with those between two hydrogen atoms.

The interaction antimuonium–hydrogen differs only in this respect: The state ψ_0 appearing in section 2.2.1 and its Hamiltonian are the same, but V has a negative sign because the charges in antimuonium are reversed. This means that all odd powers appearing in the perturbation scheme now have the opposite sign. So far as V_{00} is concerned, this is uninteresting, for it vanishes at large separations. The second perturbation remains unaltered, and since it is the predominant effect, it is clear that the main part of the van der Waals forces is the same for matter and antimatter particles.

The difference in question, E (muonium–hydrogen)—E (antimuonium–hydrogen), is therefore dominated by $\Delta_3 E$, and it is twice $\Delta_3 E$, as given by (92).

2.6. Eccentricity Effects

The distance R which occurs in the preceding calculations is normally taken to extend from the geometric center of one molecule to that of the other. If the geometric center does not coincide with the center of charge or the center of mass, special consideration must be given to that fact. In the gas phase the rotation of molecules takes place about the center of mass. Therefore all interactions must be expressed in a coordinate system having

† Details are clearly discussed in Morgan (1966).

this point as its origin. For many molecules (methane, benzene, chlorine, ethane, i.e. all those with a center of symmetry) the center of mass coincides with the center of inter- and intramolecular forces. But for heteronuclear diatomic molecules the center of mass is not the center of charge nor the center of polarizable charge.

The interaction of a spherical and a *homonuclear* diatomic molecule is, from (53),

$$\Delta E = -\frac{C_1}{R^6}(1+\gamma P_2(\cos\theta))$$

with

$$\gamma = \frac{C-D}{C+2D} = \frac{P_{\parallel}-P_{\perp}}{P_{\parallel}+2P_{\perp}}.$$

$P_{\parallel}$ and $P_{\perp}$ are the two polarizabilities of the diatomic molecule (parallel to and perpendicular to the long axis). P_2 is the second Legendre polynomial written as a function of the angle between R and the long axis of the molecule.

We now derive an expression for the interaction of a heteronuclear diatomic molecule with an atom using the *center of charge* as the origin of the diatomic molecule's coordinate system.† Our analysis will be applicable to interactions between inert gas and polar molecules. If the spherical system has a large quadrupole moment one also obtains dipole–quadrupole interactions which have been discussed by Anderson (1950).

We label the polar molecule by the numeral 1, the inert gas atom by the numeral 2. If the axis of the diatomic molecule makes an angle θ with respect to the internuclear axis, and we denote by x_{0i}, y_{0i}, z_{0i} the coordinates fixed to the diatomic molecule (z_{0i} is along the diatomic axis), then these coordinates are related to those fixed by having the z-axis coincident with the internuclear axis, namely the set $\{x_i, y_i, z_i\}$, as follows

$$\begin{aligned} x_i &= x_{0i}\cos\theta + z_{0i}\sin\theta \\ y_i &= y_{0i} \\ z_i &= z_{0i}\cos\theta - x_{0i}\sin\theta \end{aligned}$$

† This section follows the work of Roger Herman; it includes material from his dissertation (Department of Physics, Yale University, 1962), and from several papers: Herman (1962, 1963, 1966).

Using the fixed coordinates and including only dipole–dipole and dipole–quadrupole effects, the interaction, with neglect of any quadrupole contributions from the inert gas atom, is obtained from (7):

$$V = \frac{1}{R^3}\sum_{ij} q_i q_j (x_i x_j + y_i y_j - 2z_i z_j)$$

$$+\frac{3}{2R^3}\sum_{ij} q_i q_j [r_j^2 z_j + (2x_i x_j + 2y_i y_j - 3z_i z_j) z_i].$$

As before, the subscript i refers to electrons in the molecule, j to those in the atom. The second-order perturbation energy consists of two parts

$$\Delta E = -\sum_{\{n_1\}\{n_2\}} \frac{V_{00,\,n_1 n_2} V_{n_1 n_2,\,00}}{E_1(n_1)+E_2(n_2)-E_i(0)-E_2(0)}$$

$$-\sum_{n_2} \frac{V_{00,\,0n_2} V_{0n_2,\,00}}{E_2(n_2)-E_2(0)}.$$

The second term appears because of the permanent dipole moment of molecule 1. The R^{-6} terms arise from the dipole–dipole part of V. Using the mean excitation energy approximation with the same average value for the two principal axes in the diatomic molecule ($h\nu_{\parallel} = h\nu_{\perp} = I_1$, where I is the ionization potential), and the definition of polarizability we find†

$$\Delta E(R^{-6}\text{ terms}) = -\sum_{ij} \frac{q_i^2 q_j^2}{R^6(I_1+I_2)} \langle 00 | (x_{0i}\cos\theta + z_{0i}\sin\theta)^2 x_j^2$$

$$+ y_{0i}^2 y_j^2 + 4(z_{0i}\cos\theta - x_{0i}\sin\theta)^2 z_j^2 | 00\rangle$$

$$-\frac{1}{2R^6}\sum_{ij} q_i^2 q_j^2 [\langle 0 | (x_{0i}\cos\theta + z_{0i}\sin\theta) | 0\rangle^2 \langle 0 | x_j^2 | 0\rangle$$

$$+\langle 0 | y_{0i} | 0\rangle^2 \langle 0 | y_j^2 | 0\rangle$$

$$+4\langle 0 | z_{0i}\cos\theta - x_{0i}\sin\theta | 0\rangle^2 \langle 0 | z_j^2 | 0\rangle]$$

$$= -\frac{I_1 I_2}{4(I_1+I_2)R^6} P_2[P_{1,\perp}\cos^2\theta + P_{1,\parallel}\sin^2\theta + P_{1,\perp}$$

† The lengthy expressions between ⟨⟩ symbols which follow are matrix elements in Dirac bracket form.

$$+4P_{1,\parallel}\cos^2\theta+4P_{1,\perp}\sin^2\theta]$$

$$-\frac{P_2}{2R^6}\sum_i q_i^2[\langle 0|z_{0i}|0\rangle^2\sin^2\theta+4\langle 0|z_{0i}|0\rangle^2\cos^2\theta]$$

$$=-\frac{I_1I_2}{4(I_1+I_2)R^6}P_2[(5P_{1,\perp}+P_{1,\parallel})$$

$$+3(P_{1,\parallel}-P_{1,\perp})\cos^2\theta]-\frac{1}{2R^6}P_2p[3\cos^2\theta+1].$$

Here we have denoted the permanent dipole moment of the diatomic molecule by p. The first term is a special case of the equation for the interaction of asymmetric molecules, (53) with $\theta_2=\varphi=0$.

The R^{-7} interaction arises from the cross terms between the dipole–dipole and the dipole–quadrupole parts of V. After evaluating matrix elements of the noble gas atom, we obtain

$$\Delta E(R^{-7}\text{ terms})=-\frac{3P_2I_2}{R^7(I_1+I_2)}\sum_i q_i^2[\langle 0|z_i(r^2-3z_i^2)|0\rangle$$

$$-\langle 0|x_i(x_iz_i)|0\rangle-\langle 0|y_i(y_iz_i)|0\rangle]$$

$$-\frac{3P_2}{R^7}\sum_i q_i^2[\langle 0|z_i|0\rangle\langle 0|r^2-3z_i^2|0\rangle$$

$$-\langle 0|x_i|0\rangle\langle 0|x_iz_i|0\rangle-\langle 0|y_i|0\rangle\langle 0|y_iz_i|0\rangle]$$

$$=-\frac{6P_2I_2}{R^7(I_1+I_2)}(3\langle 0|x_{0i}^2z_{0i}|0\rangle(1-\cos^2\theta)$$

$$+\langle 0|z_{0i}^3|0\rangle\cos^2\theta)\cos\theta-\frac{3P_2pQ}{R^7}\cos^3\theta.$$

Here Q denotes the permanent quadrupole moment of the molecule. It is formed from the diagonal elements of the quadrupole moment tensor,†

$$Q=2Q_{zz}-Q_{xx}-Q_{yy}.$$

† Approximately, $Q=q^2(2\langle 0|\sum_i r_i^2|0\rangle-I_1(P_{1,\parallel}-P_{1,\perp})/q^2)$ which for HCl yields about 4.7×10^{-26} e.s.u.

The R^{-8} contributions are also easily derived. Considering only the permanent quadrupole moments we find

$$\Delta E(R^{-8} \text{ terms}) = -\frac{9P_2}{8R^8}\langle 0|3z_{0i}^2-r_i^2|0\rangle^2(1-2\cos^2\theta+5\cos^4\theta)$$

$$= -\frac{9P_2}{8R^8}Q^2(1-2\cos^2\theta+5\cos^4\theta).$$

Of the R^{-7} and R^{-8} terms, only the second R^{-7} contribution is important for most cases. The R^{-8} term involves contributions from Q^2 and from electronic octupole moments. These effects have been mentioned by Buckingham (1965).

Summarizing the principal terms calculated with the center of electric charge placed at the origin we thus find

$$\Delta E = -\frac{a_0}{R^6}-\frac{a_2\cos^2\theta}{R^7}-\frac{a_3}{R^8}\cos^3\theta$$

with

$$a_0 = \frac{3}{2}\,\frac{I_1I_2}{I_1+I_2}P_2\left(\frac{P_{1,\parallel}+5P_{1,\perp}}{6}\right)+\frac{P_2p^2}{2}$$

$$a_2 = \frac{3}{2}\,\frac{I_1I_2}{I_1+I_2}P_2\left(\frac{P_{1,\parallel}-P_{1,\perp}}{2}\right)+\frac{3P_2p^2}{2}$$

$$a_3 = 3P_2pQ.$$

To express this result in the center of mass coordinates, defined by R' and θ', the following well-known relations are used:†

$$\frac{1}{R} = \frac{1}{R'}\sum_{n=0}^{\infty}\left(\frac{d}{R'}\right)^n P_n(\cos\theta')$$

where d is the separation of electric and mass centers (both lie

† From Fig. 2

$$\frac{1}{R} = \frac{1}{\sqrt{[(R')^2+d^2-2dR'\cos\theta']}} = \frac{1}{R'\sqrt{([1+(d/R')^2-2(d/R')\cos\theta']}}$$

$$= \frac{1}{R'}\sum_{n=0}^{\infty}\left(\frac{d}{R'}\right)^n P_n(\cos\theta')$$

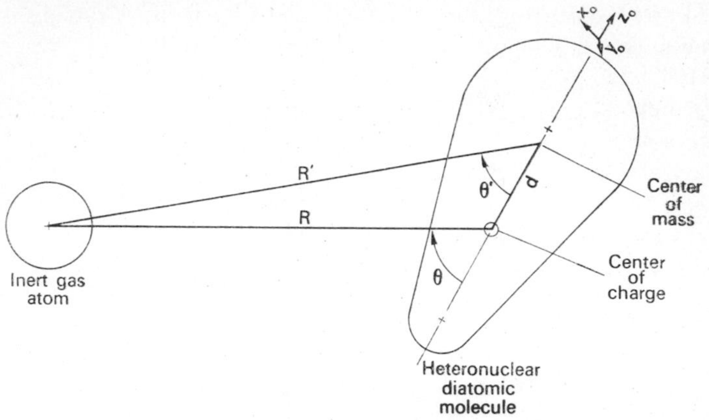

FIG. 2. Coordinate system for the interaction of an inert gas atom and a heteronuclear diatomic molecule.

on the axis of the diatomic molecule) and†

$$\cos\theta = \cos\theta' - \left(\frac{d}{R'}\right)\sin^2\theta' - \frac{3}{2}\left(\frac{d}{R'}\right)^2\cos\theta'\sin^2\theta' + \dots$$

Retaining only the leading terms, we find

$$\Delta E(R', \theta') = -\frac{(a_0 + a_2/3)}{(R')^6} - \frac{(6a_0 + 14a_2/5 + 3a_3/5)\, d\, P_1(\cos\theta)}{(R')^7}$$

$$-\frac{[16(d/R')^2 a_0 + (26/7)(d/R')^2 + 2a_2/3]P_2(\cos\theta')}{(R')^6}$$

The interaction now contains a strong R^{-7} attraction with an angular dependence given by the first Legendre polynomial $P_1(\cos\theta') = \cos\theta'$. While this contribution vanishes if the molecules rotate freely, Herman has shown that it can be very important in calculating the effect of inert gas atmospheres on the vibrational–rotational spectra of HCl.

The size of d is difficult to determine. Experimentally line shapes are not sensitive enough to define d precisely. Theoretically

† Note that

$$R\cos\theta = R'\cos\theta' - d$$

$$\cos\theta = \frac{1}{R'}(R'\cos\theta - d)\sum_{n=0}^{\infty}\left(\frac{d}{R'}\right)^n P_n(\cos\theta')$$

our knowledge of HCl state functions is not adequate either. Some general features are obvious. The internuclear distance in HCl is 1.28 Å, the center of mass is about 0.04 Å from the chlorine nucleus. Since the HCl bond is about 17% ionic and since some of the chlorine core electrons also contribute to the polarizable charge, d is less than 0.6 Å. Various estimates yield values between 0.1 and 0.35 Å. (See, for example, Herman, 1966.) Even the smaller value leads to important R^{-7} interactions in the gas phase. Of course, the long time average of this effect is very small, proportional to R^{-14}. For spectral problems, however, the time scale is very short and these effects can be important.

CHAPTER 3

Short-range Interaction

3.1. General Theory of Repulsive Forces

3.1.1. Definition of the Intermolecular Potential

The perturbation theory and the simple variational method sketched in the foregoing sections are applicable only when the perturbation is small, that is for large values of R. Under these conditions the exclusion principle, which requires the state function to be antisymmetric with respect to a transposition of electrons from one molecule to the other, may be disregarded because the overlap of charge distributions is negligible for large R.

At small distances of separation, however, a different approach becomes necessary. This will now be discussed in detail.†

† In this context it is more convenient to label the molecules a and b, the electrons by numerals. We shall thus depart from our earlier practice in which the molecules were denoted by 1 and 2.

At this point we also find it useful to employ atomic units very freely. Hence we insert a word concerning them. First of all we note that the unit of energy is equal to twice the energy of an "infinite mass Rydberg" or twice the energy of a hydrogen atom in the ground state with a nucleus of infinite mass. The unit of distance is the first Bohr radius in hydrogen, or $\hbar^2/m_e e^2$. In these units the constants $\hbar$, m_e, the mass of the electron (the subscript is used in this chapter to distinguish m_e from the mass of the nucleus), and e, the charge of the electron do not appear. Because of this, one can easily convert an equation from ordinary units to atomic units by simply setting all $\hbar$'s, m_e's, and e's equal to 1. Thus, for example, the kinetic energy operator instead of being $-\dfrac{\hbar^2}{2m_e}\nabla^2$ is $-\frac{1}{2}\nabla^2$. In the potential term we have only to set the electron charge equal to 1 and express all distances in terms of atomic units (Bohr radii).

Atomic units will be used in this chapter and the next and in several other places in this book. From the above remarks it should always be clear when these units are being used. One needs only to see if $\hbar$, m_e, or e appears.

Atomic units should not be confused with another set in which the unit

Let us consider two molecules *a* and *b*, H_a being the Hamiltonian for molecule *a*, H_b for molecule *b*, a distance *R* apart. If *a* and *b* contain *m* and *n* electrons respectively, the total Hamiltonian H_T of the system is the sum of an electronic *H* and a nuclear Hamiltonian H_N,

$$H_T = H + H_N. \tag{1}$$

$$H = H_a(1\ldots m) + H_b(m+1\ldots m+n) + V(1,\ldots, m+n, R), \tag{2}$$

where

$$H_a = \sum_{i=1}^{m}\left(-\frac{1}{2}\nabla_i^2\right) - Z_a\sum_{i=1}^{m}\frac{1}{r_{ai}} + \sum_{i>j}^{m}\frac{1}{r_{ij}} \tag{3}$$

and the interaction *V* is defined as

$$V = -Z_a\sum_{j=m+1}^{n+m}\frac{1}{r_{aj}} - Z_b\sum_{i=1}^{m}\frac{1}{r_{bi}} + \sum_{i=1}^{m}\sum_{j=m+1}^{m+n}\frac{1}{r_{ij}} + \frac{Z_aZ_b}{R}. \tag{4}$$

It is this *V* which, when expanded for large *R*, yields the asymptotic series we studied in section 2.1. The *Z*'s are the nuclear charges and the *r*'s are the scalar distances between particles defined by subscripts. The nuclear Hamiltonian defined by (1) is the sum of nuclear kinetic energies,

$$H_N = -\frac{1}{2}\left(\frac{m_e}{M_A}\right)\nabla_A^2 - \frac{1}{2}\left(\frac{m_e}{M_B}\right)\nabla_B^2. \tag{5}$$

The potential energy due to nuclear repulsion has been incorporated into *V*; m_e is the electronic mass while M_A and M_B are the nuclear masses.

The total state function Ψ_T for the system is a function of the electronic variables r_i and r_{ij} as well as the nuclear coordinates X_N:

$$\Psi_T = \Psi_T(\mathbf{r}_i, \mathbf{r}_{ij}, X_N). \tag{6}$$

This function is in general very complicated and unknown. However, since in most applications of the theory of intermolec-

of distance is the same but the unit or energy is the Rydberg, half of our unit. Numerically 1 a.u. of energy is equal to about 27.21 ev and the first Bohr radius in hydrogen, a_0, is 0.52917 Å. The use of these units, while simplifying our equations, also makes our answers independent of experimentally determined values of $\hbar$, m_e, and *e*, a feature of some importance in extremely exact calculations.

ular forces the nuclei are moving very slowly as compared with the electrons one can make an "adiabatic" approximation in which the total function is replaced by a product, one of whose factors is calculated for a fixed position of the nuclei while the other describes the probability of finding the nuclei at the positions $\boldsymbol{X}_N$:

$$\Psi_T = \Psi(\boldsymbol{r}_i, \boldsymbol{r}_{ij}, R)\mathcal{N}(\boldsymbol{X}_N). \tag{7}$$

In the first factor R is a parameter which specifies the fixed separation selected in solving the equation

$$H\Psi = E(R)\Psi. \tag{8}$$

Using (8) we find for the nuclear function

$$[H_N + E(R)]\mathcal{N}(\boldsymbol{X}_N) = E_T\mathcal{N}(\boldsymbol{X}_N), \tag{9}$$

E_T being the total energy. We thus see that $E(R)$ is the potential which must be used in the "scattering equation", eqn. (9). Usually one modifies this equation slightly by adjusting the zero of energy so that when the molecules are infinitely far apart the potential energy is zero, i.e.

$$[H_N + \Delta E(R)]\mathcal{N}(\boldsymbol{X}_N) = \Delta E_T\mathcal{N}(X_N) \tag{10}$$

with

$$\Delta E(R) = E(R) - E_a - E_b. \tag{11}$$

E_a and E_b are the energies of the isolated molecules a and b. Equation (11) defines the intermolecular potential and will be used in this form throughout the book. It is zero when the molecules are infinitely far apart and is derived on the basis of a Hamiltonian in which the nuclei are fixed at a distance R from one another.

Using perturbation theory at large separations we found in the last chapter that

$$E(R) = E_a + E_b + \Delta_2 E \qquad \text{(large } R\text{)}$$

and thus

$$\Delta E(R) = \Delta_2 E \qquad \text{(large } R\text{)}$$

This relation provides the contact between the present and the preceding chapter.

There are small errors involved in the use of the product function, eqn. (7), but these are of the order of the fourth root of the ratio of the mass of the electron to that of the nucleus (about 10^{-1}). Other possible limitations are discussed in section 3.5.

For potentials which include bound states (molecule formation) such as the ground state of H_2, this procedure, known as the Born–Oppenheimer approximation has been thoroughly studied (Born and Oppenheimer, 1927).†

For a given value of R we introduce a function φ (1, 2... $m+n$) of all electron coordinates including spins and antisymmetrize it in accordance with the Pauli principle to obtain

$$\Psi = \sum_{\lambda}(-1)^{\lambda}P_{\lambda}\varphi \equiv \mathcal{A}\varphi, \tag{12}$$

where we have written $P_{\lambda}\varphi$ for the function which results when the permutation P_{λ} among the electron coordinates is carried out upon φ; λ is an integer numbering the various permutations and is taken to be odd or even according to the parity of the permutation. The symbol $\mathcal{A}$ is called an antisymmetrizer.

In general Ψ is extremely complicated; it contains all of the information by which we could calculate the complete intermolecular potential. The exact Ψ must be a function of the relative separations of all electrons (r_{ij}) as well as the electron-nuclear coordinates ($\mathbf{r}_i$). The contribution of the $\mathbf{r}_{ij}$ terms to the energy, however, is usually of the order of about 1 eV per doubly occupied orbital; the effect on the intermolecular potential is even less since $\Delta E(R)$ is the difference between the energy at separation R and that for infinite separation. For this reason, at small separations, where the intermolecular potential is large, one can treat each electron as independent of the positions of the other electrons. At larger separations there are substantial errors in such an assumption.

In this chapter we shall restrict consideration to this particular approximate form of state function. In the next chapter we improve these functions so that the intermolecular potential can be calculated over the range where it is large as well as where it is small.

There are two commonly used approximate methods: (1) the Heitler–London, valence-bond or atomic function method, and (2) the molecular orbital method. The former emphasizes the

† For a detailed treatment of the errors involved in H_2, see Kolos and Wolniewicz (1963, 1965).

atomic character of the intermolecular interaction; the latter treats all of the electrons and nuclei together and forgets about their assignments at large separations.

We shall consider each of these methods in turn, comparing and contrasting them. Finally, we review some major variations of these two methods specifically suited to intermolecular force calculations.

3.2. Heitler–London Methods

We construct the state function of the complete n-electron system Ψ in terms of the molecular state functions Ψ_a and Ψ_b. This procedure is referred to as the Heitler–London method since it was first used by these authors to calculate the energy of the hydrogen molecule (Heitler and London, 1927). Our use of the term, Heitler–London method, generalizes their procedure to more complex cases.

Let the state function of the two molecules be $\varphi_a\ (1\ldots m)$ and $\varphi_b\ (m+1\ldots m+n)$, where these contain all intramolecular electron coordinates including electron spin. Spin must be included since many of the results depend on it critically. The antisymmetrized independent functions are

$$\Psi_a = \sum_{\lambda_a} (-1)^{\lambda_a} P_{\lambda_a} \varphi_a \equiv \mathcal{A}^a \varphi_a \tag{13}$$

and

$$\Psi_b = \sum_{\lambda_b} (-1)^{\lambda_b} P_{\lambda_b} \varphi_b \equiv \mathcal{A}^b \varphi_b, \tag{14}$$

where P_{λ_a} and P_{λ_b} are, respectively, the intramolecular permutations of all electrons in a and all electrons in b, exclusive of the intermolecular permutations, which will be described by $P_{\lambda_{ab}}$.

In writing the approximate total state function we assume that the electrons in one molecule move independently of those in the other except for the Pauli correlations, i.e.

$$\Psi = \sum_{\lambda_{ab}} (-1)^{\lambda_{ab}} P_{\lambda_{ab}} \Psi_a \Psi_b = \mathcal{A}^{ab} \mathcal{A}^b \mathcal{A}^a \varphi_a \varphi_b. \tag{15}$$

This is a generalized Heitler–London state function. We note first that (15) can not be used as an "unperturbed" function in the sense of perturbation theory because even if

$$H_a \Psi_a = E_a \Psi_a \quad \text{and} \quad H_b \Psi_b = E_b \Psi_b$$

(15) is not an eigenfunction of $H_a + H_b$. Furthermore, if one were to compute as a first approximation $\bar{V} = \langle \Psi | V | \Psi \rangle$ one would find, in general, that it remains finite even where $R \to \infty$, nor would $\bar{H}_a$ and $\bar{H}_b$ bear close relations to the atomic energies E_a and E_b.†

A quantity which does resemble an intermolecular potential energy, at least in so far as it vanishes as $R \to \infty$, is

$$\tilde{V} = \langle \varphi_a \varphi_b | \mathcal{A}(V\varphi_a\varphi_b)\rangle / \langle \varphi_a\varphi_b | \mathcal{A}(\varphi_a\varphi_b)\rangle. \tag{16}$$

Its meaning arises from the following consideration. Suppose that φ_a is an *exact* eigenstate of H_a, so that $H_a\, \varphi_a = E_a \varphi_a$ and similarly $H_b \varphi_b = E_b \varphi_b$. We then find

$$H\Psi = H\mathcal{A}(\varphi_a\varphi_b) = \mathcal{A}[H\varphi_a\varphi_b] \tag{17}$$

since H is invariant to any electron permutation, and further

$$\begin{aligned} H\Psi &= \mathcal{A}(E_a\varphi_a\varphi_b) + \mathcal{A}(E_b\varphi_a\varphi_b) + \mathcal{A}(V\varphi_a\varphi_b) \\ &= (E_a + E_b)\Psi + \mathcal{A}(V\varphi_a\varphi_b). \end{aligned} \tag{18}$$

Therefore

$$\begin{aligned} \bar{H} &= \langle \Psi | H | \Psi \rangle / \langle \Psi | \Psi \rangle \\ &= E_a + E_b + \tilde{V} \end{aligned} \tag{19}$$

with $\tilde{V}$ given by (16). To get this result we note that for two functions, u and v,

$$\langle \mathcal{A}u | \mathcal{A}v \rangle = (n+m)! \langle u | \mathcal{A}v \rangle$$

or

$$\mathcal{A}^2 = (n+m)!\mathcal{A}. \tag{20}$$

Thus according to (11), $\tilde{V}$ appears as the difference between the variational energy $\bar{H}$ and the energy at infinite separation, $E_a + E_b$, and may therefore be regarded as a measure of the intermolecular potential energy. It was thus employed in earlier calculations (Slater, 1928; Mayer, 1934; Kunimune, 1950; Rosen, 1931). This identification breaks down, however, when $H_a\varphi_a \neq E_a\, \varphi_b$, and such is universally true in practice except for atomic hydrogen, the only instance where the use of exact eigenfunctions is practical.

† These facts are further illuminated by H. Margenau (Slater volume, 1966). Similar problems have continually plagued valence bond calculations. See Simpson (1962, Chap. 3).

When the exact atomic or molecular functions are not available, the best approximation to the intermolecular energy is the difference between the total variational energy of the interacting system computed with Ψ defined in (15) for a finite value of R and the same variational energy computed for infinite R. Thus

$$\Delta E = \overline{H(R)} - \overline{H(\infty)}. \tag{21}$$

One can show that $\overline{H(\infty)} = \overline{h_a} + \overline{h_b}$, where

$$\overline{h_a} = \langle \Psi_a | H_a | \Psi_a \rangle,$$
$$\overline{h_b} = \langle \Psi_b | H_b | \Psi_b \rangle. \tag{22}$$

To see this we expand

$$\langle \Psi | H\Psi \rangle = \langle \mathcal{A}(\varphi_a \varphi_b) | H | \mathcal{A}(\varphi_a \varphi_b) \rangle = (n+m)! \langle \varphi_a \varphi_b | H \mathcal{A}(\varphi_a \varphi_b) \rangle.$$

Now we consider among the permutations included in $\mathcal{A}$, one which exchanges an electron between a and b; we shall call it P_{ij}. The integral

$$I = \langle \varphi_a \varphi_b | H P_{ij} \varphi_a \varphi_b \rangle = \langle P_{ij} \varphi_a \varphi_b | H \varphi_a \varphi_b \rangle$$

becomes in the limit of $R \to \infty$

$$\langle P_{ij} \varphi_a \varphi_b | (H_a + H_b) \varphi_a \varphi_b \rangle \quad \text{because} \quad \lim_{R \to \infty} V = 0.\dagger$$

However $\langle P_{ij} \varphi_a \varphi_b | H_a \varphi_a \varphi_b \rangle$ vanishes because its integral contains the coordinates of one electron which do not appear in H_a, in two places: once attached to nucleus a and once to nucleus b. If the overlap of φ_a and φ_b is negligible, and it certainly is at large separations, this term must vanish. For a similar reason the other part is zero. This means that, in the evaluation of $\overline{H(\infty)}$, we may ignore all intermolecular exchanges:

$$\lim_{R \to \infty} \langle \Psi | H\Psi \rangle = (m+n)! \langle \varphi_a \varphi_b | (H_a + H_b) \mathcal{A}^a \varphi_a \mathcal{A}^b \varphi_b \rangle$$
$$= \frac{(m+n)!}{m!n!} \langle \mathcal{A}^a \varphi_a \mathcal{A}^b \varphi_b | (H_a + H_b) \mathcal{A}^a \varphi_a \mathcal{A}^b \varphi_b \rangle$$
$$= \frac{(m+n)!}{m!n!} (\overline{h_a} + \overline{h_b}). \tag{23}$$

Since $\lim \langle \Psi | \Psi \rangle = (m+n)!/m!n!$ for normalized Ψ_a and Ψ_b, we have established the result

$$\lim_{R \to \infty} \overline{H} = \overline{h_a} + \overline{h_b}. \tag{24}$$

† This is true even though $\langle \Psi | V\Psi \rangle \neq 0$ as $R \to \infty$.

Hence we may use for our intermolecular potential energy the difference [in place of the exact result (11)]

$$\Delta E = \overline{H} - \overline{h_a} - \overline{h_b}. \tag{25}$$

This is not equal to $\tilde{V}$ defined in (16). Indeed, the two differ appreciably at small values of R.† Formula (25) will form the basis of most of the subsequent analysis.

An important question concerns the bounds on ΔE. Even though $\overline{H}$ is an upper bound to the total energy of the dual system, ΔE does not have this desirable quality. All one can say is this. If E_a' and E_b' are the actual atomic energies, and $\Delta E'$ is the true interaction energy, our $\overline{H}$ satisfies

$$\overline{H} \geqslant E_a' + E_b' + \Delta E'. \tag{26}$$

Hence by (25)

$$\Delta E + \overline{h_a} + \overline{h_b} \geqslant E_a' + E_b' + \Delta E' \tag{27}$$

whence

$$\Delta E \geqslant \Delta E' + (E_a' - \overline{h_a}) + (E_b' - \overline{h_b}). \tag{28}$$

But

$$E_a' - \overline{h_a} \leqslant 0 \quad \text{and} \quad E_b' - \overline{h_b} \leqslant 0.$$

Thus the inequality (28) says, disconcertingly, that our ΔE is equal to or greater than a quantity which is smaller than or equal to the true ΔE. Hence ΔE is neither an upper nor a lower bound.

The simplest and most frequently discussed specific calculation of these repulsive exchange forces applies to the hydrogen–hydrogen interaction (Heitler and London, 1927; Pauling and Wilson, 1935; Slater, 1963). The simple choice of $\varphi_a(1)\ \varphi_b(2)$ yields upon antisymmetrization the repulsive triplet curve for two hydrogen atoms (the $^3\Sigma_u$ state). Here φ_a and φ_b are the $1s$-hydrogen-atom functions with the proper spin function attached, centered about nuclei a and b respectively,

$$\varphi_a(1) = a(1)\alpha(1)$$

$$a(1) = \frac{2}{\sqrt{(4\pi)}} e^{-r_{a1}} \tag{29}$$

† For a quantitative evaluation of these differences in the helium–helium interaction see Margenau and Rosen (1953).

and α is the spin function for $S_z = 1/2$, β for $S_z = -1/2$. We recall that α and β are orthonormal with respect to integration over the spin coordinate.

$$\Psi = \mathcal{A}[a(1)\alpha(1)b(2)\alpha(2)]. \tag{30}$$

This triplet state has two α-spins; at large separations it represents two atoms with the same spin. Equation (30) is an eigenstate of S^2 and S_z corresponding to $S^2 = 3/4\,\hbar^2$ and $S_z = +\frac{1}{2}\hbar$. The normal hydrogen-molecule state function (singlet) involves both α and β spins and cannot be written as a one-term antisymmetrized expression such as (30). Expanding (30) we have

$$\Psi = \frac{1}{\sqrt{2}}[a(1)b(2)-a(2)b(1)]\alpha(1)\alpha(2). \tag{31}$$

Since we know a exactly we can calculate the energy for function (31) using (16), where

$$V_H = -\frac{1}{r_{a2}}-\frac{1}{r_{b1}}+\frac{1}{r_{12}}+\frac{1}{R} \tag{32}$$

and

$$\Delta E = \tilde{V} = \frac{\langle\Psi|\, V_H a(1)\alpha(1)b(2)\alpha(2)\rangle}{\langle\Psi|\, a(1)\alpha(1)b(2)\alpha(2)\rangle}$$

$$= \frac{\langle a(1)b(2)|\, V_H a(1)b(2)\rangle - \langle b(1)a(2)|\, V_H a(1)b(2)\rangle}{1-S^2}, \tag{33}$$

where

$$S = \langle a|b\rangle, \tag{34}$$

$$-\frac{1}{R}+\Delta E =$$

$$\frac{-\left\langle b(2)\left|\frac{1}{r_{a2}}\right|b(2)\right\rangle-\left\langle a(1)\left|\frac{1}{r_{b1}}\right|a(1)\right\rangle+\left\langle a(1)b(2)\left|\frac{1}{r_{12}}\right|a(1)b(2)\right\rangle}{1-S^2} \tag{35}$$

$$+\frac{S\left\langle b(2)\left|\frac{1}{r_{a2}}\right|a(2)\right\rangle+S\left\langle b(1)\left|\frac{1}{r_{b1}}\right|a(1)\right\rangle-\left\langle a(1)b(2)\left|\frac{1}{r_{12}}\right|b(1)a(2)\right\rangle}{1-S^2},$$

$$= \frac{-2(J-SK)+J'-K'}{1-S^2}, \tag{36}$$

with the customary definitions,

$$J = \left\langle a(1) \left| \frac{1}{r_{b1}} \right| a(1) \right\rangle = (a\beta a)$$

$$J' = \left\langle a^2(1) \left| \frac{1}{r_{12}} \right| b^2(2) \right\rangle = (aa\varrho bb)$$

$$K = \left\langle a(1) \left| \frac{1}{r_{b1}} \right| b(1) \right\rangle = (a\beta b)$$

$$K' = \left\langle a(1)b(2) \left| \frac{1}{r_{12}} \right| b(1)a(2) \right\rangle = (ab\varrho ba), \qquad (36b)$$

The last parentheses symbols represent a convenient explicit notation which will be explained (see p. 165) and used later in the book. At short distances the potential goes to infinity because of the $1/R$ electrostatic term which arises from nuclear repulsion. At large separations the results go exponentially to zero. Using well-known integrals (Slater, 1963) we find the results listed in Table 3.1.

TABLE 3.1

$R(a_0)$	E(a.u.)
0	∞
0.5	1.8989
1.0	1.7048
1.5	1.3191
2.0	1.1540
2.5	1.0760
3.0	1.0374
4.0	1.0088

At large distances J' and J behave as R^{-1}. Here it is useful to write

$$\Delta E = \frac{1}{R} + \frac{1}{(1-S)^2}(-2J+J') + \frac{1}{(1-S^2)}(2SK-K') \qquad (37)$$

$$= \frac{1}{R} + \frac{1}{1-S^2}\left[-\frac{2}{R} + e^{-2R}\left(2+\frac{2}{R}\right) + \frac{1}{R} - e^{-2R}\left(\frac{1}{R}+\frac{11}{8}\right.\right.$$

$$+\frac{3R}{4}+\frac{R^2}{6}\Big)\Big]+\frac{1}{1-S^2}(2SK-K') \tag{38}$$

$$= -\frac{S^2}{R(1-S^2)}+\frac{e^{-2R}}{1-S^2}\left(\frac{1}{R}+\frac{5}{8}-\frac{3R}{4}-\frac{R^2}{6}\right)+\frac{1}{1-S^2}(2SK-K'). \tag{39}$$

Since S behaves as e^{-R}, the entire ΔE is dominated by e^{-2R} or S^2 at large distances [remember: K behaves as e^{-R} and K' as e^{-2R}; see Slater (1963) or Pauling and Wilson (1935)]. Each term in (39), whether it is an exchange term or not, is of comparable magnitude.

Using the exact formula, (39), we obtain the results shown in Table 3.2.

TABLE 3.2

$R(a_0)$	E (a.u.)
2.5	0.07590
3.0	0.03844
3.5	0.01829
4.0	0.008789
4.5	0.004149
5.0	0.0019223
6.0	0.00039181
7.0	0.000075326
8.0	0.000013815
10.0	0.00000052059

The main reason for writing ΔE in the form of (39) is this. At large separation the actual value may be in the neighborhood of 10^{-6}. If (37) is used, the first term is of the order of 10^{-1} and thus important cancellations must take place. Unless care is taken to keep many significant figures in the computation of each term the numerical value could be meaningless. In (39) all terms are of similar size since the cancellations have been performed explicitly. For practical purposes this is very important.

The interaction we have calculated here is only an upper limit to the true interaction ($\tilde{V}$ has bounds) for we have neglected all distortion and polarization effects which "soften" the atoms by decreasing electron density between the atoms. Although our choice of atomic functions is quite primitive, it yields a reasonable interaction except for the polarization effects which lead to the long-range van der Waals' attractive interactions. At moderate distances the errors are not serious, but at very short distances severe distortions occur since the two atoms must behave together as an excited helium atom at $R = 0$.† We shall discuss this aspect in more detail for helium interactions in connection with molecular orbital theory.

In larger systems where φ_a and φ_b are not known exactly further approximations must be made. One is to assume that electrons in the interacting molecules move in orbitals independent of one another. For example, to calculate the interaction of two helium atoms by the present method is a more complex problem since both φ_a and φ_b now involve the radius vectors of two electrons as well as their relative positions:

$$\varphi_a = \varphi_a(\boldsymbol{r}_1, \boldsymbol{r}_2, \boldsymbol{r}_{12}).$$

Usually one assumes that

$$\Psi_a = \mathcal{A}^a[\phi_a(1)\alpha(1)\phi_a(2)\beta(2)], \tag{40}$$

where ϕ is a one electron function depending *only* on r_1 *or* r_2. The best possible function of this type, using energy as a criterion, is the Hartree Fock atomic function or orbital.‡ To obtain it we minimize the energy, $\langle \Psi_a \mid H_a \mid \Psi_a \rangle$, subject to the condition $\langle \phi_a \mid \phi_a \rangle = 1$, i.e. for helium we minimize the functional

$$I = \langle \mathcal{A}^a\phi_a(1)\phi_a(2) \mid H_a \mid \mathcal{A}^a\phi_a(1)\phi_a(2)\rangle - \lambda\langle\phi_a(1) \mid \phi_a(1)\rangle - \lambda\langle\phi_a(2) \mid \phi_a(2)\rangle \tag{41}$$

the constant λ being a Lagrangian multiplier.* Since

$$H_a = -\frac{1}{2}\nabla_1^2 - \frac{1}{2}\nabla_2^2 - \frac{2}{r_1} - \frac{2}{r_2} + \frac{1}{r_{12}} \tag{42}$$

we find

$$\left(-\frac{1}{2}\nabla_1^2 - \frac{1}{r_1}\right)\phi_a(1) + \int \frac{\phi_a^2(2)}{r_{12}}\, d\tau_2 \phi_a(1) - \frac{\lambda}{2}\phi_a(1) = 0$$

† The triplet state cannot collapse to the ground state of helium due to spin restrictions. The singlet can.

‡ Henceforth, Hartree-Fock will be abbreviated to HF.

* For a discussion of Lagrangian multipliers see Margenau and Murphy (1956, p. 209). Only one Lagrangian multiplier is used here because we discuss the ordinary or restricted HF theory in which two electrons, one of α and one of β spin, are placed into the same spatial orbital.

or, upon recognizing this as a Schrödinger equation for ϕ_a with eigenvalue $\lambda/2$,

$$\left[-\frac{1}{2}\nabla_1^2-\frac{1}{r_1}+\int\frac{\phi_a^2(2)}{r_{12}}\,d\tau_2\right]\phi_a(1) = \varepsilon\phi_a(1). \qquad (43)$$

ε is called the orbital energy. The HF orbital is the solution of the Schrödinger equation for an electron moving in the field of the nucleus and the average field of the other electron, which is represented by the integral term of (43). The equation is non-linear and a self-consistent procedure must be used in solving it. A ϕ_a is chosen and used in the integral. The equation is then solved for a new ϕ_a which is used to calculate a new potential. This process is continued until no further changes in ϕ_a occur. For a discussion of HF theory see Slater (1960), and the later sections of this chapter.

The best approximations to ϕ_a in (43) are generally written as linear combinations of Slater orbitals obtained by using the Roothaan procedure, (Roothaan, 1951, 1960; Slater, 1963, 1964—Appendix 7 and several articles) although some numerical solutions are also available.

In Table 3.3 we list three approximate orbitals for helium as

TABLE 3.3. *Approximate Atomic Orbitals for Helium and Corresponding Energies*

$$\phi_a = \sum_i c_i(n_i)\phi_{an_i}(\zeta_i)$$

$$\phi_{an_i} = (2\zeta_i)^{n_i+1/2}[(2n_i)!]^{-1/2}r_a^{n_i-1}e^{-r_a\zeta_i}$$

$c_1(1) = 1.00000$	$c_1(1) = 0.18159$ $c_2(1) = 0.84289$	$c_1(1) = 1.36211$ $c_2(2) = 0.10724$ $c_3(2) = 0.28189$
$\zeta_1 = 1.6875$	$\zeta_1 = 4.573$ $\zeta_2 = 2.448$	$\zeta_1 = 1.450$ $\zeta_2 = 2.641$ $\zeta_3 = 1.723$
$\varepsilon = -0.89648$	$\varepsilon = -0.91792$	$\varepsilon = -0.91795$
$E_a = -2.847656$	$E_a = -2.861673$	$E_a = -2.861680$

determined by the Roothaan procedure [Bagus, Gilbert, Roothaan, Cohen (to be published)] and state, at the bottom of each column, the energies to which they correspond.

We now turn to the calculations of ΔE for helium. The simplest of these did not use HF orbitals. Margenau and Rosen (1953) chose a $1s$ orbital

$$\phi_a(1) = \frac{\sqrt{(4\delta^3)}}{\sqrt{(4\pi)}} e^{-\delta r} \cdot (\text{spin function}) \tag{44}$$

and used (37) with $\delta = 1.6875$ (the "Wang function"). Rosen also calculated the interaction employing for φ_a an open shell state function which places electrons 1 and 2 in slightly different orbitals by permitting two different values of δ, i.e.

$$\varphi_a = [\phi_a(1)\phi_a'(2)+\phi_a(2)\phi_a'(1)]\,[\alpha(1)\beta(2)-\alpha(2)\beta(1)]$$

where ϕ_a and ϕ_a' are $1s$ orbitals of hydrogen type with exponents δ and δ', respectively. At short distances his values are now known to be too large (Phillipson, 1962) suggesting that some terms he neglected become important at small separations.

Calculations using more elaborate forms of φ_a, such as the HF functions, will be discussed later since they are equivalent to an elementary form of molecular orbital theory (Slater, 1963, Appendix 10).

More approximate calculations have been made on the interaction of neon and argon atoms (Bleick and Mayer, 1934; Kunimune, 1950a, b). In these φ_a was taken to be the product of very simple one-electron orbitals. They aim at $\tilde{V}$ (or a form similar to it) instead of ΔE and are thus subject to serious errors. In addition, the approximate state functions employed have a tendency to yield too small a repulsion since poor wave functions generally undervalue the electron density in the fringes of the atom. This effect is most serious for moderate and large separations. At small separations with large overlap ($\langle \phi_a | \phi_b \rangle > {}^1/_{10}$) such details are less important than distortion of the atoms themselves. An example of this is provided by the helium results to be discussed in the next section. At larger separations the use of approximate state functions leads to considerable errors in the repulsive interaction—using (44) instead of the best HF orbital leads to an error of a factor of two in $\Delta E(R)$ at $5.8a_0$—while for separations less than $3a_0$, the very approximate result, (44), is very adequate (see Table 3.4, later in this chapter, for example).

A welcome feature of these atomic orbital calculations, aside from the simplicity of the functions employed is, of course, their obvious physical meaning. There are many disadvantages as well. First, it is difficult to find a simple way to improve these calculations, especially at the smaller separations. Secondly, when many electrons are involved and φ_a and φ_b are written as products of one-electron orbitals, the matrix elements appear multiplied by numerous overlap factors. The work then becomes cumbersome and requires great care to make sure that all effects are included. A variation of this procedure in which orthogonal atomic orbitals (OAO) or localized orbitals are used is to be studied in section. 3.4.

Numerical results for ΔE based on functions of which (44) is typical will be given later (Table 3.4), when they can be compared with values obtained by more elaborate procedures.

3.3. Molecular Orbital (MO) Theory

When the electron densities of two atoms† overlap significantly it is no longer reasonable to consider the interacting species as separate atoms and we must return to the more general problem of two nuclei and $m+n$ electrons. Strictly speaking, this is always the case for atoms interacting at small separations. In general, therefore, we may not assign electrons to one atom or nucleus. Instead, they are to be described by orbitals which have the symmetry required by the nuclear or molecular framework. For a homonuclear system we allow functions which are even or odd with respect to the interchange of nuclei. Also, we classify these molecular orbitals and states by their angular momentum about the internuclear axis‡ m_λ, a good quantum number: σ orbitals are cylindrically symmetric while π, δ, etc., have various lesser types of symmetry. The orbitals corresponding to increased electron density (over that of one atom alone) between the nuclei are bonding orbitals while those with decreased density are antibonding. In a homonuclear system the orbitals, each

† Here we treat only the interaction of two atoms, purposely limiting our discussion because there are no comparably accurate calculations on the interaction of molecules.

‡ The internuclear axis is also the z-axis.

containing two electrons, one of α and one of β spin normally have increasing energies in the following order:

$$1\sigma_g(1s),\ 1\sigma_u(1s),\ 2\sigma_g(2s),\ 2\sigma_u(2s),\ \{1\pi_u(2p),\ 2\pi_u(2p)\},\ 3\sigma_g(2p),$$
$$\{1\pi_g(2p),\ 2\pi_g(2p)\},\ 3\sigma_u(2p). \quad (45)$$

In this notation the symbols following the molecular orbitals specify the atomic states which form the orbital, e.g. $2\sigma_g(2s)$ is essentially the sum of $2s$ atomic orbitals on each center. Often, however, we shall not specify the orbital composition but reserve the parentheses to indicate the electron assigned to that orbital. Of the above list the σ_u and π_g orbitals are antibonding. The ground state of the hydrogen molecule ($^1\Sigma_g$) has the form

$$\mathcal{A}[1\sigma_g(1)\alpha(1)1\sigma_g(2)\beta(2)]. \quad (46)$$

The repulsive triplet state ($^3\Sigma_u$) is written

$$\mathcal{A}[1\sigma_g(1)\alpha(1)1\sigma_u(2)\alpha(2)]. \quad (47)$$

The state function of the helium–helium system ($^1\Sigma_g$) is:

$$\mathcal{A}[1\sigma_g(1)\alpha(1)1\sigma_g(2)\beta(2)1\sigma_u(3)\alpha(3)1\sigma_u(4)\beta(4)]. \quad (48)$$

A full discussion of molecular orbital theory, may be found in Ballhausen and Gray (1965), Slater (1963, vol. 1), Kotani, Ohno, and Kayama (1961, vol. 17/2), or, in a more elementary form in Gray (1964).

Our intention here is to outline those aspects which are important for the calculation of intermolecular forces. Since much of the work leading to that end is the determination of accurate orbitals—computation of ΔE is a simple matter when orbitals are known—a large part of the present section is given to a discussion of methods for finding proper molecular orbitals.

With equal numbers of bonding and antibonding electrons, there is no molecular bonding—no stable molecules—but a repulsive interaction results. Then van der Waals' attraction does not appear in the MO theory. This important point is discussed in Appendix A.

All one-electron MO's are orthogonal and therefore matrix elements and energy calculations are rather simple. The main problem lies in the circumstance that these functions do not reduce to atomic wave functions when the nuclei are infinitely far apart. The molecular orbitals obey a symmetry requirement which forces them to produce equal electron density on both nuclei. The total electron density $\Psi^*\Psi$ does indeed resemble that of two atoms with almost no electron density between their centers, but individual parts such as $(1\sigma_g(1))^*(1\sigma_g(1))$ or $(1\sigma_u(3))^*(1\sigma_u(3))$ distribute density on both centers. We will later see that this inconvenience tends to be serious only when improvements are attempted on functions (47) and (48). It is very serious in the ground state of H_2, given by equation (46), since at large separations this state decomposes into $H+1/2\ (H^+ + H^-)$ and not into 2H. Such complications do not occur if we treat systems with equal numbers of bonding and antibonding electrons.

To calculate the MO energy we first need to determine the MO functions. For a homonuclear molecule containing N electrons,

$$\Psi = \mathcal{A}\left(\prod_{i=1}^{N} \Phi_i\right). \tag{49}$$

Each orbital has a definite symmetry and can be written in the form

$$\Phi_i = \sum_j c_{ji}\left(\frac{\phi_{aj}+\lambda_i\phi_{bj}}{\sqrt{2}}\right)\cdot s, \tag{50}$$

where ϕ_{aj} and ϕ_{bj} are "atomic" functions centered about nuclei a and b, and s is the spin function (α or β). While the ϕ's are usually Slater orbitals, they could also be Gaussians or any other localized function. To satisfy the g and u symmetry requirements, $\lambda_i = (-1)^{m_\lambda}$ for g symmetry and $\lambda_i = (-1)^{m_\lambda+1}$ for u symmetry, m_λ being the quantum number for the component of angular momentum around the z-axis; for σ orbitals $m_\lambda = 0$, for π orbitals $m_\lambda = \pm 1$, etc.

The Slater orbitals are

$$\phi_{aj} = (2\zeta_j)^{n_j+1/2}[(2n_j)!]^{-1/2} r_a^{n_j-1} e^{-r_a\zeta_j} Y_l^m(\theta, \varphi). \tag{51}$$

Due to symmetry restrictions only certain types of atomic orbitals can contribute to (50). These orbitals must have the same magnetic quantum numbers as the molecular orbital. Thus a σ orbital is composed of atomic functions which are symmetric about the z-axis of the atom ($m = 0$).

The molecular orbitals Φ_i, can be obtained by solving either a differential or a matrix equation. After the pioneering work of Roothaan (1951) the matrix method has been used almost exclusively; an outline of his version of the HF procedure for closed shell molecules (two electrons per molecular orbital) follows. It is particularly simple for this case, which fortunately is the one of greatest usefulness in calculating intermolecular forces. For a treatment of the open shell case, the works of Roothaan (1951, 1960) or the reviews by Berthier (1964) and Simonetta and Gianinetti (1964, pp. 83, 57) should be consulted.

In the case of a closed-shell, diatomic molecule it is convenient to define a unique HF operator,

$$h = I+g; \tag{52}$$

I represents the one-electron operators; for any electron i

$$I = -\frac{1}{2}\nabla_i^2 - \frac{Z_a}{r_{ai}} - \frac{Z_b}{r_{bi}}, \tag{53}$$

while g is the total two-electron contribution which gives rise to the matrix elements

$$g_i = \sum_j \langle \Phi_j(1)|\, R(1i)\, |\Phi_j(1)\rangle \tag{54}$$

$$= \sum_j (J_j - K_j). \tag{55}$$

J_j and K_j the direct Coulomb operator and the Coulomb exchange operator, are defined as the two parts of

$$R(1i) = \frac{1}{r_{1i}}(1-P_{1i}) \tag{56}$$

where P_{1i} is the permutation operator which interchanges electrons 1 and i in the functions it acts upon. Notice that g_i contains the terms in which $j = i$, so that this operator is the same for all orbitals.

Each molecular orbital Φ_i shall be expanded in terms of the set of functions X_j (defined by reference to (50)) which carry coefficients c_{ji}. A suitable set of functions $\{X\}$ is chosen such that each Φ_i is adequately represented by the smallest possible number of X_i. If we also define the row vector $\boldsymbol{X}$

$$\boldsymbol{X} = (X_1 X_2 \ldots X_m) \tag{57}$$

and the column vector

$$\boldsymbol{c}_i = \begin{pmatrix} c_{1i} \\ c_{2i} \\ \vdots \\ c_{mi} \end{pmatrix} \tag{58}$$

the molecular orbitals of (50) are

$$\Phi_i = Xc_i s. \tag{59}$$

Since the purpose of this method is to find the best antisymmetrized product function, we must minimize the HF energy $\langle \Psi | H | \Psi \rangle$ subject to the condition that $\langle \Psi | \Psi \rangle = 1$.

Using (52) we find

$$\langle \Psi | H | \Psi \rangle = \sum_i I_i + \sum_i \sum_{\substack{j \\ (i>j)}} (J_{ij} - K_{ij}), \tag{60}$$

where J_{ij} and K_{ij} are, as before, the integrals

$$J_{ij} = \langle \Phi_i(1)\Phi_j(2) \left| \frac{1}{r_{12}} \right| \Phi_i(1)\Phi_j(2) \rangle, \tag{61}$$

$$K_{ij} = \langle \Phi_i(1)\Phi_j(2) \left| \frac{1}{r_{12}} \right| \Phi_i(2)\Phi_j(1) \rangle. \tag{62}$$

Representing the operators I, J_{ij}, and K_{ij} by their matrices, we can write†

$$\langle \Psi | H | \Psi \rangle = \sum_i c_i^+ \overleftrightarrow{I} c_i + \tfrac{1}{2} \sum_i \sum_j c_i^+ (\overleftrightarrow{J}_j - \overleftrightarrow{K}_j) c_i. \tag{63}$$

For orthonormal molecular orbitals

$$\langle \Phi_i | \Phi_i \rangle = c_i^+ \overleftrightarrow{S} c_j = \delta_{ij}, \tag{64}$$

where $\overleftrightarrow{S}$ is the overlap matrix with elements:

$$(S)_{kl} = \langle X_k | X_l \rangle.$$

Minimizing (63) with respect to variations in the coefficients while writing $-2\varepsilon_{ij}$ as a Lagrangian multiplier, we have

$$\delta \{ \langle \Psi | H | \Psi \rangle - 2\varepsilon_{ij} c_i^\dagger \overleftrightarrow{S} c_j \} = 0 \tag{65}$$

for every i, j, and this leads to the minimizing condition (Roothaan, 1951)

$$\overleftrightarrow{h} c_i = \sum_j \overleftrightarrow{S} c_j \varepsilon_{ij} \tag{66}$$

where $\overleftrightarrow{h} = \overleftrightarrow{I} + \sum_j (\overleftrightarrow{J}_j - \overleftrightarrow{K}_j)$. We can always force ε into diagonal form (Slater, 1960, vol. 2) by a unitary transformation of the

† Remember, c_i^+ is a row matrix. See, for instance, Margenau and Murphy, (1956, vol. 1, chap. 10).

orbitals within these closed shell systems. Hence we may write

$$\overleftrightarrow{\boldsymbol{h}}\, \boldsymbol{c}_i = \varepsilon_i \overleftrightarrow{\boldsymbol{S}}\, \boldsymbol{c}_i, \qquad i = 1, 2 \ldots n. \tag{67}$$

Equation (67) is non-linear since $\boldsymbol{J}_j$ and $\boldsymbol{K}_j$ which appear in $\boldsymbol{h}$ also depend on the c_i. For a typical coefficient we have

$$\sum_{k=1}^{m} \Bigg\{ \langle X_l | I | X_k \rangle + \sum_{j=1}^{n} \sum_{r} \sum_{s} c_{rj}^{+} c_{sj} \left[\left\langle X_l(1) X_r(2) \left| \frac{1}{r_{12}} \right| X_k(1) X_s(2) \right\rangle \right.$$
$$\left. - \left\langle X_l(1) X_r(2) \left| \frac{1}{r_{12}} \right| X_s(1) X_k(2) \right\rangle \right] - \varepsilon_i S_{lk} \Bigg\} c_{ki} = 0. \tag{68}$$

To solve this non-linear problem one first selects a set of basis functions $\boldsymbol{X}$ and assumes a set of coefficients for each orbital from which are then constructed the set of matrices $\boldsymbol{J}_j$ and $\boldsymbol{K}_j$. These matrices are used in (67) to solve for new coefficients. These in turn define a new set of J_j and K_j which determine a new set of $\boldsymbol{c}_i$. When after several cycles the input set agrees to within some specified tolerance with the output set, the process has converged and the orbitals are said to be self-consistent.

The HF differential equation for each orbital is very similar to (67); it is

$$h\Phi_i = \varepsilon_i \Phi_i. \tag{69}$$

In fact, (67) and (69) would be identical if the set $\{X\}$ were complete. Since this is usually not the case we shall refer to orbitals obtained from (67) as self-consistent field (SCF) orbitals and not as HF orbitals. The energy and orbitals of the SCF procedure can be pushed arbitrarily close to the HF values by choosing sufficiently large, sufficiently complete basis sets.

In practice two simplifications can be made in solving (67). First, in a closed shell atom each orbital is occupied by one α and one β electron. Therefore we have only to solve $n/2$ equations in an n electron system. The two-electron term in h is

$$\sum_{\substack{\text{different spatial} \\ \text{orbitals}}}^{n/2} (2J_j - K_j).$$

Secondly, the set of basis orbitals $\{X_p\}$ can be divided into various symmetry types and an equation like (67) holds for each type separately ($\sigma_g, \sigma_u, \pi_g, \pi_u, \delta_g \ldots$).

The many symmetry equations must nevertheless be solved together because of coupling which is present in J and K_j.

For more details on the operations involved in finding the orbitals the works of Roothaan and Bagus (1963), and Wahl (1964) should be studied.

The SCF equations have been solved for many systems to various degrees of approximation. Until recently most calculations involved a minimum amount of exponent variation. [ζ_j in eqn. (51).] Today the more elaborate calculations perform the self-consistent procedure for a variety of exponents and select that basis set whose exponents yield the best energy. An example of early calculations of this sort is the work of Fraga and Ransil (see Ransil, 1960). We owe some of the best orbitals obtainable for diatomic molecules with today's computers to the effort at the University of Chicago led by Roothaan.† Publications by members of this group (among them Wahl, Bagus, Cade, Gilbert, Huo) represent the present limits of the SCF method. [Huo (1965) CO and BF and other work in press]; Wahl [(1964) (F_2)]; Cade, D. Sales, and Wahl [(1966) (N_2 and N_2^+]; [Das and Wahl 1966 extended HF calculations on H_2, Li_2, and F_2]. In comparison with these results, SCF calculations on polyatomic molecules still seem primitive. Best results have been obtained by Pitzer and Lipscomb, (1963) and Moskowitz and Harrison, (1965).

Once the orbitals are known the total energy can be calculated in accordance with (60):

$$E_{\text{HF}} = \sum_{i=1}^{n} \varepsilon_i - \sum_{i>j}^{n} (J_{ij} - K_{ij}). \tag{70}$$

After a little algebra which involves adding (60) and (70) we can also write

$$E_{\text{HF}} = \tfrac{1}{2} \sum_{i=1}^{n} (I_i + \varepsilon_i). \tag{71}$$

It is important to understand the physical significance of the symbols ε_i, which appear throughout this work. ε_i is called an orbital energy. Its meaning becomes clear when we calculate the energy of an ion of the same system, formed by removing electron k. If the orbitals in the ion are the *same* as those in the neutral system, the energy of the ion is

$$E_{\text{HF}}\,(\text{ion}) = \sum_{i=1(\neq k)}^{n} I_i + \tfrac{1}{2} \sum_{i\neq k} \sum_{j} (J_{ij} - K_{ij})$$

† Another large group including Clementi, Nesbet, McLean, and Yoshimine is active at the IBM laboratory in San Jose, California.

$$= \sum_{i=1}^{n} I_i - I_k + \tfrac{1}{2} \sum_i \sum_j (J_{ij} - K_{ij}) - \sum_{i(\neq k)} (J_{ik} - K_{ik})$$

$$= E_{\mathrm{HF}}\,(\text{neutral atom}) - \langle \Phi_k | I + \sum_i (J_i - K_i) | \Phi_k \rangle$$

$$= E_{\mathrm{HF}}\,(\text{neutral atom}) - \varepsilon_k. \tag{72}$$

Thus the orbital energy ε_k represents the energy that would be required to ionize the system by removing electron k, providing all the other orbitals remained unchanged. In many cases ε_k is therefore almost exactly the ionization potential.

From our definition, eqn. (11), the intermolecular potential in the HF or SCF approximation is

$$[\Delta E(R)]_{\mathrm{HF}} = E_{\mathrm{HF}} - (E_a)_{\mathrm{HF}} - (E_b)_{\mathrm{HF}}. \tag{73}$$

To obtain it we subtract from E_{HF} the energy at $R \to \infty$, namely the sum of the HF energies of atoms a and b. Just as the energy calculated by the Heitler–London method is not a bound on the interaction energy if approximate atomic functions are used (see (25) and the following paragraphs), (73) likewise does not usually possess any bounds.

The procedure outlined here at some length is not useful when open shell atoms are involved. In that case, however, the ordinary HF or molecular orbital method is unsuited to define $\Delta E(R)$ anyway, since the requirement that all obitals be doubly occupied forces the system to dissociate into *ions* at large separations. Only in certain excited states with higher spin multiplicity does the Pauli principle cause atoms to be formed upon dissociation. The $^3\Sigma_u$ state of H_2 is the simplest example; it does not dissociate into an H^- ion since there can not be two $1s$ electrons with α spin in the same ion.

We now turn to specific results. Accurate values of $[\Delta E(R)]_{\mathrm{SCF}}$ have been determined for the hydrogen, helium, and neon interactions, in addition to those systems where chemical bonding is involved. We shall consider primarily the helium interaction.

The state function for two helium atoms is composed of two molecular orbitals, $1\sigma_g$ and $1\sigma_u$. Each of these in the SCF procedure is a linear combination of basis functions X_p, but the functions need not be the same for both orbitals since, as mentioned previously, each orbital has its own equation (67). Each X_p is the sum or difference of one atomic orbital on center *a* and one on center *b*. Since at large separations the electron density should resemble that of an isolated atom, a good first choice for the atomic orbitals in X_p would be those which give the best energy for the atom. Exactly what changes take place as the atoms are brought together we can not predict in *a priori* fashion but we do know that there must be some distortion of the electron distribution between the nuclei. Griffing and Wehner (1953) calculated the intermolecular potential using only one basis function for $1\sigma_g$ and $1\sigma_u$. The atomic function used was a 1*s* orbital with exponent equal to 1.6875. This is the best simple helium atom wave function (Pauling and Wilson, 1935; Kauzman, 1957; see also Table 3.3). Their calculation, which will be considered in more detail later, made clear the need for extreme care in evaluating (68). The energy difference is a very small part of the total energy, smaller, in fact, than the error in the well-known Hylleraas calculations on the helium atom, hence the total energy must be known precisely if the difference is to be meaningful. Because of small errors in some of their integrals the results of Griffing and Wehner meet this test only partially. In the region of small overlap ($R > 3a_0$) supplementary calculations (Griffing and Wehner, 1953) showed that the best exponent for use in the atomic orbital was 1.6875. This is the same value one finds for the isolated atom, hence distortion of the electron density is small at such separations. Griffing and Wehner were therefore justified in using as exponents the atomic values at all separations.

However, Huzinaga (1957) pointed out that at very small distances the $1\sigma_g$ orbital will contract and $1\sigma_u$ will expand. This is because in the limit of zero separation the system must reduce to the beryllium atom. The $1\sigma_g$ orbital becomes a 1*s* orbital of the beryllium atom, and this has a smaller radius and larger exponent than the helium 1*s* orbital. The $1\sigma_u$ orbital, on the other hand, must become a larger orbital of beryllium which has an exponent smaller than 1.6875. Thus Huzinaga suggested the use of different exponents, i.e. different ζ in (51), for the $1\sigma_g$ and $1\sigma_u$ orbitals. This method, often referred to as the "MO-ζ method", led to a significant improvement in energy for distances less than 3 Bohr radii. At $R = 1$, for example, Huzinaga found ζ to be 2.25 for $1\sigma_g$ and 1.25 for $1\sigma_u$, while at $R = 0$ Slater (1963, vol. 1, p. 116) obtained 3.68 for $1\sigma j$ and 0.379 for $1\sigma_u$.

Ransil (1961) extended this work by using a three-term basis set for each orbital. While the exponents of the $1s$, $2s$, and $2p$ functions in each orbital were taken to be equal, a different exponent was used for the $1\sigma_g$ than for the $1\sigma_u$ orbital. Below $3a_0$ his results were significantly lower than the earlier work. However, the $[\Delta E]_{SCF}$ obtained for this basis set with the best variationally determined exponents exhibited an attractive well almost as deep as is indicated by the experimental data for the total potential curve. Since, as shown in Appendix A, the van der Waals' attraction at large distances is *not* contained in the MO treatment, this feature of his results was difficult to understand. We now know that this minimum is a spurious result occasioned by the fact that the choice of equal exponents caused the $2s$ function to contribute to the energy in the molecule (finite R) whereas it cannot do so in the atoms (infinite R) (Roothaan and Weiss, 1960, table I).† Relaxing the equal exponent constraint removed the attractive well.

Recently two detailed studies have been made of the SCF energy for this system (Kestner, 1968; Gilbert and Wahl, 1967). The results of Kestner are listed in Table 3.4. The basis set consisted of three terms whose linear combination was found to yield accurate energies for the helium atom (Bagus, *et al.* to be published),‡ plus such other atomic functions $2p\,(m = 0)$, $3s$ and $3d$ $(m = 0)$ as would contribute significantly. Exponents were arbitrarily varied from their atomic values until the total energy was a minimum at each internuclear separation.

From a detailed study of the calculated orbitals we observe that the deviation of exponents from their atomic values for $R > 3a_0$ is very small. Two types of changes take place in the state function when atomic overlap is small. There are small symmetric distortions about each nucleus (the coefficients of the less extended basis functions with larger ζ become more important as R decreases) and some $2p(m = 0)$-type distortions (the coefficient of this basis function increases almost exponentially as R decreases.) Both changes result in decreased electron density between the nuclei. Basis functions containing atomic d and f orbitals contribute negligibly.

† We wish to thank Professor Roothaan for pointing this out to us.
‡ In this work the most accurate orbital in Table 3.3 was used.

TABLE 3.4. *Energy of Interaction between Helium Atoms with no Electron Correlation Included*

R (a_0)	Interaction energy (a.u.) LCAO–MO: Single Slater orbitals	LCAO–MO: Atomic orbitals	Complete SCF: Calculations
0.0			
0.5			3.0328
1.0			0.92957
2.0			0.1207
3.0	0.013131	0.015445	0.013537
4.0	0.0010752	0.0014753	0.0013563
4.7			0.000256_5
5.0	0.000077499	0.00013284	
5.2			0.000075_4
5.5	0.00001996	0.000038818	0.000036_8
6.0	0.0000050235	0.000011227	0.000010_6
7.0	2.993×10^{-7}	0.9138×10^{-6}	
8.0	1.679×10^{-8}	7.2139×10^{-8}	

At very short distances ($R < 3a_0$) the SCF procedure indicates a very significant change in the nature of the molecular orbitals. The $1\sigma_g$ orbital loses almost all of the $2p_z$ mixing while the $1\sigma_u$ orbital acquires a good deal of it. Consider a simple SCF–MO at very small R:

$$\sigma_g \sim \frac{1}{\sqrt{[2(1+S)]}}\,(e^{-\delta r_{a1}}+e^{-\delta r_{b1}})$$

$$\sigma_u \sim \frac{1}{\sqrt{[2(1-S)]}}\,(e^{-\delta r_{a1}}-e^{-\delta r_{b1}}) \tag{74}$$

when $R \to 0$, $S \to 1$ as $e^{-\delta R}$ (S is the overlap of orbitals $e^{-\delta r_{a1}}$ and $e^{-\delta r_{b1}}$) and $\sigma_g \to e^{-\delta r_{a1}}$, which is a $1s$ united-atom orbital. However, since

$$\sigma_u \to (1-S)^{-1/2}e^{-\delta r_{a1}}(1-e^{-\delta(r_{b1}-r_{a1})})$$

and $r_{b1}-r_{a1} \sim -R\cos\theta$ according to the cosine law, we find as $R \to 0$,

$$\sigma_u \to e^{-\delta r_{a1}}\cos\theta. \tag{75}$$

This represents a sort of "$1p$" united-atom orbital. Actually the united-atom solution should be that corresponding to a beryllium atom with its configuration $1s^2\ 2s^2$. The $1\sigma_g$ orbital goes into the $1s$ form naturally, but a simple $1\sigma_u$ does not become a $2s$ orbital. The strong $2p$-mixing found in the SCF result partially corrects the $1\sigma_u$ behavior at small distances.

Table 3.4 presents a comparison of the complete SCF results and various approximations to it. The first two columns refer to calculations in which the basis set consisted of one- and two-term approximations to the atomic HF orbital. No orbital exponents were varied to obtain these results, the atomic values being used for all separations.

We summarize the results in Table 3.4 by fitting simple exponential expressions to the numerical data using atomic units.

$$\Delta E_{\text{LCAO-MO (one term)}}(R) = 55.81e^{-2.720R} \quad (3-8a_0)$$

$$\Delta E_{\text{LCAO-MO(HF atomic orbitals)}}(R) = 28.13e^{-2.462R} \quad (3-8a_0)$$

$$\Delta E_{\text{SCF}}(R) = 22.42e^{-2.424R} \quad (4-6a_0)$$

or

$$\Delta E_{\text{SCF}}(R) = e^{-2.906R}(7.764+4.490R + 3.693R^2+1.046R^3) \quad R \geqq 1a_0. \tag{76}$$

These fits are good to 2% over the ranges indicated. There is, however, no formal justification for the form of the potentials chosen. The large exponent for the one-term result is again representative of the poor asymptotic properties of a one-term atomic wave function.

There have been several attempts to calculate the interaction at short distances by a perturbation expansion starting from beryllium orbitals, that is, a united-atom expansion. The latest work by Miller and Present (1963) suggests that such expansions are useful only for distances less than $0.6a_0$. Brown and Steiner (1966) [see also Brown, 1966] and Levine (1964) have also shown that in one-electron problems an expansion of the energy in powers of R is not possible. There is, at least, a term in $R^5 \ln R$. Such small separations are of little physical interest at present since the interaction energy is extremely large (about 80 ev at $0.5a_0$) and is thus important only in extremely high-energy collisions.

One particular approximate form of the MO procedure is of such wide use that it deserves special consideration. This is the

LCAO (linear combination of atomic orbitals) procedure. As the name implies, the basis functions used are those which yield the best energy for the atom. Exponents and coefficients, except for changes due to symmetry, are not allowed to vary from their atomic values. Since the term LCAO has been employed in the literature for molecular orbitals formed from various types of atomic orbitals, we shall often feel the need to qualify our notation by specifying the atomic orbitals involved, e.g. one-term Slater orbitals, HF orbitals, etc. If there were no distortion or overlap these would be the correct molecular orbitals at large separations for the interaction of closed shell atoms.

We shall now turn to the detailed calculations, applying the LCAO–MO method to hydrogen and helium interactions. The results will prove identical with those of Heitler–London, described in the previous section.

Consider $H_2(^3\Sigma_u^+)$ in the LCAO approximation. The normalized molecular orbitals are

$$1\sigma_g = \frac{1}{\sqrt{[2(1+S)]}}(a+b) \qquad (77)$$

$$1\sigma_u = \frac{1}{\sqrt{[2(1-S)]}}(a-b)$$

$$S = \langle a | b \rangle \qquad (78)$$

where a and b are the ordinary hydrogen atom functions ($\zeta = 1$) localized about nuclei a and b, respectively.

The antisymmetrized wave function, eqn. (47), is

$$\Psi_{H_2}(^3\Sigma_u) = \mathcal{A}\,(1\sigma_g(1)\alpha(1)1\sigma_u(2)\alpha(2))$$

$$= \frac{1}{2\sqrt{(1-S^2)}} \begin{vmatrix} a(1)+b(1) & a(2)+b(2) \\ a(1)-b(1) & a(2)-b(2) \end{vmatrix} \alpha(1)\alpha(2)$$

$$= \frac{1}{\sqrt{(1-S^2)}} \begin{vmatrix} a(1) & a(2) \\ b(1) & b(2) \end{vmatrix} \alpha(1)\alpha(2)\,. \qquad (79)$$

Except for normalization, this is the Heitler–London result, eqn. (30), for which the energy is given by (33).

The energy of interaction in the general SCF procedure can not be written in compact form, but if we assume an LCAO–MO state function we can write all results in terms of integrals over atomic orbitals. We now demonstrate this explicitly for helium.

The molecular orbitals (LCAO–MO) needed to calculate the helium interaction are

$$1\sigma_g = \frac{1}{\sqrt{[2(1+S)]}}(A+B) \tag{80}$$

$$1\sigma_u = \frac{1}{\sqrt{[2(1-S)]}}(A-B), \tag{81}$$

where A and B are *atomic HF orbitals* on nuclei A and B† respectively. The HF energy of the *atom* is

$$E_{\text{HF}} = 2\langle A| -\frac{1}{2}\nabla^2 - \frac{2}{r}|A\rangle + \left\langle A(1)A(2)\left|\frac{1}{r_{12}}\right|A(1)A(2)\right\rangle. \tag{82}$$

To obtain the molecular energy we use the entire determinantal function

$$\Psi = \mathcal{A}[1\sigma_g(1)\alpha(1)1\sigma_g(2)\beta(2)1\sigma_u(3)\alpha(3)1\sigma_u(4)\beta(4)]$$
$$= \frac{1}{\sqrt{4!}}\begin{vmatrix} 1\sigma_g(1)\alpha(1) & 1\sigma_g(2)\alpha(2) & 1\sigma_g(3)\alpha(3) & 1\sigma_g(4)\alpha(4) \\ 1\sigma_g(1)\beta(1) & 1\sigma_g(2)\beta(2) & 1\sigma_g(3)\beta(3) & 1\sigma_g(4)\beta(4) \\ 1\sigma_u(1)\alpha(1) & 1\sigma_u(2)\alpha(2) & 1\sigma_u(3)\alpha(3) & 1\sigma_u(4)\alpha(4) \\ 1\sigma_u(1)\beta(1) & 1\sigma_u(2)\beta(2) & 1\sigma_u(3)\beta(3) & 1\sigma_u(4)\beta(4) \end{vmatrix}. \tag{83}$$

By combining rows with like spins we obtain the Heitler–London function based on the atomic HF orbitals as discussed earlier in this chapter:

$$\Psi = \frac{2}{\sqrt{(4!)(1-S^2)}}\begin{vmatrix} A(1)\alpha(1) & A(2)\alpha(2) & A(3)\alpha(3) & A(4)\alpha(4) \\ A(1)\beta(1) & A(2)\beta(2) & A(3)\beta(3) & A(4)\beta(4) \\ B(1)\alpha(1) & B(2)\alpha(2) & B(3)\alpha(3) & B(4)\alpha(4) \\ B(1)\beta(1) & B(2)\beta(2) & B(3)\beta(3) & B(4)\beta(4) \end{vmatrix} \tag{84}$$

This correspondence between the two methods holds for any homonuclear interaction which has an even number of alpha and beta spins.

The total energy is

$$E = \langle\Psi|H|\Psi\rangle \tag{85}$$

since $\langle\Psi|\Psi\rangle = 1$.

† The letters A and B are used in two senses, but when they denote the orbital they will usually carry a numerical argument, and so no confusion should arise.

In the notation of Griffing and Wehner (1953) we write

$$E_{\text{LCAO-MO}} = 2H_g+2H_u+J_{11}+4J_{12}+J_{22}-2K_{12}+4/R$$

$$H_g = \left\langle 1\sigma_g(1) \left| -\frac{1}{2}\nabla^2-\frac{2}{r_{a1}}-\frac{2}{r_{b1}} \right| 1\sigma_g(1) \right\rangle$$

$$= 2N_g^2\left[\left\langle A \left| -\frac{1}{2}\nabla^2-\frac{2}{r_{a1}}-\frac{2}{r_{b1}} \right| A \right\rangle + \left\langle A \left| -\frac{1}{2}\nabla^2-\frac{2}{r_{a1}}-\frac{2}{r_{b1}} \right| B \right\rangle\right].$$

The J and K symbols are

$$J_{11} = \left\langle 1\sigma_g(1)1\sigma_g(2) \left| \frac{1}{r_{12}} \right| 1\sigma_g(1)1\sigma_g(2) \right\rangle$$

$$= N_g^4[2(AA\varrho AA)+8(AA\varrho AB)+4(AA\varrho BB)+2(AB\varrho AB)]$$

$$J_{12} = N_g^2N_u^2[2(AA\varrho AA)+2(AB\varrho AB)-4(AA\varrho BB)]$$

$$J_{22} = N_u^4[2(AA\varrho AA)-8(AA\varrho AB)+4(AA\varrho BB)+2(AB\varrho AB)]$$

$$K_{12} = N_g^2N_u^2[2(AA\varrho AA)-2(AB\varrho AB)], \tag{86}$$

where

$$N_g = \frac{1}{\sqrt{[2(1+S)]}}, \quad N_u = \frac{1}{\sqrt{[2(1-S)]}}, \quad S = \langle A\,|\,B)\rangle$$

$$(pr\varrho gt) = \left\langle p(1)r(2) \left| \frac{1}{r_{12}} \right| g(1)t(2) \right\rangle. \tag{87}$$

In the same notation

$$(E_{\text{HF}})_a = 2\left\langle A \left| -\frac{1}{2}\nabla^2-\frac{2}{r} \right| A \right\rangle+(AA\varrho AA). \tag{88}$$

$(AA\varrho AA)$ is a one-center Coulomb integral, $(AB\varrho AB)$ is a two-center Coulomb integral, $(AA\varrho AB)$ is a two-center hybrid or ionic integral, and $(AA\varrho BB)$ is a two-center exchange integral. Similarly,

$$(\Delta E)_{\text{LCAO}} = [4(N_g^2+N_u^2)-4]\left\langle A \left| -\frac{1}{2}\nabla^2-\frac{2}{r_{a1}} \right| A \right\rangle$$

$$+4(N_g^2+N_u^2)\left\langle A \left| -\frac{2}{r_{b1}} \right| A \right\rangle$$

$$+4(N_g^2-N_u^2)\left\langle A \left| -\frac{1}{2}\nabla^2-\frac{2}{r_{a1}}-\frac{2}{r_{b1}} \right| B \right\rangle$$

$$+[2(N_g^4+2N_g^2N_u^2+N_u^4)-2](AA\varrho AA)$$
$$+8(N_g^4-N_u^4)(AA\varrho AB)$$
$$+2(N_g^4+6N_g^2N_u^2+N_u^4)(AB\varrho AB)$$
$$+4(N_g^4-4N_g^2N_u^2+N_u^4)(AA\varrho BB)+4/R. \qquad (89)$$

When normalization constants are introduced we obtain

$$(\Delta E)_{\text{LCAO}} = \left[\frac{4S^2}{1-S^2}\right]\left\langle A\left|-\frac{1}{2}\nabla^2-\frac{2}{r_{a1}}\right|A\right\rangle$$
$$-\frac{4S}{1-S^2}\left\langle A\left|-\frac{1}{2}\nabla^2-\frac{2}{r_{a1}}-\frac{2}{r_{b1}}\right|B\right\rangle$$
$$-\frac{8S}{(1-S^2)^2}(AA\varrho AB)-\frac{(2-6S^2)}{(1-S^2)^2}(AA\varrho BB)$$
$$+\frac{4S^2-2S^4}{(1-S^2)^2}(AA\varrho AA)$$
$$+\frac{4}{1-S^2}\left\langle A\left|-\frac{2}{r_{b1}}\right|A\right\rangle$$
$$+\frac{4-2S^2}{(1-S^2)^2}(AB\varrho AB)+4/R. \qquad (90)$$

Inspection shows that all but the last three terms are of order S^2 or higher, while the latter are individually proportional to $1/R$ at large separations. This observation, however, is misleading since these terms combine to approach zero in the manner of S^2/R, as we showed for the hydrogen interactions, cf. (39). It is tempting to separate the energy into those parts which have coefficients proportional to overlap and those which do not, as Löwdin did (1950). For example, if S is zero, (90) reduces to

$$(\Delta E)_{S=0} = -2(AA\varrho BB)+4\left\langle A\left|-\frac{2}{r_{b1}}\right|A\right\rangle$$
$$+4(AB\varrho AB)+4/R. \qquad (91)$$

But this is not a very useful expression since *all* terms, when properly treated, behave roughly as S^2. Löwdin called $\Delta E-(\Delta E)_{S=0}$ the S energy. It should be observed, however, that both terms, which are usually of comparable magnitude, vanish if the atoms do not overlap.

The last three terms in (90), even though they individually behave as $1/R$, collectively contribute less than the energy resulting from electron exchange, even at large separations.† The Coulomb and other long-range parts are collectively screened off quite rapidly as R increases. When combined, the last three terms in (90) yield

$$-\frac{2S^2(1-2S^2)}{(1-S^2)^2R}-G, \tag{92}$$

where G is a function which is obtained in a manner analogous to the way in which we combined the first two terms in (38) to obtain the second last term in (39). The specific form of G for any purpose can be obtained from the integral formulas of Roothaan (1951). It goes to zero approximately as S^2. When the orbitals A and B are one-term Slater orbitals with exponent ζ, one finds that (Kestner and Sinanoglu, 1966)

$$\begin{aligned} e^{2\zeta R}G = & -\frac{8}{1-S^2}\left(\zeta+\frac{1}{R}\right) \\ & +\frac{4-2S^2}{(1-S^2)^2}\left(\frac{1}{R}+\frac{11}{8}\zeta+\frac{3}{4}\zeta^2R+\frac{1}{6}\zeta^3R^2\right). \end{aligned} \tag{93}$$

For other types of orbitals G is a more complicated function.

The total interaction in the form of (90), but with expression (92) in place of the last three terms of (90), has been evaluated for a one-term atomic orbital with $\zeta=1.6875$ and for a two-term orbital (Kestner and Sinanoglu, 1966) proposed by Green *et al.* (1954).‡ The various results are collected in Table 4.4 together with the complete SCF results. It will be seen that the one-term atomic orbital leads to an underestimate of the repulsion at large separations but is rather satisfactory at smaller distances for reasons discussed before, namely the low electron density in the tail of the atomic wave function. The LCAO–MO involving atomic HF wave functions compares favorably with the complete SCF results for separations of more than $3a_0$ (1.6 Å). The maximum error of only 12% (for $R > 3a_0$) is within the range generally acceptable for practical applications. It should be empha-

† The same point has been made concerning the H_2 interaction by Ellison (1961). This could have been seen from (39) if individual terms had been evaluated.

‡ This orbital is very similar to the atomic SCF results of Bagus *et al.* (to be published). See Table 3.3.

sized that the LCAO–MO calculation even involving HF atomic orbitals is very simple and capable of high precision, whereas the complete SCF results, although more accurate, are very elaborate and time consuming even with the new programs and fast computers. What we desire is a compromise, a simple method to correct these differences for the many distortion effects found in the true orbital. We shall return to this point in the next chapter.

3.4. Alternate Approaches to the State Function

Several zero-order approaches to the interactions have been tried aside from the Heitler–London, the MO, and the united-atom expansions discussed above. Some of these are statistical (Abrahamson, 1963, 1964) and not within the scope of our discussion, for while yielding reasonable answers there seems to be no way of assessing their accuracy.

We likewise omit at this point discussion of Gaussian or one-center basis sets since these become increasingly inadequate at large separations and are thus in principle not well suited for treating intermolecular forces. (See, however, Kotani, Ohno, and Kayama, 1961, vol. 37/1, p. 78.) Qualitative use of them will be made in Chapter 5.

More interesting is the prospect of calculating the force and not the energy. The cancellations which concerned us above should be automatically cared for in such calculations since at large separations the force must vanish. The potential can then be obtained by an integration. Beginning with the Hellman–Feynman theorem (Hellmann, 1937, p. 285; Feynmann, 1939), which features an electrostatic approach to quantum chemistry, Hurley (1954, 1956) used this procedure to study interactions involving bonding. Salem (1961) has treated repulsive interactions such as the He_2 problem. A basic fault of this procedure is, however, that it is only as accurate as the state function employed, and that to extract a potential curve is quite tedious. Whereas energy calculations are correct to *second order* in the relative error of the function, these force calculations are correct, in general, only to the *first order*, and this can lead to large errors in ΔE. Nevertheless, this approach is interesting and likely to provide an area of continuing research. We shall consider next the use of localized orthogonal orbitals and elliptic coordinate expansions.

3.4.1. *Localized Orthogonal Orbitals*

As already noted, one of the defects of molecular orbital theory from the point of view of intermolecular forces is the delocalization of electron density in the orbital functions. Even at large separations each electron spends half of its time about each nucleus, yet we know that a reasonable approximation results if we put one electron around one particular nucleus. In fact, this was done in the Heitler–London theory which, in the case of H_2 ($^3\Sigma_u$) and He_2, gave the same energy as the molecular orbital theory in LCAO–MO form. We showed that the wave functions of the LCAO–MO and the approximate Heitler–London forms are identical whenever an equal number of bonding and antibonding electrons are involved. On the other hand, the Heitler–London theory was cumbersome because the atomic orbitals were not orthogonal.

To remedy these difficulties, Coulson and Fisher (1949) suggested using functions of the form

$$A = \frac{a+\lambda b}{\sqrt{(1+2\lambda S+\lambda^2)}}, \tag{94}$$

$$B = \frac{b+\lambda a}{\sqrt{(1+2\lambda S+\lambda^2)}} \tag{95}$$

for the H_2 and He_2 systems. The H_2 ($^3\Sigma_u$) function would be

$$\Psi = [A(1)B(2)-A(2)B(1)]\alpha(1)\alpha(2). \tag{96}$$

The parameter λ could be chosen to give the lowest energy for the ground state of H_2 ($^1\Sigma_g$), thus providing for a variation in the relative proportions of the covalent $a(1)\ b(2)$, and ionic $a(1)\ b(1)$ components in the MO function. For our systems with equal numbers of bonding and antibonding electrons this ratio is always fixed.

However we can also choose λ so that A and B are orthogonal, $\langle A \mid B \rangle = 0$. For this purpose we require that

$$(\lambda_0^2+1)S+2\lambda_0 = 0$$

or

$$\lambda_0 = \frac{-1\pm\sqrt{(1-S^2)}}{S}. \tag{97}$$

The energy obtained with this function is that of the equivalent LCAO–MO or Heitler–London calculation for H_2 ($^3\Sigma_u$), He_2, or any interaction of closed-shell atoms with an equal number of alpha and beta spins and an equal number of bonding and anti-bonding electrons.

To show the equivalence of the LCAO–MO and the orthogonalized atomic orbital state function, we write in place of A and B

$$X_a = [(l_1+l_2)a+(l_1-l_2)b]\alpha$$

and

$$X_b = [(l_1-l_2)a+(l_1+l_2)b]\alpha. \tag{98}$$

Since we require that

$$\langle X_a | X_a \rangle = \langle X_b | X_b \rangle = 2(l_1^2+l_2^2)+2(l_1^2-l_2^2)S = 1$$
$$\langle X_a | X_b \rangle = 2(l_1^2-l_2^2)+2(l_1^2+l_2^2)S = 0, \tag{99}$$

we find

$$l_1 = \frac{1}{2\sqrt{(1+S)}} \quad \text{and} \quad l_2 = \frac{1}{2\sqrt{(1-S)}}.$$

Thus

$$l_1+l_2 = K_1 = \frac{1}{2}\left(\frac{1}{\sqrt{(1+S)}}+\frac{1}{\sqrt{(1-S)}}\right), \tag{100}$$
$$l_1-l_2 = K_2 = \frac{1}{2}\left(\frac{1}{\sqrt{(1+S)}}-\frac{1}{\sqrt{(1-S)}}\right).$$

Comparing with the LCAO–MO approximation, (77) and (78), we notice that

$$X_a = \frac{1}{\sqrt{2}}(1\sigma_g+1\sigma_u)\alpha,$$
$$X_b = \frac{1}{\sqrt{2}}(1\sigma_g-1\sigma_u)\alpha, \tag{101}$$

and therefore, except for normalization,

$$\psi_{\text{LCAO-MO}} = \begin{vmatrix} 1\sigma_g(1)\alpha(1) & 1\sigma_g(2)\alpha(2) \\ 1\sigma_u(1)\alpha(1) & 1\sigma_u(2)\alpha(2) \end{vmatrix} \tag{102}$$

is the same as

$$\psi_{\text{OAO}} = \frac{1}{2}\begin{vmatrix} (1\sigma_g+1\sigma_u)\alpha(1) & (1\sigma_g+1\sigma_u)\alpha(2) \\ (1\sigma_g-1\sigma_u)\alpha(1) & (1\sigma_g-1\sigma_u)\alpha(2) \end{vmatrix} = \begin{vmatrix} X_a(1) & X_a(2) \\ X_b(1) & X_b(2) \end{vmatrix} \tag{103}$$

The latter we obtain by manipulating the appropriate columns. This in turn is the same state function and thus has the same energy as the Heitler–London function obtained from the same atomic orbitals.

For small S

$$K_1 = 1+\tfrac{3}{8}S^2+\ldots$$
$$K_2 = -\tfrac{1}{2}S(1+\tfrac{5}{8}S^2+\ldots). \tag{104}$$

These relations can be written more compactly. For example, the MO eigenfunctions can be related to the orthogonal atomic orbital functions by a matrix $\overleftrightarrow{t}$.

$$\begin{pmatrix} X_a \\ X_b \end{pmatrix} = \overleftrightarrow{t} \begin{pmatrix} 1\sigma_g \\ 1\sigma_u \end{pmatrix}, \tag{105}$$

where

$$\overleftrightarrow{t} = \frac{1}{\sqrt{2}}\begin{pmatrix} 1 & 1 \\ 1 & -1 \end{pmatrix}. \tag{106}$$

Clearly, $\overleftrightarrow{t}$ is hermitian and unitary

$$\overleftrightarrow{t} = \overleftrightarrow{t}^{\dagger} = \overleftrightarrow{t}^{-1}. \tag{107}$$

The total energy of the system is not affected by such a unitary transformation. By virtue of (71)

$$\begin{aligned} E &= \sum_i \sum_\varrho t_{i\varrho}^{-1}(I_i+\varepsilon_i)t_{i\varrho} \\ &= \sum_\varrho (I_\varrho+\varepsilon_\varrho), \end{aligned} \tag{108}$$

where I_ϱ and ε_ϱ are calculated with the use of the new transformed orbitals: $\langle X_\varrho(1)\,|I|\,X_\varrho(1)\rangle$. Thus the energy of orthogonal atomic orbitals (OAO), LCAO–MO, and the Heitler–London calculations are all the same if the same atomic state functions are used. This proof holds when equal numbers of bonding and antibonding electrons and an even number of alpha and beta spins are involved, as mentioned.

The orthogonalization in the case of larger systems such as Ne_2 can best be achieved by using a method of Löwdin (1956). If our basis set is a column vector of atomic orbitals

$$\Psi = \begin{pmatrix} \phi_{A1} \\ \phi_{B1} \\ \phi_{A2} \\ \phi_{B2} \\ \vdots \\ \phi_{AM} \\ \phi_{BM} \end{pmatrix} \tag{109}$$

we can transform to an orthonormal set by writing

$$\Psi^0 = \overleftrightarrow{T}\Psi \tag{110}$$

where

$$\overleftrightarrow{T}^{\dagger}\,\overleftrightarrow{S}\overleftrightarrow{T} = \overleftrightarrow{1} \tag{111}$$

and S is again the overlap matrix $\overleftrightarrow{S} = \langle \Psi \mid \Psi \rangle$, $\overleftrightarrow{1}$ the identity matrix.

Löwdin suggested a symmetrical form of T, namely $T = (\overleftrightarrow{S})^{-1/2}$. His procedure is called symmetric orthogonalization. For $H_2\,(^3\Sigma_u)$ this leads to results equivalent to our (100)–(101) since

$$(S)^{-1/2} = \begin{pmatrix} K_1 & K_2 \\ K_2 & K_1 \end{pmatrix}. \tag{112}$$

In general the transformation matrix $\overleftrightarrow{T}$ is not unitary.†

Our technique of transforming an MO to a localized orbital in (105) is obviously more general than the LCAO–MO method. We can, in fact, apply $\overleftrightarrow{t}$ or its related form to any MO system. While we have only discussed H_2, the helium system is a trivial extension of this approach since there is no coupling between like and unlike spins. The localized helium orbitals are those of (101), and $\overleftrightarrow{t}$ is simply

$$\overleftrightarrow{t} = \frac{1}{\sqrt{2}}\begin{pmatrix} 1 & 0 & 1 & 0 \\ 0 & 1 & 0 & 1 \\ 1 & 0 & -1 & 0 \\ 0 & 1 & 0 & -1 \end{pmatrix}. \tag{113}$$

The MO-column vector is

$$\Psi = \begin{pmatrix} 1\sigma_g\alpha(1) \\ 1\sigma_g\beta(2) \\ 1\sigma_u\alpha(3) \\ 1\sigma_u\beta(4) \end{pmatrix}. \tag{114}$$

$\overleftrightarrow{t}$ matrices are also available for more elaborate molecules.‡

If the exact MO is transformed the result will be called an "equivalent localized orbital" (ELO), to distinguish it from other localized forms (LO). Alternate procedures have been proposed

† For further elaboration see Kotani, Ohno, and Kayama (1961), vol. 37/2, pp. 138–40.

‡ *Ibid.*, and papers of J. Lennard-Jones and J. Pople, e.g. *Proc. Roy. Soc.* **A202**, 155, 166 (1950).

to localize the orbitals. For example, a minimum exchange criterion has been used with great success by Ruedenberg and Edmiston (1963); for interacting atoms it is equivalent to our procedure involving $\vec{t}$. Others are discussed in detail by Gilbert (1964, p. 405). For intermolecular force calculations only the very simple forms of $\vec{t}$ recorded here are sufficient.

The transformed molecular orbitals, the ELO, are defined by a modified HF equation:

$$[I_\varrho+\sum_\delta(J_\delta-K_\delta)]X_\varrho = \varepsilon_\varrho X_\varrho+\sum_\nu \gamma_{\varrho\nu}X_\nu \tag{115}$$

which follows from applying $\vec{t}$ to (66). *All sums* are again taken over all spin-orbitals (spatial times spin functions) X_ϱ. The coefficient $\gamma_{\varrho\nu}$ appears because the X_ϱ are not basis vectors for the irreducible representations of our diatomic systems, i.e. they are not symmetry orbitals:

$$\gamma_{\varrho\nu} = \langle X_\varrho \,|\, \sum_\delta(J_\delta-K_\delta)+I_\varrho \,|\, X_\nu\rangle . \tag{116}$$

These quantities, often called "off diagonal energies", are usually small even in molecular systems (Hall, 1951; McKoy unpublished; Sinanoglu, 1964). They do not contribute to the HF energy, as given by (108). While little effort has been directed toward the calculation of orbitals via (115), work along these lines should prove fruitful in the future. Again, the energy yielded by an ELO function is identical with that of an MO function and represents an improvement over approximate forms such as those involved in the OAO, LCAO–MO, or Heitler–London atomic orbital approaches.

3.4.2. Elliptical Coordinate Expansions

For the hydrogen-molecule ion (H_2^+) the substitution of spheroidal or elliptical coordinates leads to a separable Hamiltonian and thence to an exact solution. These coordinates are defined by:†

$$\xi_1 = \frac{r_{a1}+r_{b1}}{R} \qquad 1 \leqslant \xi \leqslant \infty, \tag{117}$$

$$\eta_1 = \frac{r_{a1}-r_{b1}}{R} \qquad -1 \leqslant \eta < 1,$$

† They are also referred to as prolate spheroidal coordinates (see Margenau and Murphy, 1956, p. 180).

where φ_1 is the azimuthal angle about the molecular (internuclear) axis ($0 \leqslant \varphi < 2\pi$).

Orbitals involving these coordinates are appropriate for all diatomic systems. A general form would be

$$\phi_j(r_j) = \sum_i \exp(-\delta_i \xi_j - \zeta_i \eta_j) \xi_j^{n_j} \eta_j^{m_j} \exp(i m_\lambda \varphi_j)$$
$$\times [(\xi_j^2 - 1)(1 - \eta_j^2)]^{1/2 |m_\lambda|}, \qquad (118)$$

where n_j and m_j are parameters which must be positive integers, δ and ζ are unrestricted except that δ must be positive, and m_λ is the axial angular momentum (see section 3.3). The total wave function is the antisymmetrized product of these orbitals.

Sakamoto and Ishiguro (1956) used a simple form for the orbitals

$$\phi_1^E(r_j) = e^{-\delta \xi_j - \zeta \eta_j}$$
$$= e^{-(\delta+\zeta) r_{aj}/R} e^{-(\delta-\zeta) r_{bj}/R} \qquad (119)$$

and

$$\phi_2^E(r_j) = e^{-\delta \xi_j + \zeta \eta_j}$$
$$= e^{-(\delta-\zeta) r_{aj}/R} e^{-(\delta+\zeta) r_{bj}/R} \qquad (120)$$

with a total wave function for the helium–helium system of the following form:

$$\Psi = \mathscr{A}[\phi_1^E(1)\alpha(1)\phi_1^E(2)\beta(2)\phi_2^E(3)\alpha(3)\phi_2^E(4)\beta(4)] \qquad (121)$$

except for normalization constants. At large separations these become atomic orbitals with $\zeta = \delta = \frac{27}{16}(R/2)$.† At smaller separations the orbitals become distorted from their atomic form. Numerical results based on (121) are very similar to those of the LCAO–MO and Heitler–London calculations described before for separations greater than $3a_0$.

Taylor and Harris (1964) improved the above function in two ways. First, they used two terms of (118) for each orbital (with $m_j = n_j = m_\lambda = 0$) and, secondly, they introduced different orbitals for α and β spins. The total state function is now complicated by the need to have the proper spin symmetry, but this difficulty is easily resolved. In the limit of infinite separation the latter authors found a helium atom energy of -2.87558 a.u.

† The best single Slater atomic orbital has an exponent of 1.6875 or 27/16.

(99.03% of the exact energy −2.90372). The SCF energy is listed in Table 3.3. For $\Delta E(R)$ they found values 0.14 a.u. higher than the SCF results at $1a_0$, but around $3a_0$ their results are very close to the LCAO–MO values with HF atomic orbitals [$\Delta E(R) =$ 0.0162 a.u. at $3a_0$ vs. 0.0154 by the LCAO–MO procedure, and 0.0135 by the SCF method].†

While such procedures as these are useful, it is not certain at present if they offer great advantages over even a simple procedure like that involving LCAO–MO's. Here again more work is needed, especially in the van der Waals' attractive region.

3.5. Possible Limitations of the Adiabatic Approximations

One of the reasons for the great interest in the interaction of helium atoms at short distances is that experimental data are available. Amdur and co-workers‡ have used high energy total cross-section measurements to evaluate an intermolecular potential in the region around $R = 0.5$ Å. Their values differ markedly (about 9 ev at 0.5 Å) from theoretical values in this region. True, the theoretical values are not bounded, but as the state functions become more flexible the calculated potential must approach the true potential as a consequence of the general variational principle. For this reason many theoretical studies were undertaken to improve the calculations of helium interactions. The latest values for the interaction given in Table 3.4, and especially the work of Phillipson (1962) indicate that the discrepancy can not be resolved within the standard theoretical calculations, those based on eqn. (11). It had been suggested (Berry, Kestner, and McKoy, 1963; Erginsoy, 1965, vol. 2) that the adiabatic or Born–Oppenheimer approximation, eqn. (7), may fail at high velocities used in the experiments. The nuclei might be moving so rapidly that the electrons can not adjust themselves to their motion. In that case calculations based on fixed nuclei would not be applicable. It also means that the concept of a potential would no longer be valid.

† The values quoted are based on our definition of $\Delta E(R)$, equation (11), where $E(R)$ and E_a and E_b are calculated with the same state function. Taylor and Harris calculate $\Delta E(R)$ using their best-calculated E_a and E_b.

‡ Amdur and Bertrand (1962), Amdur (1949), Amdur and Harkness (1954), Amdur, Jordan, and Colgate (1961), summarized by Amdur and Jordan (1966, vol. 10).

Thorson and Bandrauk (1964) and Thorson (1964) studied not only the deviations from the Born–Oppenheimer approximations but the entire collision process. They found that although there may be small errors in the adiabatic approximation, there are compensating features in the collision process which remove these almost entirely. Thus, the adiabatically calculated potentials should be correct even for collisions involving energies of a few thousand volts.

To resolve the difficulty Amdur (1965) has recently re-examined† the interpretation of his experimental results. For a cylindrical or square-beam profile, the experimental results were found to be very sensitive to the exact shape and size of the detector. It was with such sensitive beam geometry that the measurements discussed above were made. With a triangular or trapezoidal beam profile, the size of the detector is not so critical. All of these statements apply only when the potential does not vary rapidly with separation, i.e. at small separations. When $R > 1$ Å the potential varies more rapidly and such critical sensitivities disappear. It is for this reason that experimental and theoretical results around and beyond 1 Å are in reasonable agreement.

Amdur has also corrected his results in the neighborhood of 0.6 Å, in view of faulty gas pressure determinations, and within experimental error they now agree with Phillipson's theoretical curves.

Other experiments provide further indication that the adiabatic approximation is applicable even for 1000 ev helium atom collisions. For example, scattering of high energy helium atoms from room temperature argon atoms yields within 8% the same potential as the scattering of high energy argon atoms from room temperature helium atoms.

At very high energies, to be sure, the adiabatic approximation must fail. Current evidence suggests that the error is serious for several thousand volt helium atoms but the concept of an intermolecular potential, i.e. (11), is valid even at energies of 1000 ev. Non-adiabatic effects are also likely to be important in head-on collisions (Berry, Kestner, and McKoy, 1963). They should, therefore, be looked for in large-angle scattering.

† A comprehensive account of relevant matters is found in vol. 8 of *Methods of Experimental Physics*, Chap. 3. Academic Press, 1967, Ed. B. Bederson and W. L. Fete.

Corrections to the Born–Oppenheimer approximation have been of interest in another context. Strictly speaking it should be said that the concept of an intermolecular potential, i.e. as a function V of the intermolecular vector $\boldsymbol{R}$ such that $-\nabla_R V$ can be interpreted as the force acting between the two molecules, is meaningful only under the assumption that the motion of the nuclei can be neglected. However, even if the coupling between nuclear and electronic motions is taken into account, the concept of an intermolecular potential is a valid one if certain coupling terms are small.

Dalgarno and McCarroll (1956, 1957) calculated the coupling terms between electronic and nuclear motions for a diatomic molecule whose atoms are at large internuclear separations R and concluded that these give rise to an additional R^{-2}-dependent interatomic potential. Such a term would eventually dominate the other long-range constituents of V and would cause surprising and strange effects in the values of a number of gas-kinetic quantities. The coupling terms in question are matrix elements between electronic states of the atoms for which the nuclei are at rest. Nevertheless, these electronic states depend on $\boldsymbol{R}$ because the electron cloud accompanies the nuclei. Dalgarno and McCarroll chose a set of "rotating" electron states in their calculation, i.e. a set of electron states describing an electron cloud whose configuration with respect to $\boldsymbol{R}$ remains unchanged as $\boldsymbol{R}$ rotates. However, for large internuclear separations the electron clouds of two atoms in the course of passing each other do not influence each other very much,† and, hence, their configurations do not remain stationary with respect to the internuclear axis, but with respect to a space-fixed coordinate system instead. Realizing this, Hans Laue (1967) showed that, if one chooses "non-rotating" electron states, i.e. states describing electron clouds whose configuration remains the same with respect to a space-fixed coordinate system as $\boldsymbol{R}$ rotates, the coupling coefficients between nuclear and electronic motions are constants of the order of magnitude of the isotope effect. The R^{-2}-dependence

† Criteria for the "adiabatic" behavior (constant alignment of electronic angular momentum with respect to $\boldsymbol{R}$) of electron clouds are given by Klein and Margenau (1959). For the meaning of adiabaticity in this context see Margenau and Lewis (1959). The former reference also estimates the distances of separation R at which the electron clouds begin to retain their alignment with $\boldsymbol{R}$.

of the coupling coefficients found by Dalgarno and McCarroll therefore indicates the presence of a centrifugal potential, accompanying the revolution of the electron cloud as its configuration maintains its alignment with the revolving internuclear line. The mathematical analysis shows that for large R the coupling coefficients for rotating states form a matrix with large off-diagonal elements and that as a consequence the concept of a single interatomic potential describing the relative motion of the atoms fails. The R^{-2}-dependence of the coupling coefficients for rotating states has therefore no physical significance for large R. On the other hand, if nonrotating states are used, the matrix of coupling coefficients has only small elements, of the order of the isotope effect, which are independent of R and entail a corresponding, indeed negligible, contribution to the interatomic potential.

For the reader's convenience we list in the following table the rather confusing notation used in defining molecular orbitals.

TABLE 3.5. *Notation Commonly Used in Defining Molecular Orbitals*

Designation	Description
HF	Orbitals determined by the HF self-consistent field procedure
AO	*Atomic orbitals*
MO	*Molecular orbitals*
LCAO–MO	*Molecular orbitals* formed by taking *linear combinations* of *atomic orbitals*. The best such combination uses HF atomic orbitals
OAO	*Orbitals* in a molecule formed by making the *atomic orbitals* on one center orthogonal to atomic orbitals on other centers
SCF	*Self-consistent field* orbitals found by using the matrix equation formulation of HF theory with a small basis set. As the basis set is made larger these orbitals approach the ideal of the HF orbitals
LO	*Localized orbitals.* These are approximate molecular orbitals localized about the various nuclei in a polyatomic molecule
ELO	*Equivalent localized orbitals.* These are localized Hartree–Fock orbitals related to one another by elements of the molecular symmetry. They can be derived by a self-consistent field treatment, or by transforming the HF orbitals

CHAPTER 4

THE INTERACTION OF SMALL ATOMIC SYSTEMS AT INTERMEDIATE DISTANCES

IN THE previous chapter we employed various simple wave functions in computing atomic interactions, restricting attention primarily to He_2 and H_2, and we found a purely repulsive intermolecular potential. We know, however, that the long range interaction of two neutral species is proportional to R^{-6} and attractive.

The study of real gases yields evidence of both the attractive and repulsive parts of the potential. The variation of density of a gas with pressure or temperature, and the variation of viscosity with density and temperature are but two examples in which the assumption of a purely repulsive potential leads to incorrect conclusions.

As a prime example of the relationship between the intermolecular potential and the properties of a real gas,† we consider the virial expansion which relates the pressure p of a gas to the density ϱ and the temperature T:

$$\frac{p}{kT} = \varrho + B_2(T)\varrho^2 + B_3(T)\varrho^3 + \dots . \quad (1)$$

Here B_n is the nth virial coefficient and k is Boltzmann's constant. B_2 depends only on the intermolecular potential, $\Delta E(R)$, between two spherical molecules in a vacuum (ter Haar, 1954; Hill, 1960):

$$B_2(T) = -\frac{1}{2}\int_0^\infty [e^{-\Delta E(R)/kT}-1]4\pi R^2\, dR, \quad (2)$$

when quantum effects can be neglected. From this it might appear that, knowing $B_2(T)$ experimentally over a very large range of temperatures, one could invert the relation to obtain $\Delta E(R)$. This is, in fact, not possible since this inversion is unique only when

† A complete discussion of the theory of real gases is found in Hirschfelder, Curtiss, and Bird (1954). Recent determinations of intermolecular potentials from experimental data are discussed in Appendix B of this book.

$\Delta E(R)$ is a monotonic function of the internuclear separation (Keller and Zumino, 1959). Also, the temperature range over which we know B_2 is usually rather restricted. Thus we cannot obtain a unique potential from virial (pressure versus density) data. The current procedure is therefore to assume a form of $\Delta E(R)$, usually containing two variable parameters, and to adjust these until they give the best fit to $B_2(T)$.

Several standard but artificial potential forms are commonly used to fit the gas data.

The Lennard-Jones 6–12 potential

$$\Delta E(R) = \varepsilon_0 \left[\left(\frac{r_0}{R}\right)^{12} - 2\left(\frac{r_0}{R}\right)^{6} \right], \tag{3a}$$

where ε_0 is the depth and r_0 the position of the minimum.

Another is a three-parameter extension of (3a) known as the Kihara potential (Kihara, 1953, 1958):

$$\Delta E(R) = A\left(\frac{m}{n}\right) \varepsilon_0 \left[\left(\frac{1-\gamma}{\frac{R}{\sigma}-\gamma}\right)^{n} - \left(\frac{1-\gamma}{\frac{R}{\sigma}-\gamma}\right)^{m} \right], \tag{3b}$$

where σ is the collision diameter of the molecule and

$$A(x) = \frac{1}{1-x}\, x^{-\frac{x}{1-x}}.$$

Usually the values $m = 6$ and $n = 12$ are selected.

Lastly we list the exp-6 potential:

$$\begin{aligned} \Delta E(R) &= \frac{\varepsilon_0}{1-(6/\alpha)} \left[\frac{6}{\alpha} \exp\left(\alpha\left[1-\frac{R}{r_0}\right]\right) - \left(\frac{r_0}{R}\right)^{6} \right] & R \geq r_m \\ &= \infty & R < r_m \end{aligned} \tag{4}$$

where α is a dimensionless parameter whose value ordinarily lies between 12 and 15, and r_m is the distance at which $\Delta E(R)$ becomes a maximum. Equation (4) has to be cut off at r_m since it goes to $-\infty$ as $R \rightarrow 0$, and the maximum at r_m makes (4) very unrealistic.

The concept of an exponential and an R^{-6} term was first employed by Slater and Kirkwood (1931). Margenau (1931) calculated and added an R^{-8} term. Because it introduces another parameter, this procedure is often avoided and the burden of the R^{-8} term is put upon the coefficient of R^{-6}, which is thereby easily falsified. The additive combination of an exponential and an R^{-6} term was common in the early literature; the specific form of (4) in terms of ε_0 and α was used by Buckingham.†

One can employ a similar procedure to obtain a potential from the transport properties of a gas, viscosity, self-diffusion, etc. However, the simple two-parameter potentials obtained from different properties of the gas do not agree with one another. The potential forms selected are too simple. The high temperature transport data, for example, tend to place excessive weight upon the repulsive regions. They also do not yield satisfactory solid-state properties; the error here, however, may be caused by many-body effects which will be considered in Chapter 5.

But even if we restrict our attention to the gas, our "experimental" potentials listed above have large uncertainties. We do not know the position or depth of the potential minimum within an accuracy better than 10% in most cases. Thus thorough theoretical studies are needed to provide guidance in the determination of "experimental" potentials.‡

The problem of calculating intermolecular forces at all distances with a single state function is very difficult indeed. At large R it is legitimate to neglect atomic overlap and the intermolecular antisymmetry of the state function. At short distances one can calculate the repulsion by the devices of the last chapter. However, in the intermediate region it is necessary to include all effects, and the net result is numerically small. For example, the energy of two helium atoms is about 5.8 a.u. (158 ev) while the depth of the potential is about 3.2×10^{-5} a.u., 8.7×10^{-4} ev, or 7 cm^{-1}. Even the best *a priori* calculations of molecular energies suffer from errors of the order of volts. Clearly, then, we must calculate ΔE in such a way that the potential does not appear as a difference between large terms. Yet we must include all effects. At large separations this was done easily since the state of one atom is practically unchanged by the presence of the other (i.e. the atoms did not overlap). Therefore the energy of the atoms themselves, which is very large, could simply be subtracted out. When overlap is in-

† The specific form listed for $R < r_m$ is due to Rice and Hirschfelder (1954).

‡ One such study of the adjustment of two-body potentials to fit low pressure gas data may be found in the *Discussions of the Faraday Society*, No. 40 (Bristol, 1966). See especially the papers of Rowlinson and Munn. Also, see Appendix B.

volved, the cancellations are not so obvious but, as was shown with the LCAO–MO calculations, even there we can explicitly write ΔE in a form which is not a delicate difference between large energies.

In studying the region of intermediate separations, where the minimum occurs, we shall again consider mainly hydrogen and helium interactions, devoting most space to the helium case since this is more typical of larger systems. We review simple perturbation methods as well as very elaborate calculations and then summarize several general approaches which have been or will soon be used in the calculation of the interaction of larger systems.

4.1. Hydrogen Atom Interactions

The repulsive region is reasonably well represented by molecular orbital theory, and the Heitler–London approach yields a rather similar potential. To compare the two theories we follow Donath and Pitzer (1956) and write the molecular orbitals for the bonding and antibonding states as the sum of hydrogen $1s$ (respectively $1s_a$ and $1s_b$) and $2p$ states (the X_i in eqn. (57) of Chapter 3):

$$1\sigma_g\colon \Phi_1 = (1/\sqrt{2})[(1-\delta_+^2)^{1/2}(1s_a+1s_b)+\delta_+(2p_{z,a}+2p_{z,b})], \quad (5)$$

$$1\sigma_u\colon \Phi_2 = (1/\sqrt{2})[(1-\delta_-^2)^{1/2}(1s_a-1s_b)-\delta_-(2p_{z,a}-2p_{z,b})]. \quad (6)$$

In (5) and (6) δ_+ and δ_- are variable parameters, determined by minimizing the total energy. Overlap has been neglected in the normalization as it is a small correction at separations where these equations will be used. In this function the positive lobes of the two p-orbitals extend toward one another along the z or internuclear axis.

On expanding the triplet wave function, using the above orbitals, we find

$$\begin{aligned}\Psi = {}& (1-\delta_+^2)^{1/2}(1-\delta_-^2)^{1/2}\mathcal{A}(1s_a(1)1s_b(2)\alpha(1)\alpha(2))\\ &+\tfrac{1}{2}\{\delta_+\sqrt{(1-\delta_-^2)}+\delta_-\sqrt{(1-\delta_+^2)}\}\{\mathcal{A}(1s_a(1)2p_{z,a}(2)\alpha(1)\alpha(2))\\ &\qquad+\mathcal{A}(2p_{z,b}(1)1s_b(2)\alpha(1)\alpha(2))\}\\ &+\tfrac{1}{2}\{\delta_+\sqrt{(1-\delta_-^2)}-\delta_-\sqrt{(1-\delta_+^2)}\}\{\mathcal{A}(1s_a(1)2p_{z,b}(2)\alpha(1)\alpha(2))\\ &\qquad+\mathcal{A}(2p_{z,a}(1)1s_b(2)\alpha(1)\alpha(2))\}\\ &-\delta_+\delta_-\mathcal{A}(2p_{z,a}(1)2p_{z,b}(2)\alpha(1)\alpha(2)). \end{aligned} \quad (7)$$

The first term is the Heitler–London state function. The second term introduces an "ionic polarizing" effect, since both atomic orbitals appearing in it extend about the same center. Inspection will show that the third term represents a "covalent polarizing" contribution, while the last determinant allows for polarization of both atomic orbitals simultaneously.

An even more general and therefore better function of the same type would insure independent variation of the four functions. To define it we introduce coefficients C_0, C_1, C_2, and C_3 in place of the expressions involving the two parameters δ_+ and δ_- of the MO function. This general function then allows the orbitals to suffer distortion along the internuclear axis. In particular, the orbitals can readjust themselves to lower the electron density between the two nuclei and thereby lower the energy in the variational expressions.

There is, of course, no reason to restrict the basis set to only $1s$ and $2p$ functions. In fact, since we desire our approximate function to approach the exact function as closely as possible, we should introduce all excited states. From analytic function theory we know that a well-behaved function can be expanded in terms of a complete basis set. Hydrogen atom functions including the continuum form such a complete set. But in practice one cannot use the infinite set, and one is forced to select only a few easily tractable functions. Further, one does not use the true hydrogen atom functions, but either a set in which the exponential behavior is reflected by an extra, variable parameter or a modification of the hydrogen set which does not include the continuum.

The latter choice yields the following two radial functions:†

$$(2\eta r)^l e^{-\eta r} L_{n+l}^{2l+1}(2\eta r) \tag{8}$$

or

$$(2\eta r)^l e^{-\eta r} L_{n+l+1}^{2l+2}(2\eta r).$$

The normal hydrogen atom solutions are, for comparison,

$$(2\eta r/n)^l e^{-\eta r/n} L_{n+l}^{2l+1}(2\eta r/n), \tag{9}$$

where n is the principal and l the angular momentum quantum number, while L_p^q is the associated Laguerre polynomial.

If we use a complete basis set $\{\phi_{ai}, \phi_{bj}\}$, i.e. the union of sets centered about each nuclei, we could expand the exact state func-

† More details are given by Löwdin (1959) and in the references cited there.

tion and write

$$\Psi' = \sum_{ij} d_{ij}\mathcal{A}(\phi_{ai}(1)\alpha(1)\phi_{bj}(2)\alpha(2))$$
$$+\sum_{ij} f_{ij}[\mathcal{A}(\phi_{ai}(1)\alpha(1)\phi_{aj}(2)\alpha(2))+\mathcal{A}(\phi_{bi}(1)\alpha(1)\phi_{bj}(2)\alpha(2))]. \tag{10}$$

The dominant term in the first summation must be the Heitler–London determinant.

The function above is a variant of a state representation often characterized by the labels "configuration–interaction" (CI) or "superposition-of-configurations" state function. To the original determinant, the Heitler–London term according to our procedure, one adds additional configurations, all of the appropriate total molecular symmetry. The mixing coefficients d_{ij} and f_{ij} are found by minimizing the total energy ($\overline{H}$ of Chapter 3). This is usually done by matrix methods which produce as a byproduct the higher energy solutions, which are approximate functions for the excited states of the total system.

That procedure which specifically uses as its dominant term the SCF result is customarily called the CI procedure. It requires all additions to the dominant term to be of definite symmetry, and changes involving only one orbital can not contribute in the first order to corrections of the energy. Because the function defined by (10) differs in some of these respects, we shall speak of it as resulting from a "superposition-of-configurations".

Equation (10) contains an infinite number of terms and as such is useless. Therefore we must discover which terms are most important. We do this by considering several studies which have been made at large and small separations.

The types of configurations needed to produce the correct potential at large separations can be found by examining the function used by Slater and Kirkwood to obtain the R^{-6} attraction (see section 2.4) namely

$$1s_a(1)1s_b(2)\left[1+\frac{C}{R^3}(x_1x_2+y_1y_2-2z_1z_2)\right]. \tag{11}$$

For hydrogen this yielded a London energy of $-6/R^6$ and the constant $C=1.00$. (The correct value of the energy is $-6.499\ R^{-6}$.) We can rearrange (11) in the form of (10) for intermediate separations where the Pauli principle becomes vital:

$$\Psi_{SK} = \mathcal{A}[1s_a(1)1s_b(2)\alpha(1)\alpha(2)] + d_{22}\mathcal{A}[2p_{x,b}(1)2p_{x,a}(2)\alpha(1)\alpha(2)]$$
$$+ d_{22}[2p_{y,a}(1)2p_{y,b}(2)\alpha(1)\alpha(2)] + d_{33}\mathcal{A}[2p_{z,a}(1)2p_{z,b}(2)\alpha(1)\alpha(2)]. \quad (12)$$

The last term is one of the components of the Donath–Pitzer state function, eqn. (7). The last three terms represent the instantaneous correlations of the induced dipoles in the two atoms. The coefficients d_{ii} in (12) are simple functions of R only at large separations when overlap is negligible; they are then proportional to R^{-3}.

Equation (12) is likewise incomplete because, as mentioned earlier, the R^{-8} terms are also important at large separations. Such effects involve d orbitals; thus various $2p_a3d_b$, $3d_a2p_b$, and $3d_a3d_b$ terms should be added to (12) to include all of the important long-range attractive parts in the potential. Again such contributions have coefficients proportional to R^{-4} or R^{-5} when overlap can be neglected. At smaller separations their coefficients are very complicated functions of R. It is misleading, therefore, to consider a $2p_a3d_b$ term as simply an R^{-8} contribution, since this is only true when R is very large.

There are also procedures which force the function in (10) to converge rapidly, allowing one to use very few terms. In the most common procedure these best orbitals have been given the name natural spin orbitals (Löwdin, 1959). They have been used in connection with the helium atom and the hydrogen molecule, particularly in studying the repulsive triplet of hydrogen at large separations.†

There have been several general "superposition-of-configuration" studies of the repulsive H_2 system at small separations. We shall consider three typical calculations, one at small separations and two at intermediate separations.

For small separations Cade (1961) used a function like (10) composed of six configurations involving hydrogen atom state functions: $1s^2$, $1s2s$, $1s2p_z$, $2s^2$, $2s2p_z$, and $2p_z^2$. Only the covalent terms were introduced, i.e. only one orbital was placed on each nucleus in each configuration corresponding to the first summation in (10). Table 4.1 presents his results.

† Hirschfelder and Löwdin (1959 and erratum 1965). Their revised potential is $-\dfrac{6.499026}{R^6} - \dfrac{124.395}{R^8}$.

TABLE 4.1. *Calculations by Cade on the Hydrogen Interaction*

R (a_0)	Cade value $\Delta E(R)$ a.u.
0.81	0.6120
1.37	0.2806
1.93	0.1398
2.45	0.0704
2.97	0.0346
3.99	0.0082
5.00	0.0018

Comparing these values with those of the LCAO–MO or Heitler–London functions in Chapter 3 (following formula (39) of Chapter 3) we note that they yield energies differing by less than 5% for $R > 3a_0$. As seen from the coefficients in his wave function, only for $R < 2.5$ do the additional covalent terms used by Cade become greater than 10% of the primary (Heitler–London) term. Therefore, for separations at which the potential is less than 1.8 ev, the Heitler–London function represents the repulsive potential with reasonable accuracy.

Donath and Pitzer compared the molecular orbital theory, which is based on (7), with that employing the general superposition of configurations and using the same terms as (7). The appropriate function is

$$\begin{aligned}\Psi_{\mathrm{Dp}} = {} & C_0\mathcal{A}[1s_a(1)1s_b(2)\alpha(1)\alpha(2)] \\ & + C_1\{\mathcal{A}[1s_a(1)2p_{z,a}(2)\alpha(1)\alpha(2)]+\mathcal{A}[2p_{z,b}(1)1s_b(2)\alpha(1)\alpha(2)]\} \\ & + C_2\{\mathcal{A}[1s_a(1)2p_{z,b}(2)\alpha(1)\alpha(2)]+\mathcal{A}[2p_{z,a}(1)1s_b(2)\alpha(1)\alpha(2)]\} \\ & + C_3\mathcal{A}[2p_{z,a}(1)2p_{z,b}(2)\alpha(1)\alpha(2)]. \end{aligned} \tag{13}$$

The $2p_z$ orbital adopted was that of a hydrogen-like atom of nuclear charge $Z = 2$. With this choice the exponents of the $1s$ and $2p_z$ functions are identical. Thus any admixture of the $2p_z$ function primarily affects the angular distribution of electrons about each nucleus with very little change in the radial distribution.† Since C_1, C_2, and C_3 are much less than C_0, a variation–perturbation procedure was used to find the contribution of each term separately. No cross terms involving C_1, C_2, and C_3 in the

† See Hirschfelder and Löwdin (1954) for a further discussion.

TABLE 4.2. *Energy Contributions from Terms in the Donath–Pitzer State Function, Equation (13)*

			ΔE in a.u. $\times 10^3$			
$R(a_0)$	ΔE_0	ΔE_1	ΔE_2	ΔE_3	ΔE Eqn. (13)	$(\Delta E)_{MO}$ Eqn. (7)
6	0.3918	−0.02398	−0.00043	−0.0733	0.2946	0.3664
7	0.07533	−0.00462	−0.00002	−0.03173	0.03897	0.07057
8	0.01382	−0.00083		−0.01482	−0.00183	0.01297
10	0.000421	−0.000024		−0.003983	−0.003586	0.000395
12	0.0000123	−0.0000006		−0.001338	−0.001326	0.0000116

energy were considered. In Table 4.2 we list the energy contribution ΔE_i corresponding to the configuration represented by the coefficient C_i.

Several interesting features appear. (a) The covalent polarizing terms are of little importance at these separations. They become important in Cade's wave function for $R < 2.5a_0$. (b) The attraction (negative ΔE) arises from the $2p_{z,a}2p_{z,b}$ term. If the other terms in (12) (x, y terms) had been included, the potential would have been even more negative. (c) The molecular-orbital function (7) is highly inadequate to render account of the complete potential. This is discussed in detail in Appendix A. The energy obtained by using the molecular orbital function remains positive at all separations. (d) All coefficients, listed in Table 4.3, are small compared with C_0.

In a much earlier, important publication, Hirschfelder and Linnett (1950) presented a similar calculation in which they neglected ionic contributions but concentrated attention on the potential in the region of the van der Waals minimum. Their trial function, though not written in the form of (10), is equivalent to it. It has the form

$$\varphi_{\mathrm{HL}} = \mathcal{A}\{\phi_a(1)\phi_b(2)[1+(\alpha\delta^2)(x_{a1}x_{b2}+y_{a1}y_{b2}) + \beta\delta^2 z_{a1}z_{b2}]\alpha(1)\alpha(2)\}. \tag{14}$$

In this work the z-axis about each nucleus is directed to the right. This choice is opposite of that in the work of Donath and Pitzer

TABLE 4.3. *Coefficients of Terms in the Donath–Pitzer State Functions, Equations (7) and (13)*

$R(a_0)$	C_0	C_1	C_2	C_3	$\delta_+ = \delta_-$
6	0.9993	0.005973	–0.000897	–0.008482	0.004175
7	0.9998	0.002553	–0.000190	–0.005597	0.001796
8	1.0000	+0.001063	–0.000037	–0.003861	0.000749
10	1.0000	+0.000175	–0.000011	–0.001997	0.000123
12	1.0000	+0.0000271	–0.00000003	–0.001157	0.0000192

or Cade. ϕ_a and ϕ_b are simple Slater orbitals given by eqn. (44) of Chapter 3 with an effective nuclear charge δ which is varied to give the lowest energy ($\bar{H}$ of section 3.2). The parameter α provides for the mixing of the p_x and p_y contributions; it is negative because of electronic repulsion. β, another parameter, is positive and indicates that electrons tend to stay as far apart as possible. Except for sign (due to choice of coordinates), β is equivalent to C_3 of the Donath–Pitzer function. Any difference in numerical values arises from the variation of exponents in the Hirschfelder–Linnett results. At large separations $\delta \to 1.0$ and comparison with (11) shows that

$$\alpha = -\beta/2 = -\frac{C}{R^3} = -\frac{1}{R^3}. \tag{15}$$

Any deviation of δ and αR^3 from constant values indicates a failure of the multipole expansion, while deviations of β/α from 2.0 indicate p_z-distortions other than those of dispersion effects. The magnitude of these departures is shown in Table 4.4.

The minimum of ΔE comes at $R = 8.45a_0$ with a depth of -0.0074 kcal/mole or 2.59 cm^{-1}. The potential is zero at $7.15a_0$. Thus even at 6–7a_0 the errors in the multipole expansion and the effects of distortion ($\beta/\alpha \neq 2$) are very small, suggesting many simplifications. For distances around the minimum (even to $6a_0$) one can use the atomic functions ($\delta = 1$) and multipole expansions with no distortion. The exchange and overlap effects are important only for separations less than 7–8a_0. Since exchange effects are small they can be approximated if one is interested

TABLE 4.4. *Parameters of the Hirschfelder–Linnett State Function, Equation (14)*

$R(a_0)$	δ	$-\alpha R^3$	(β/α)	$\Delta E(R)_{HL}$
1.23	0.8140	0.0769	10.98	0.3696
1.72	0.8698	0.1703	7.77	0.17126
1.97	0.8870	0.2281	7.199	0.11980
2.19	0.9132	0.2786	6.693	0.08884
2.64	0.9511	0.3861	6.058	0.04851
3.07	0.9758	0.4931	5.480	0.02605
3.54	0.9893	0.5982	4.825	0.01346
4.02	0.9956	0.6965	4.142	0.006850
4.51	0.9981	0.7822	3.525	0.003199
5.01	0.9977	0.8458	3.014	0.001446
6.01	0.9982	0.9425	2.415	0.0002482
7.01	0.9986	0.9810	2.146	0.00002219
8.03	0.9960	1.0037	2.046	-0.00000933
10.02	0.9979	1.0053	2.003	-0.000006054

only in the region about the potential minimum. But before discussing approximate methods, we turn to an alternate type of wave function which is capable of attaining any desired degree of accuracy.

The most accurate calculations of the hydrogen interaction have been made with the use of functions of elliptic coordinates. The forms considered in the last chapter led to repulsion only since they neglected the motions of electrons relative to one another. This can be introduced by a superposition of configurations provided some of the configurations allow electrons to stay away from each other. It is also included when the interelectronic coordinate r_{12} is used explicitly in the state function. This can be made to introduce the correlation effects which lead to the van der Waals' attraction. To be sure, most of the original work in this area was concerned with the ground state singlet (Coolidge and James, 1933; Kolos and Roothaan, 1960; Kolos and Wolniewicz, 1963, 1964), but some calculations have been performed on the triplet state of the H–H combination [Kolos and Roothaan, 1960; James, Coolidge, and Present, 1936; Kolos and Wolniewicz, 1965; for work on other excited states see Present, 1935; Coolidge and

James, 1938; James and Coolidge, 1938); W. Kolos and L. Wolniewicz, 1965] and these will now be outlined.

The total state function of the ${}^3\Sigma_u^+$ state is

$$\Psi = \sum_i C_i \Phi_i \cdot [\alpha(1)\alpha(2)], \tag{16}$$

where

$$\Phi_i = \Psi_{p_i q_i r_i s_i \mu_i} - \Psi_{r_i s_i p_i q_i \mu_i} \tag{17}$$

$$\Psi_{pqrs\mu} = e^{-\delta(\xi_1+\xi_2)} \xi_1^p \eta_1^q \xi_2^r \eta_2^s r_{12}^\mu, \tag{18}$$

and $q_i + s_i$ is odd. While James, Coolidge, and Present used only 15 terms in this expansion, Kolos and Roothaan employed 34, including several involving $r_{12}{}^2$. Recently, Wolniewicz and Kolos (1965) extended this work to large distances with 25 carefully chosen terms and were able to obtain the complete potential curve with a depth of 4.3 cm^{-1} at $7.85a_0$ (4.15 Å). Their results are given in Table 4.5.

These energies lie considerably below all those previously calculated. For $R < 4$ they are about 50% below the Heitler–London values. However, at large separations ($R > 10a_0$) even these results may be in error by 10–20% as may be judged from comparisons with perturbation theory results. They have uncertainties of about 2–5% between 8 and $10a_0$, the region of the potential minimum, solely because of computer limitations. All this points to the general limitations of this procedure: after very extensive calculations the depth of the potential is known to only two or three significant figures. In larger systems such methods become even less efficient since energy values must be known accurately to 7–8 significant figures.

Both for ease of calculation and even for attaining higher accuracy in limited regions, less elaborate methods will now be suggested. While these methods have restricted use, they can be accurate as well as simple near the potential minimum where distortion of the atomic functions, i.e. electron overlap, is not great.

The simplest approximate potential of this sort which one can write combines the Heitler–London approach of the last chapter with the second-order non-overlap calculations of the dispersion interactions

$$\Delta E = Ae^{-\mu R} - \frac{C_1}{R^6} - \frac{C_2}{R^8}. \tag{19}$$

TABLE 4.5. *Kolos–Wolniewicz Results for the Lowest Triplet State of Hydrogen*

R (a_0)	Number of terms used in Eq. (16)	Potential—$\Delta E(R)$	
		Atomic units	Wave numbers
1.0	53	0.3784773	83066.2
1.5	53	0.1903905	41785.9
2.0	53	0.1029364	22591.9
2.5	53	0.0545537	11973.2
3.0	53	0.0279896	6143.0
3.5	53	0.0138721	3044.6
4.0	53	0.0066219	1453.3
4.5	53	0.0030285	664.7
5.0	35	0.0013151	288.6
5.5	35	0.0005297	116.3
6.0	35	0.0001875	41.1
6.5	35	0.0000479	10.5
7.0	25	−0.0000030	−0.7
7.2	25	−0.0000117	−2.6
7.4	25	−0.0000167	−3.9
7.6	25	−0.0000191	−4.2
7.8	25	−0.0000197	−4.3
8.0	25	−0.0000196	−4.3
8.2	25	−0.0000191	−4.2
8.4	25	−0.0000173	−3.8
8.7	25	−0.0000152	−3.3
9.0	25	−0.0000127	−2.8
9.5	18	−0.0000095	−2.1
10.0	18	−0.0000067	−1.5

The first term is a first-order perturbation result, the expectation value of the total Hamiltonian calculated with the zero-order Heitler–London function. The last two terms are calculated in the second order without using a properly antisymmetrized state function.

This formula is in error in several respects: (a) neglect of exchange in second-order energy, and (b) neglect of changes in the zero-order function for small R. The results of Cade, and Hirschfelder and Linnett indicate that the error due to the failure of the multipole expansion is minor for $R > 7a_0$. Near the potential minimum the error is less than 5%. However, the Donath and

Pitzer results (Table 4.2) indicate that the actual contribution of terms which yield the long range attraction (their C_3 term in (13)) differs from the value expected from the pure R^{-6} result by 15% at $6a_0$. Almost all of this is due to the exchange contribution. The values at $8a_0$ differ by about 3%.

In view of these findings we review now an older approach to the problem of finding the potential at intermediate distances.

Margenau (1939) suggested a simple method to correct for exchange and applied it to both the H–H and the He–He interactions. Using the Slater–Kirkwood function, (11), in a variational calculation of $\overline{H}$ (cf. Chapter 3), he writes

$$\Psi = \mathcal{A}[(1+\lambda D)1s_a(1)1s_b(2)\alpha(1)\alpha(2)], \tag{20}$$

where $D = H-E_0 = V+H_0-E$ (H_0 and E_0 refer to the separated atoms). Then

$$\Delta E = \langle\Psi|D|\Psi\rangle/\langle\Psi|\Psi\rangle. \tag{21}$$

If one now uses for the parameter λ the value previously determined as best without exchange, one obtains (neglecting higher powers of λ in the denominator)

$$\Delta E \cong \frac{D_{00}(DP)_{00}+\lambda'[(D^2)_{00}-(D^2P)_{00}]}{1-P_{00}+2\lambda'[D_{00}-(DP)_{00}]}, \tag{22}$$

where $(D^n)_{00} = \langle 1s_a(1)1s_b(2)|D^n|1s_a(1)1s_b(2)\rangle$ and P is an operator which exchanges electrons 1 and 2. Here λ', the non-exchange value, is $-(D^2)_{00}/(D^3)_{00}$, the Slater–Kirkwood value discussed in Chapter 2. When λ' is zero we have $\Delta E = \tilde{V}$ [cf. Chapter 3] the Heitler–London result [eqn. (19) of Chapter 3]. The Slater–Kirkwood result in our notation is

$$E_2 = \lambda'(D^2)_{00} \tag{23}$$

and therefore

$$\Delta E \cong \frac{\tilde{V}+(1-P_{00})^{-1}[E_2-\lambda'(D^2P)_{00}]}{1+2\lambda'\tilde{V}}. \tag{24}$$

At $R \simeq 7a_0$, $\tilde{V}$ and E_2 are about 0.002 ev and λ' is typically of the order of the reciprocal of an excitation energy, so that $2\lambda'\tilde{V}<0.004$. This value and $P_{00} = S^2$, where S is the atomic overlap integral, suggest the following approximate result:

$$\Delta E \simeq \tilde{V}+(1-S^2)^{-1}[E_2-\lambda'(D^2P)_{00}], \tag{25}$$

$$\simeq \tilde{V}+E_2-\lambda'(D^2P)_{00}+S^2E_2. \tag{26}$$

The first term is the Heitler–London interaction, the second the dispersion interaction as given by (19), and the last two are correction terms. The third is the "second-order exchange term" (Margenau, 1939). The last term in (26) is very small (for $R = 6a_0$, $S = 0.047$ and $S^2 = 0.0022$).

The third term in (26) may be expanded:

$$(D^2P)_{00} = \langle D1s_a(1)1s_b(2)|DP1s_a(1)1s_b(2)\rangle, \tag{27}$$

$$= \langle 1s_a(1)1s_b(2)|(r_{12}^{-1}+R^{-1}-r_{b1}^{-1}-r_{a2}^{-1})$$
$$(r_{12}^{-1}+R^{-1}-r_{a1}^{-1}-r_{b2}^{-1})1s_b(1)1s_a(2)\rangle. \tag{28}$$

The evaluation of this integral requires great care as it contains many cancellations at moderate separations. All terms in the potential seem to behave as R^{-1} individually, but are proportional to R^{-3} collectively at large separations. Margenau originally evaluated (28) using formulas of Eisenschitz and London (1930). However, some of the individual integrals, particularly the one involving r_{12}^{-2} contained in that paper, were approximate, and as Dalgarno and Lynn (1956) point out, the numerical values of (28) obtained with their use are inaccurate. The latter authors suggest an alternative form. If V^m is the first term of a multipole expansion of V, i.e. $(X_1X_2+Y_1Y_2-2Z_1Z_2)R^{-3}$, X_1 is the sum of the x-coordinates of all electrons in atom 1, as in Chapter 2, they propose to use

$$(D^2P)_{00}$$
$$= \langle 1s_a(1)1s_b(2)\,|\,V^mPV1s_a(1)1s_b(2)\rangle \frac{\langle 1s_a(1)1s_b(2)\,|\,PV1s_a(1)1s_b(2)\rangle}{\langle 1s_a(1)1s_b(2)\,|\,PV^m1s_a(1)1s_b(2)\rangle}, \tag{29}$$

where the fraction is meant to correct for the inaccuracies of the multipole expansion.

If we use only the dipole–dipole term of the multipole series and neglect corrections, we may simply write

$$(D^2P)_{00} \sim \left\langle 1s_a(1)1s_b(2) \left| \frac{X_{a1}X_{b2}+Y_{a1}Y_{b2}-2Z_{a1}Z_{b2}}{R^3} \right.\right.$$
$$\left.\left. \times \frac{X_{a2}X_{b1}+Y_{a2}Y_{b1}-2Z_{a2}Z_{b1}}{R^3} \right| 1s_b(1)1s_a(2) \right\rangle$$
$$\sim \frac{1}{R^6}\,[2\langle 2p_{x,a}\,|\,2p_{x,b}\rangle^2+4\langle 2p_{z,a}\,|\,2p_{z,b}\rangle^2]. \tag{30}$$

An even better value for this contribution can be obtained from the Hirschfelder–Linnett (1950) procedure. Their equivalent term, with *no* approximations concerning multipole expansion and the like, is

$$\lambda'(D^2P)_{00} = \Big\langle 1s_a(1)1s_b(2)\big|[\alpha\delta^2(X_{a1}X_{b2}+Y_{a1}Y_{b2})+\beta\delta^2(Z_{a1}Z_{b2})] \times\left[\frac{1}{r_{12}}-\frac{1}{r_{b1}}-\frac{1}{r_{a2}}+\frac{1}{R}\right]1s_b(1)1s_a(2)\Big\rangle. \quad (31)$$

Their paper gives details useful in evaluating the necessary integrals.

To examine the value of the second-order exchange term in relation to the London energy E_2 and the repulsive energy $\tilde{V}$, we shall use $\lambda' = [p/E_0]^{-1}$, where E_0 is the ionization potential of the hydrogen atom and p is a correction factor, taken to be 0.92 for hydrogen in view of earlier considerations (see Chapter 2). At $7a_0$, near the zero of the potential, we find

$$E_2 = -\frac{C_1}{R^6}-\left(\frac{C_2}{R^8}+\frac{C_3}{R^{10}}\right) = -2.407\times10^{-3} \text{ ev}$$

$$\tilde{V} = 2.05\times10^{-3} \text{ ev}$$

while the second-order exchange term is

$$\lambda'(D^2P)_{00} = 0.030\times10^{-3} \text{ ev} \qquad \text{according to (30)},$$
$$= 0.020\times10^{-3} \text{ ev} \qquad \text{according to (31)}.$$

At $8a_0$ the exchange corrections are even smaller relative to E_2† and beyond $10a_0$ they can be neglected. The potential minimum according to Kolos and Wolniewicz (1965) occurs at $8.45a_0$ (4.15 Å). The second-order exchange effect is small when compared with the non-overlap dispersion result. At $7a_0$ this correction is large compared with the total potential, but this is only because the total potential is practically zero at this distance. Near the potential minimum the correction term is a small part, less than 1%, of the total potential. At separations smaller than $5a_0$ this entire procedure is not useful as the correction terms become very large.

† From (30) we obtain

$$\lambda'(D^2P)_{00} = 0.0052\times10^{-3} \text{ ev}$$

versus

$$\tilde{V} = 0.376\times10^{-3} \text{ ev and } E_2 = -0.96\times10^{-3} \text{ ev for } 8a_0.$$

For reference, the total potential at $5a_0$ according to Kolos and Wolniewicz (Table 4.5) is 0.036 ev.

Dalgarno and Lynn (1956) adopted a modification of Margenau's second-order perturbation procedure to obtain the correction terms for H_2^+ and H_2. They used the approximation involving a mean excitation energy which was discussed in Chapter 2, replacing the second-order perturbation energy by

$$\varepsilon_2 = \frac{1}{\bar{E}}\,[\langle\Psi|\,V^2|\Psi\rangle - \langle\Psi|\,V|\,\Psi\rangle^2]. \tag{32}$$

Here Ψ is the zero-order wave function in the previous LCAO–MO form of Chapter 3. $\bar{E}$ is the mean excitation energy which is adjusted to yield the correct result when $R \to \infty$. As a further correction Dalgarno and Lynn (1956) suggest using

$$\bar{E} = \bar{E}(R \to \infty) + CS, \tag{33}$$

where C is a constant obtained by fitting ε_2 to the correct result at some small separation and S is again the overlap integral between the two atoms. For H_2^+, where the exact values are known, the error in assuming a constant $\bar{E}$ was found to be only 4% at $5a_0$. Also for H_2^+ at $5a_0$ the error in using

$$\Delta E = \tilde{V} + \varepsilon_2 \tag{34}$$

was only 2.4%, while the second-order exchange procedure, (26), was in error by 4%. The approximation of Margenau consists of using only the first term in (32) and $\lambda = \bar{E}^{-1}$, where $\bar{E}$ is the value variationally obtained when overlap is neglected.

Dalgarno and Lynn's procedure for H_2 is of uncertain rigor since Ψ is not an eigenfunction of H_0, the sum of the Hamiltonians of atoms a and b (see section 3.2 for a complete discussion). The Margenau procedure is not subject to this difficulty.

Other approximate techniques have been proposed,† but they are perhaps not of interest in the present context.

As a summary of the various approximate calculations on the triplet state of H_2, we note that the primary contributions to the intermolecular potential were

$$\Delta E = \tilde{V} + \Delta E_1 + \Delta E^{\text{attract}}, \tag{35}$$

† Musher (1963) employs a perturbation method which partially sums the second-order energy. Frost and Braunstein (1951) write $\Psi = \mathcal{A}[1\sigma_g(1)\alpha(1)\,1\sigma_g(2)\beta(2)(1+\gamma r_{12})]$ for the ground state singlet of H_2. This has not been extended to the triplet state.

namely, the Heitler–London repulsion, the singly substituted configurations in the Donath and Pitzer function, the second term in (10), and the doubly substituted configurations which lead to energy terms behaving as R^{-6}, R^{-8}, and R^{-10} at large separations. For separations around the potential minimum, one can use the second-order exchange approximation in which

$$\Delta E^{\text{attract}} = E_2 + \lambda'(D^2P)_{00} \tag{36}$$

provided E_2 is the result calculated with no overlap and the second term is the correction for overlap. The best values of E_2 are probably those of Chan and Dalgarno (1965) and Bell (1965):†

$$E_2 = -\frac{6.499}{R^6} - \frac{124.4}{R^8} - \frac{3286}{R^{10}}. \tag{37}$$

Some of the previous values of R^{-10} are in error because they included only the quadrupole–quadrupole contribution and not the dipole–octupole portion.

In Table 4.6 we list the values obtained from (35), with (36) and (37), together with some of the other approximate calculations. The results suggest that (35) is correct with an accuracy better than 10% for $R > 8a_0$. For similar precision one could also add the higher-order E_2 terms and ΔE_1 to the Hirschfelder and Linnett values.

TABLE 4.6 *Summary of Approximation Values for Hydrogen Atom Interactions*

R (a_0)	$\Delta E \times 10^3$ ev				
	E_2	ΔE [eqn. (35)][a]	Hirschfelder–Linnett	Donath–Pitzer	Kolos–Wolniewicz
6	−7.284	2.876	6.758	8.016	5.1019
7	−2.407	−0.4531	0.604	1.0604	−0.082
8	−0.9596	−0.6014	−0.254	−0.0498	−0.533
10	−0.2196	−0.2061	−0.1647	−0.09758	−0.182
12	−0.06854	−0.06822	−0.05427	−0.03608	—

[a] $\lambda'(D^2P)_{00}$ is calculated using (30).

† More recently W. Kolos (*Int. J. Quantum Chem.* **1**, 169, 1967) has given even more precise values.

Hydrogen is a very special case and knowledge of its interactions does not help in the treatment of larger systems. Each hydrogen atom has but one electron and the problem of intra-atomic electron repulsions never enters. Hydrogen is the only atom for which exact solutions are available. In addition, there is little experimental knowledge of hydrogen atom interactions because of their tendency to form molecules. For these reasons our attention will now be directed toward the interaction of helium atoms and larger systems.

4.2. Helium Interactions

For helium there exists ample experimental information. Many adjustments of trial potentials to virial and transport data have been performed and in Fig. 3 we present some of the latest "experimental" potentials.† The parameters for these curves which are of greatest interest are σ, the separation at which the potential changes from attractive to repulsive, and ε_0 and r_0, the depth and position of the potential minimum. Specific results are summarized in Table 4.7.

TABLE 4.7. *"Experimental" Potentials for Helium Interactions*

Potential[a]	σ (a_0)	ε_0 (°K)	r_0 (a_0)
Lennard–Jones 6–12	4.830	10.22	5.422
Mason Rice (1954) exp-6 (MR–1)	5.178	9.16	5.92 ($\alpha = 12.4$)
Kilpatrick, Keller, and Hammel (MR–5)	5.280	7.563	6.027 ($\alpha = 12.4$)

[a] All from Kilpatrick, Keller, and Hammel (1955).

It is interesting to note that in spite of the disparity in these potentials the first and third fit the second virial coefficient equally well.

† L.W. Bruch and I. I. McGee [*J. Chem. Phys.* **46,** 2959 (1967)], using experimental and theoretical data, find $\sigma = 5.00 \pm .03 a_0$, $r_0 = 5.62 \pm .03 a_0$ and $\varepsilon_0 = 12.5 \pm 1.2$°K. This implies that overlap corrections may be less serious than suggested later in this chapter. It also suggests that there may be a very weakly bound He_2 molecule.

The helium atom interactions are complicated by the presence of intra-atomic electron repulsions. Even the long-range interactions of the atoms are not accurately known and continue to be the object of increasingly accurate calculations (Davison, 1966). There are available good approximate state functions, the most complicated of which lead to complete agreement with atomic

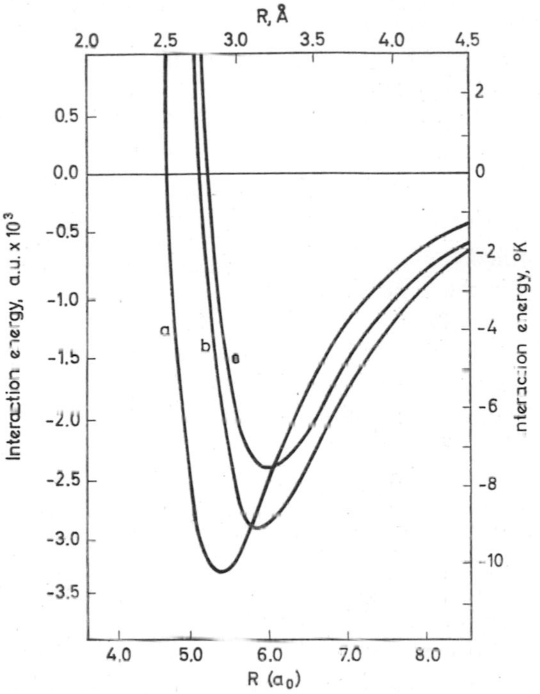

FIG. 3. "Experimental" helium–helium interactions. *a*, Lennard-Jones 6–12 potential fit by Kilpatrick, Keller, and Hammel (1955). *b*, Exp-6 potential fit by Mason and Rice (1954) (MR–1). *c*, Exp-6 potential fit to new experimental data by Kilpatrick, Keller, and Hammel (1955) (MR–5).

experimental data. In calculating the helium interactions, however, one must take into account the electron repulsions within the atoms as well as the electron interactions between atoms which were considered for hydrogen atoms in the first part of this chapter.

The first calculations of this interaction employed the Heitler–London method with an atomic function which was the product of two one-electron orbitals for the first-order energy, and added to this the calculated long-range R^{-6} attraction (Slater and Kirkwood, 1931). Next, an improved value of the R^{-6} term was computed and the R^{-8} interaction as well as the second-order exchange term was added (Margenau, 1939). The seond-order exchange effect was found to be very small. From approximate calculations by Margenau it is likely that this effect is even smaller in neon, argon, and other systems with a rare gas structure and moderate polarizability. The Slater result (Slater and Kirkwood, 1931) leads to a repulsion which is much too weak. It has an exponential behavior very similar to the LCAO–MO with HF atomic orbitals and simulates the complete SCF results discussed in Chapter 3, but its magnitude is small. This reflects the techniques of integral evaluation and the approximations used. In particular, Slater identified $\tilde{V}$ and ΔE, and this error accounts for much of the low value for the repulsion.

Hirschfelder, Curtiss, and Bird attempted in their book to correct some of these faults and to compound what seemed at the time the best theoretical potential for helium. They used Rosen's (1950) calculated repulsion, Margenau's (1939) second-order exchange and R^{-6} attraction, and Page's (1938) R^{-8} attraction to obtain in atomic units

$$\Delta E = 21.22e^{-2.33R} - 12.85e^{-2.82R} - \frac{1.45}{R^6} - \frac{11.19}{R^8}. \tag{38}$$

The second-order exchange term employed here (the second on the right) is now known to be of the wrong sign. However, it is very small. Also, the last two terms are a few per cent too small. The recent values of Davison (1966) are

$$-\frac{1.47}{R^6} - \frac{14.2}{R^8}. \tag{39}$$

We also know that the Rosen repulsion is too large at very small separations (Phillipson, 1962). Apparently, several bond functions were neglected in its calculation. On the whole, however, the potential (38) has been found servicable in some respects. It has a depth of 6.48°K at $6a_0$.

Later we shall improve (38) by using more recent values for its several parts. We note, however, that such a procedure is not fundamentally satisfying since it is forced to employ many different functions and techniques. Very little is known about how these effects interfere with one another, nor is it easy to tell whether all important features have been included. We therefore review other ways of obtaining the complete potential within one consistent mathematical scheme.

Phillipson (1962), realizing that a simple LCAO–MO was not adequate to describe an SCF orbital, let alone the van der Waals attraction, made an elaborate machine computation based on a function involving a sixty-four-term configuration interaction:

$$\Psi = \sum_{k=1}^{64} C_k \mathcal{A} \left(\prod_{i=1}^{4} \Phi_{ki} \right). \tag{40}$$

Each orbital contained only a single function (one X_i of eqn. (57) of Chapter 3), but many types of orbitals were included: a $1\sigma_g$ and $1\sigma_u$ orbital formed from $1s$ Slater atomic functions with exponent ζ, a $1\sigma_g'$ and $1\sigma_u'$ orbital formed from $1s$ Slater functions with exponent ζ', a $\sigma_g(2s)$ and a $\sigma_u(2s)$ orbital formed from $2s$ Slater orbitals; a $\sigma_g(2p)$ and $\sigma_u(2p)$ orbital formed from $2p_z$ Slater orbitals; and two $\pi_u(2p)$ and two $\pi_g(2p)$ orbitals formed from $2p_x$ and $2p_y$ Slater atomic orbitals. Included in (40) were only those configurations which had the proper overall symmetry. Phillipson employed the MO–ζ method in which the σ_g and σ_u orbitals are allowed to have different exponents. At large separations he obtained a helium atom energy in error by only 0.72 ev, while the HF value is in error by about 1.1 ev.

His function is not as accurate in the united-atom region. However, he finds that even at 0.5 Å the proper united atom configurations contribute very little. Therefore we expect his results to be accurate even at separations of less than 1 Å. Results for a ten-, a sixty-four-configuration state function, and the SCF function discussed in Chapter 3 are listed in Table 4.8. The ten-term function contains only the $1s$ orbitals while the sixty-four-term result contains $1s$, $2s$, and $2p$ atomic orbitals. It includes some of the van der Waals' attraction and therefore lies lower than the SCF result.

Phillipson claims to have found a small minimum when only one term in (40) was used and exponents were varied. This minimum occurs at 3.3 Å ($6.24a_0$) and has a depth of 0.33°K or 0.29×10^{-4} ev. This minimum was not found in the complete SCF treatments (Chapter 3). Its small magnitude and strange location might raise the suspicion that it is spurious because of

TABLE 4.8. *Phillipson Configuration Interaction Calculations of the Helium Interactions*

R (Å)	E (calc.) ev		
	ψ_{10}	ψ_{64}	ψ_{SCF}
0.5	28.76	27.76	28.26
0.625	18.06	17.33	17.58
0.75	11.36	10.78	10.92
1.0	4.39	4.059	4.137
2.0	0.074	0.0329	0.06193

a failure of several large quantities to cancel one another properly in the calculation.

Phillipson's calculations emphasize two general facts: (1) to obtain very good results a great many determinants must be included and the functions must be judiciously chosen, and (2) to obtain the van der Waals' minimum by such a procedure is a very complex problem. When one uses sets of molecular orbitals, much of one's physical intuition is lost, and an appraisal of the relevance of different functions becomes difficult. In particular the distinction between inter- and intra-atomic effects is not simply preserved.

Moore (1960) and Kim (1962) have used a trial function similar to that of Hirschfelder and Linnett for hydrogen. They chose

$$\Psi = \mathcal{A}\left[\left(1+\alpha\delta^2 \sum_{i=1}^{2}\sum_{j=3}^{4} Z_{ai}Z_{bj}+\beta\delta^2 \sum_{i=1}^{2}\sum_{j=3}^{4}(x_{ai}x_{bj}+y_{ai}y_{bj})\right)\varphi_o\right]. \tag{41}$$

At large separations electrons one and two are on center a and electrons three and four are on center b. φ_0 is the product of four atomic functions and $\mathcal{A}\varphi_0$ is the Heitler–London expression. The atomic wave functions used were single Slater orbitals, given by eqn. (44) of Chapter 3 with exponent $\delta = 1.6875$ for $R > 3a_0$.

This function can also be written as a superposition of configurations in the form

$$\begin{aligned}\Psi = {} & \mathcal{A}\varphi_0+C_1\mathcal{A}[2p_{z_1a}(1)1s_a(2)2p_{z_1b}(3)1s_b(4)\alpha(1)\beta(2)\alpha(3)\beta(4)]\\ & +C_2\mathcal{A}[2p_{x_1a}(1)1s_a(2)2p_{x_1b}(3)1s_b(4)\alpha(1)\beta(2)\alpha(3)\beta(4)]\\ & +C_2\mathcal{A}[2p_{y_1a}(1)1s_a(2)2p_{y_1b}(3)1s_b(4)\alpha(1)\beta(2)\alpha(3)\beta(4)]\\ & +\dots\end{aligned} \tag{42}$$

The remaining terms contain atomic $2p$ functions for electrons one and four, electrons two and three, and electrons two and four. Written in the form of (42), we see that Ψ is in fact a generlization of the Slater–Kirkwood expression, eqn. (11). There are four pairs of interactions involved between helium atoms. Equation (42) contains thirteen determinants but only two variable parameters.

Kim and Moore used the same function but obtained slightly different results for $\Delta E(R)$. Moore minimized only the numerator of $\bar{H}$ [eqn. (19) of Chapter 3], while Kim minimized the entire $\bar{H}$. If overlap is small ($S = 0.1$ for $R = 3a_0$), this error should be insignificant. The primary differences between the two results were traced by Kim to the values of the integrals used. Small numerical errors in the individual integrals lead to sizeable errors in the potential calculated. This points up the importance of determining ΔE in such a way that large cancellations occur automatically. While Kim and Moore did not indicate what happens to ΔE at large R, Kim approximated his results by the formula

$$14.22e^{-2.34R} + 13.45e^{-3.78R} - \frac{.911}{R^{5.9}} \qquad (43)$$

in the region of 1.2 to $7.1a_0$. The last two terms do contain the second-order exchange effects, and they are positive contributions, as we found them for hydrogen interactions.

Kestner (1966), with use of a similar function, obtained the long-range behavior: $-0.984R^{-6}$.

Despite the relative simplicity of these calculations their results are significant since they have led to a reasonable form for the potential. Phillipson was not able to accomplish this with a large number of configurations.

Moore and Kim, nevertheless, made several approximations which must be considered further.

(1) They used as atomic functions single Slater orbitals. These are known to give very poor results at large distances. The overlap of two Slater orbitals at $5a_0$ is 0.718×10^{-2}, while that of two HF *atomic* orbitals is 1.42×10^{-2}. At larger separations these values differ even more. Since all repulsive contributions depend strongly on the overlap, a single Slater orbital will lead to an underestimate of such effects (see Chapter 3).

(2) They neglect contributions of *intra*-atomic electron repulsions to the state function.

(3) In assuming that (41) is a good approximation for calculating ΔE, they neglect all contributions from configurations involving d or f orbitals. As was shown for hydrogen interactions in the previous section, the other contributions which at large separations lead to the R^{-8} and R^{-10} terms are very important near the potential minimum.

(4) They introduce only two variation parameters and assume that the effects involving two electrons of like spin are the same as those involving two electrons of different spins.

Many of these assumptions are indeed valid, though often for very complex reasons, as will be apparent from a study of an alternative method which is to follow. Unfortunately, however, the approaches of Hirschfelder and Linnett and Moore and Kim are almost inapplicable to larger systems because of the great number of terms they require. It has been estimated that for neon one needs a minimum of 109 determinants in the total state function. To use such a function in computing $\overline{H}$ is practically impossible. Hence, to calculate interactions between larger systems we shall need to modify the procedures described so far.

This will be done with the aid of Sinanoglu's "Many–electron theory of atoms and molecules" (Sinanoglu, 1962, 1964; Sinanoglu and Tuan, 1964) which can be readily applied to intermolecular force calculations. Relevant details will first be sketched.† We use as our zero-order state function the HF-SCF results, writing φ_0 in either the localized (equivalent localized orbitals, ELO, of Chapter 3) or the delocalized form (MO). For helium interactions‡ we have

$$\varphi_0^{\mathrm{HF}} = \mathcal{A}(\eta_1\eta_2\eta_3\eta_4) \qquad (44)$$

or

$$\varphi_0^{\mathrm{HF}} = \mathcal{A}(\Phi_1\Phi_2\Phi_3\Phi_4), \qquad (45)$$

where the η_i are the localized orbitals and the Φ_i are the molecular orbitals. Both η_i and Φ_i are spin orbitals, which means they are the product of a spatial times a spin function.

† They are taken mostly from Sinanoglu (1964), Sinanoglu and Kestner (1966), and Kestner (1966).

‡ In this section we treat only the helium interaction; the method is much more general, as will be seen later.

Additive corrections to φ_0 are of four types:

$$\mathcal{A}(\hat{f}_1\Phi_2\Phi_3\Phi_4), \quad \mathcal{A}(\Phi_1\hat{f}_2\Phi_3\Phi_4), \text{ etc.,} \tag{46}$$

$$\mathcal{A}(\Phi_1\hat{U}_{23}\Phi_4), \quad \mathcal{A}(\hat{U}_{12}\Phi_3\Phi_4), \text{ etc.,} \tag{47}$$

$$\mathcal{A}(\Phi_1\hat{U}_{234}), \text{ etc.,} \tag{48}$$

$$\hat{U}_{1234}. \tag{49}$$

Our notation is meant to suggest that $\hat{f}_i$ is some function of the coordinates of electron i only, $\hat{U}_{ij}$ of the coordinates of electrons i and j, $\hat{U}_{ijk}$ is a function only of the coordinates of electrons $i, j,$ and k, etc. Let us call these one-, two-, three-, and four-electron substitutions, since we have replaced corresponding numbers of one-electron orbitals in (45) by functions of one, two, three, and four electrons. When these functions are taken to be products of excited individual orbitals, as in calculations involving configuration interaction, they are also referred to as one, two, three, and four electron "excitations".

It is convenient to choose general functions which are antisymmetric themselves as well as orthogonal to all occupied HF orbitals (which are identified by $\wedge$); e.g. we require that

$$\hat{U}_{ijk}(x_1,x_2,x_3) = -\hat{U}_{ijk}(x_2,x_1,x_3), \tag{50}$$

$$\langle \hat{U}_{ijk} | \mathcal{A}(\Phi_k\Phi_l\Phi_m) \rangle = 0, \tag{51}$$

$$\langle \hat{U}_{ijk}(x_1,x_2,x_3) | \Phi_l(x_1) \rangle_{x_1} = 0, \tag{52}$$

(integration over coordinates of electron one only), where Φ_k, Φ_l, and Φ_m designate any one of the four spin orbitals in φ_0^{HF}. These relations greatly simplify the procedure.

The exact state function for the helium interaction has the form

$$\Psi = \varphi_0^{\mathrm{HF}} + r_1 + r_2 + r_3 + r_4, \tag{53}$$

where r_1 is the collection of all one-electron substitutions, r_2 denotes all two-electron substitutions, and so forth. The effects included in r_1 are those in which one orbital is modified independently of all the other orbitals. r_2 includes all correlations of any two specific electrons at one time, independent of all other electron correlations. For two helium atoms there are six two-electron correlations. We shall call these pair correlations, and the functions $\hat{U}_{ij}$, pair functions. r_3 and r_4 represent all correlations involving three and four electrons simultaneously.

If $\varphi_0^{\rm HF}$ is the exact HF function, r_1 will not occur in the first- but only in the second-order correlations of that function,† i.e. it affects the fourth-order energy correction. By "order" we mean here the power of the perturbation. In our calculations this is the difference between the total Hamiltonian and our zero-order HF Hamiltonian. (It represents the instantaneous interaction over and above any average interaction between electrons. In (57), which follows, it appears as $m_{\varrho\nu}$.) First-order corrections to $\varphi_0^{\rm HF}$ contain only two-electron substitutions (see, for example, Sinanoglu, 1961). The second-order correction to $\varphi_0^{\rm HF}$ contains one-, two-, three-, and certain special four-electron substitutions, but the four-electron effects are simple products of two lower-order, two-electron substitutions (see Stanton, 1965; Kestner, 1966).

If $E_{\rm HF}$ is the energy obtained from $\varphi_0^{\rm HF}$, the difference $E-E_{\rm HF}$ is called the "correlation energy", $E_{\rm corr}$. Whereas the HF energy reflects the average interactions of electrons, the correlation energy reflects the deviations of interactions from their average values, i.e. the correlations in their motions. The intermolecular interaction receives contributions from both $E_{\rm HF}$ and $E_{\rm corr}$.

In single atoms and molecules, one-, three-, and four-electron substitutions are of minor importance since the probability, for example, of having three electrons close together is small in view of the exclusion principle (only two electrons of three will have unlike spins) (Sinanoglu, 1961, 1962). Two electrons, on the other hand, can interact strongly.

As to their effect on intermolecular forces, many-electron correlations other than binary are not necessarily small since three electrons are not kept apart any more than are two electrons at most internuclear separations. As might be expected, however, these contributions are relatively more important at moderate separations than at extremely small R, where the situation resembles that in a single atom. For helium all estimates suggest that these many-electron contributions amount to less than 15% of the interaction energy at all separations.

† This is Brillouin's theorem. An analogous result holds for the localized SCF orbitals (Kahalas and Nesbet, 1963).

Thus we can write the total state function for a pair of helium atoms as follows:†

$$\begin{aligned}\Psi = \varphi_0^{\text{HF}} &+ \mathcal{A}\{\hat{\mu}_{12}\} + \mathcal{A}\{\hat{\mu}_{34}\} + \mathcal{A}\{\hat{\mu}_{13}\} \\ &+ \mathcal{A}\{\hat{\mu}_{14}\} + \mathcal{A}\{\hat{\mu}_{23}\} + \mathcal{A}\{\hat{\mu}_{24}\} \\ &+ \mathcal{A}\{\hat{\mu}_{12}\hat{\mu}_{34}\} + \mathcal{A}\{\hat{\mu}_{24}\hat{\mu}_{13}\} + \mathcal{A}\{\hat{\mu}_{23}\hat{\mu}_{14}\}. \qquad (54)\end{aligned}$$

In this expression $\hat{\mu}_{ij}$ is the equivalent of $\hat{U}_{ij}$, but the former is based on localized orbitals while the latter is based on molecular orbitals. We abbreviate similarly the counterparts of the terms in (47) and write $\hat{\mu}_{1234}$ in its lowest order, i.e.

$$\{\hat{\mu}_{1234}\} \equiv \mathcal{A}(\hat{\mu}_{12}\hat{\mu}_{34}).$$

While (54) can be used either with the delocalized (MO) or localized form (ELO) of the HF function, we shall work here exclusively with the latter. When the overlap between atomic functions on the two centers is small, these localized orbitals are only slightly distorted from the atomic functions themselves. As before, electrons one and two with α and β spins, respectively, will be assigned to the localized orbital about center a, and electrons three and four to the orbital about center b.

There are two types of pair functions in (54): intra-orbital (intra-atomic at large separations) pairs (μ_{12} and μ_{34}) and inter-orbital pairs ($\hat{\mu}_{13}$, $\hat{\mu}_{23}$, $\hat{\mu}_{14}$, and $\hat{\mu}_{24}$). The latter yield the dispersion forces at large separations. These were previously introduced as $\lambda D\varphi_0$, e.g. in (20). The intra-pairs have very little dependence on R when the potential is not strongly repulsive, while the interpairs have a strong dependence on distance (R^{-6}, etc., at large R values).

The total variational energy for a general function

$$\Psi = \varphi_0 + \chi_{ss} \qquad (55)$$

is

$$E = \frac{\langle\Psi|H|\Psi\rangle}{\langle\Psi|\Psi\rangle} = E_{\text{HF}} + \frac{2\langle\varphi_0|H - E_{\text{HF}}|\chi_{ss}\rangle + \langle\chi_{ss}|H - E_{\text{HF}}|\chi_{ss}\rangle}{1 + \langle\chi_{ss}|\chi_{ss}\rangle} \qquad (56)$$

since $\langle\varphi_0|\chi_{ss}\rangle = 0$ and $E_{\text{HF}} = \langle\varphi_0|H|\varphi_0\rangle$ when $\langle\varphi_0|\varphi_0\rangle = 1$.

The total Hamiltonian is the sum of a HF part and a perturbation

$$H = \sum_{\varrho=1}^{4} h_p^0 + \sum_{\varrho>\nu} m_{\varrho\nu} \qquad (57)$$

† $\mathcal{A}\{\hat{\mu}_{12}\} \equiv \mathcal{A}(\hat{\mu}_{12}\Phi_3\Phi_4)$

where the perturbation

$$m_{\varrho\nu} = r_{\varrho\nu}^{-1} - J_\varrho(\varrho) - J_\varrho(\nu) - J_\nu(\varrho) - J_\nu(\nu) + K_\varrho(e) + K_\varrho(\nu) + K_\nu(e) + K_\nu(\nu) + J_{\varrho\nu} - K_{\varrho\nu} \tag{58}$$

The Coulomb and exchange potentials, J_ϱ and K_ϱ, were introduced in Chapter 3. [See eqns. (61)–(63) of Chapter 3.]

If we introduce a new operator, namely the one-electron Hamiltonian minus the orbital energies

$$e_\varrho = h_\varrho - \varepsilon_\varrho, \tag{59}$$

we can substitute the specific χ_{ss} from (54) into (56) to obtain

$$E = E_{\text{HF}} + \sum_{\varrho>\nu} \frac{\tilde{\varepsilon}''_{\varrho\nu} D_{\varrho\nu}(1+\langle\hat{\mu}^2_{\varrho\nu}\rangle)}{D} + \Xi + \Delta. \tag{60}$$

The meaning of these symbols is as follows:

$$\tilde{\varepsilon}''_{\varrho\nu} = \frac{\langle\hat{\mu}_{\varrho\nu}|r_{\varrho\nu}^{-1}|\mathcal{A}_2(\eta_\varrho\eta_\nu)\rangle + \langle\hat{\mu}_{\varrho\nu}|e_\varrho + e_\nu + m_{\varrho\nu}|\hat{\mu}_{\varrho\nu}\rangle}{1+\langle\hat{\mu}^2_{\varrho\nu}\rangle}, \tag{61}$$

where $\mathcal{A}_2$ is a two-electron, antisymmetrizing operator. Ξ is a collection of all relevant matrix elements of the $m_{\varrho\nu}$ between different terms in χ_{ss}, e.g.

$$\langle\{\hat{\mu}_{\varrho\nu}\}|m_{\varrho\delta}|\{\hat{\mu}_{\varrho\nu}\}\rangle. \tag{62}$$

The major contributions to Ξ are matrix elements involving two different pair functions. D is the normalization constant:

$$D = \langle\Psi|\Psi\rangle = 1 + \sum_{\varrho>\nu}\langle\hat{\mu}^2_{\varrho\nu}\rangle + \sum_{\varrho>\nu}\sum_{\delta>\gamma}\langle\hat{\mu}^2_{\varrho\nu}\rangle\langle\hat{\mu}^2_{\delta\gamma}\rangle \quad (\varrho,\nu \neq \delta,\gamma). \tag{63}$$

$D_{\varrho\nu}$ is D with all terms involving either electrons ϱ or ν omitted, i.e. for helium

$$D_{13} = 1 + \langle\hat{\mu}^2_{24}\rangle$$
$$D_{24} = 1 + \langle\hat{\mu}^2_{13}\rangle.$$

Typically, $\langle\hat{\mu}^2_{\varrho\nu}\rangle$ is less than one-tenth and thus $D_{\varrho\nu}(1+\langle\hat{\mu}^2_{\varrho\nu}\rangle)/D$ can be safely approximated by expanding to first order in $\langle\hat{\mu}^2\rangle$.

The quantity Δ is a small correction which is called for by the use of localized orbitals. It is proportional to overlap cubed divided by R cubed at moderate separations. It is furthermore proportional to the $\gamma_{\varrho v}$ of eqn. (109) of Chapter 3 and will be called the off-diagonal contribution.

For ease in calculation we shall rewrite (60) as follows:

$$E = E_{\rm HF} + \frac{2\tilde{\varepsilon}_{12}'' D_{12}(1+\langle \hat{\mu}_{12}^2\rangle)}{D} + \frac{2\tilde{\varepsilon}_{13}^{(a)} D_{13}^{(a)}(1+\langle \hat{\mu}_{13}^2\rangle)}{D}$$
$$+ \frac{2\tilde{\varepsilon}_{14}^{(a)} D_{14}^{(a)}(1+\langle \hat{\mu}_{14}^2\rangle)}{D} + \Delta + \frac{2\Delta\tilde{\varepsilon}_{13}^{(a)} D_{13}^{(a)}(1+\langle \hat{\mu}_{13}^2\rangle)}{D}$$
$$+ \frac{2\Delta\tilde{\varepsilon}_{14}^{(a)} D_{14}^{(a)}(1+\langle \hat{\mu}_{14}^2\rangle)}{D} + \Xi' \qquad (64)$$

a form which results on combining similar terms and dividing ε_{13} (like-spin interorbital effect) and ε_{14} (unlike spin) into two parts:

$$\tilde{\varepsilon}_{13}^{(a)} = \langle \hat{\mu}_{13}^{(a)} | r_{13}^{-1} | \mathcal{A}(\eta_1\eta_2)\rangle + \langle \hat{\mu}_{13} | e_1^{(a)} + e_3^{(a)} | \hat{\mu}_{13}\rangle \qquad (65)$$

with

$$e_1^{(a)} = h_1^0 - \varepsilon_1^0 - J_1(1) - K_1(1). \qquad (66)$$

By this division we have made $\Delta\tilde{\varepsilon}_{13}^{(a)}$ very small. When the multipole expansion of r_{13}^{-1} holds, the quantity

$$\Delta\tilde{\varepsilon}_{13}^{(a)} = \langle \hat{\mu}_{13} | m_{13}^{(a)} | \hat{\mu}_{13}\rangle \qquad (67)$$

behaves as R^{-11} (with $e_1+e_3+m_{13} = e_1^{(a)}+e_3^{(a)}+m_{13}^{(a)}$). This is part of the third-order perturbation effects discussed in Chapter 2.

Equation (60) or (64) is a variational expression with respect to any change in the function (54). To be sure (60) and (64) are very complicated even in the four-electron helium interaction. In larger systems the complexity increases rapidly with the number of electrons. But suppose that we minimize only a part of (64). There is no guarantee, of course, that this will yield an absolute minimum for *that* part; probably it does not. However, if the result of this partial minimization is substituted in the complete expression (64), and all terms are evaluated, we do have an upper limit for the total energy; the equality sign cannot hold in this situation. If we carefully minimize very large parts of (64), the error in the total energy after evaluating all terms should be small. In fact, a variation of this procedure can be used to derive the standard perturbation theories.†

† For a discussion of this form of variation–perturbation theory, see Sinanoglu (1961d).

We shall minimize ε_{12}'', $\varepsilon_{13}^{(a)}$, and $\varepsilon_{14}^{(a)}$ separately. The first quantity is of the order of atomic correlation energies, about 1 ev, while the other two are of the order of 10^{-4} ev. By minimizing these parts separately we can treat each in its own special way, using methods familiar from the theory of the helium atom on ε_{12}'', and methods applicable to dispersion forces on $\varepsilon_{13}^{(a)}$ and $\varepsilon_{14}^{(a)}$. In this way the major parts of ΔE are determined directly and not by methods involving differences.

The intermolecular potential thus obtained is, in view of (64),

$$\begin{aligned}\Delta E &= \Delta E_{\mathrm{HF}} + \Delta E_{\mathrm{corr}} \\ &= \Delta E_{\mathrm{HF}} + 2\left[\frac{\tilde{\varepsilon}_{12}'' D_{12}}{D} - \varepsilon_{12}^{\mathrm{atom}}\right] + \Delta \\ &\quad + (2/D)(\tilde{\varepsilon}_{13}^{(a)} D_{13} + \tilde{\varepsilon}_{14}^{(a)} D_{14} + \Delta\tilde{\varepsilon}_{13}^{(a)} D_{13} + \Delta\tilde{\varepsilon}_{14}^{(a)} D_{14}) + \Xi. \qquad (68)\end{aligned}$$

Here the term in square brackets is the change in the intra-atomic correlation energy with distance. It arises from two causes: normalization ($D_{12}/D \neq 1$) and distortion ($\tilde{\varepsilon}_{12}'' \neq \varepsilon_{12}^{\mathrm{atom}}$). If we express all normalization constants D to first order in matrix elements over pair functions, we obtain

$$\begin{aligned}\Delta E = \Delta E_{\mathrm{HF}} &+ 2[\tilde{\varepsilon}_{12}''(1 + 2\langle \hat{\mu}_{13}^2 \rangle - 2\langle \hat{\mu}_{14}^2 \rangle) - \varepsilon_{12}^{\mathrm{atom}}] \\ &+ [2\tilde{\varepsilon}_{13}^{(a)} + 2\tilde{\varepsilon}_{14}^{(a)}](1 - 2\langle \hat{\mu}_{12}^2 \rangle) + \Delta \\ &+ 2(\Delta\tilde{\varepsilon}_{13}^{(a)} + \Delta\tilde{\varepsilon}_{14}^{(a)})(1 - 2\langle \hat{\mu}_{12}^2 \rangle) \\ &+ 8\langle \mathcal{A}(\hat{\mu}_{13}\eta_2) \,|\, r_{12}^{-1} \,|\, \mathcal{A}(\hat{\mu}_{23}\eta_1) \rangle (1 - 2\langle \hat{\mu}_{12}^2 \rangle) \\ &+ 8\langle \mathcal{A}(\hat{\mu}_{12}\eta_3) \,|\, r_{13}^{-1} \,|\, \mathcal{A}(\hat{\mu}_{23}\eta_1) \rangle (1 - 2\langle \hat{\mu}_{12}^2 \rangle) \\ &+ 8\langle \mathcal{A}(\hat{\mu}_{12}\eta_3) \,|\, r_{23}^{-1} \,|\, \mathcal{A}(\hat{\mu}_{13}\eta_2) \rangle (1 - 2\langle \hat{\mu}_{12}^2 \rangle). \qquad (69)\end{aligned}$$

In this expression $\hat{\mu}_{23}$ and $\hat{\mu}_{14}$ have the same form and parameters; they differ only with respect to electron labels. $\varepsilon_{12}''-\varepsilon_{12}$ for the atom would be zero if the localized orbitals were not distorted from their HF atomic forms. Because these orbitals are not identical, this quantity will introduce dependence on the overlap of the atomic orbitals on the two centers.

Terms in (69) involving two different pair functions are the explicit major parts of Ξ. They are of two distinct types; one involving two interatomic (interorbital) pair functions and another coupling the interatomic pairs to the intra-atomic effects. We often abbreviate them by the symbols ∇ and ∇ (the dotted line symbolizes the r_{ij}^{-1}, the solid line the $\hat{\mu}_{\varrho\nu}$ expression). The first group was estimated by Donath and Pitzer (see Pitzer, 1959) to be large for the neon–neon interaction. Sinanoglu pointed out that the second group of terms could be even larger. Such is the case for the helium interaction.

From the perspective now gained it is easier to appraise the previous calculations. Moore, Kim, Slater and Kirkwood and essentially all other authors (to some extent even Phillipson) neglect $\hat{\mu}_{12}$ and thus miss all effects due to changes in the intra-orbital correlation and, more important, the contributions to Ξ which occur at all separations, behaving as R^{-6} at large distances.

Approximate calculations using this scheme are now available for separations over $4a_0$.† For large separations, where overlap may be neglected ($R > 8a_0$), very accurate results have been obtained. Instead of the true SCF function which has only recently become available, an LCAO–MO with HF orbitals was used. As indicated in Chapter 3 [cf. eqns. (76) of Chapter 3 and Table 3.4], this is a good approximation to the true SCF function for $R > 3a_0$. It was transformed to the localized form by eqns. (98) and (100) of Chapter 3. The trial functions are.

$$\bar{\mu}_{14} = \psi_{14/C_{14}} - \mathcal{A}(\eta_1\eta_4),$$
$$\bar{\mu}_{13} = \psi_{13/C_{13}} - \mathcal{A}(\eta_1\eta_3). \tag{70}$$

where

$$\psi_{14} = \mathcal{A}\{[1+S_{14}(x_1x_4+y_1y_4-\gamma_{14}z_1z_4)]\eta_1\eta_4\} \tag{71}$$

with a similar expression for ψ_{13}. The parameters, S_{14}, S_{13}, γ_{13}, and γ_{14} are varied to yield the lowest pair energies, $\tilde{\varepsilon}_{13}^{(a)}$ and $\tilde{\varepsilon}_{14}^{(a)}$. These parameters are functions of the internuclear separation. The form of ψ_{14} is the same as that used by Hirschfelder and Linnett for H_2 and by Kim and Moore for the helium interactions. The latter two authors, however, set $\hat{\mu}_{12} \equiv 0$ and minimized the whole of (64) at one time using a very approximate Heitler–London form (the atomic orbitals were single Slater terms) for φ_0. They also set $S_{13} = S_{14}$ and $\gamma_{14} = \gamma_{13}$. If (70) is to be used in our equations, $\bar{\mu}_{13}$ and $\bar{\mu}_{14}$ must be made orthogonal with

† Kestner and Sinanoglu (1966) and Kestner (unpublished). The latter values will be quoted here since they are based on more accurate atomic orbitals.

respect to all individual orbitals, η_k in analogy with (51) and (52), to yield $\hat{\mu}_{13}$ and $\hat{\mu}_{14}$.

As to the intra-atomic pair correlation function, $\hat{\mu}_{12}$, one extracts it from the Hylleraas (1929) six-term helium function and orthogonalizes it with respect to the orbitals in our helium-molecule system. One begins by using equations such as (70) with ψ_{12} provided by Hylleraas.

The results are shown in Fig. 4 for distances from 5.5 to $9a_0$. In these calculations all small normalization effects were neglected.

The following general statements which are not affected by the approximate quality of the wave functions used can now be made.

(1) Beyond $7.5a_0$ (4 Å) the simple nonoverlap dispersion results apply with an error less than 5%.

(2) The effects of spin symmetry ($\tilde{\varepsilon}_{13}^{(a)}$ vs. $\tilde{\varepsilon}_{14}^{(a)}$) begin at about $7a_0$ and increase as distance decreases. These effects are exhibited by the behavior of the z_1z_4 and z_1z_3 components of the pair functions. Thus $\varepsilon_{13}^{(a)}$ differs widely from $\varepsilon_{14}^{(a)}$ (like versus unlike spin pairs) for small distances, indicating differences in the $2p_z$ mixing. Moore (1960) and Kim (1963) simplified their analysis by enforcing the equalities $\gamma_{14} = \gamma_{13}$ and $S_{14} = S_{13}$, and this introduced errors. When the pairs are minimized separately, as described above, such additional parameter variation is handled easily.

(3) Overlap and exchange effects are not very large at the potential minimum. This agrees with the second-order exchange results of Margenau and with Lynn's (1958) use of a constant "mean excitation energy" after the manner of Dalgarno and Lynn (see H_2 section). An expansion in powers of overlap can be used with complete confidence, neglecting powers of S greater than 2.

(4) The relative contribution of three-electron coupling terms, the triangles, is roughly constant over the distances of greatest interest. At large separations it amounts to 22%, while at the potential minimum it is 27% of ΔE_{corr}.

(5) The change in intra-atomic correlation energy due to distortion of the atom into localized orbitals is small, roughly 2% of the correlation energy difference, near the potential minimum. For larger R it decreases rapidly; at shorter dis-

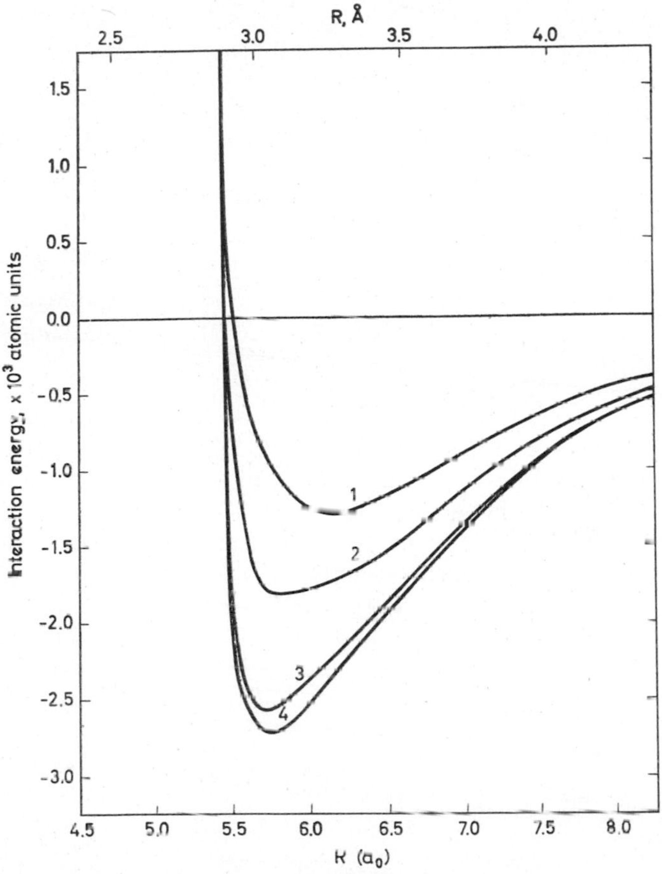

FIG. 4. The helium–helium interaction. 1, calculated curve (using approximate state functions as described in text). 2, estimated curve with proper long-range R^{-6} interaction (all atomic p orbital mixing included). 3, estimated curve with proper long-range R^{-6} and R^{-8} interaction (all atomic p and d orbital mixing estimated). 4, estimated total interaction (curve 3 plus small corrections for errors in E_{HF} and other known long-range effects). This potential probably exaggerates repulsion for $R < 5.5a_0$.

tances it increases very rapidly and causes severe problems for separations less than $4a_0$.

(6) The off-diagonal contribution Δ in (69) is small for all separations beyond $4a_0$. It varies as S^3R^{-3}, where S is the overlap integral between two atoms, since it is proportional to the off-diagonal energy $\gamma_{\varrho\nu}$ times the matrix element between inter- and intra-atomic pair functions.

(7) Even around the potential minimum the multipole expansion for Coulomb integrals can be used to an accuracy of better than 1%. The main cause of the departure from R^{-6} behavior of the pair energies lies in the exchange integrals and the normalization corrections, the quantity c in (70). This again is understandable from the standpoint of the method of second-order exchange. If we write ΔE_{corr} as $d(R)R^{-6}$, we find that $d(R)$ varies by only 12% as R varies from the potential minimum to infinity.

Elementary calculations based on (69) and with φ_0^{HF} taken as a linear combination of HF atomic orbitals (see Table 3.4) yield an intermolecular potential of depth 3.0°K at $6a_0$ (3.2 Å) (1°K = 3.1668×10^{-6} a.u. = 0.69502 cm^{-1}) as shown in Fig. 1. Before commenting further on the results, let us consider the long-range behavior in more detail.

At large separations one expands the pair function of (71) and includes other effects as follows:

$$\hat{u}_{14}(R \to \infty) = \frac{b_1}{R^3}[x_1x_4+y_1y_4-2z_1z_4]\eta_1'\eta_4'$$
$$+\frac{b_2}{R^4}M_4\eta_1'\eta_4'+\frac{b_3}{R^5}M_5\eta_1'\eta_4', \tag{72}$$

where M_4 and M_5 are higher multipole expressions (see Chapter 3). We write $\eta_i' = (1+\beta r_i+\gamma r_i^2)\phi_i^{\text{atom}}$, where ϕ_i^{atom} is the atomic SCF spin orbital. The b's as well as the β's and γ's of each term are variable parameters. For separations over $8a_0$ multipole expansions of r_{13}^{-1}, r_{23}^{-1}, etc., can legitimately be used. When all normalization terms are neglected and $\tilde{\varepsilon}_{13}^{(a)}$ and $\tilde{\varepsilon}_{14}^{(a)}$ are separately minimized (they are equal at the distances where overlap is neglected) we find the results of Table 4.9. Note that if overlap can be neglected, $\Delta E_{\text{HF}} = 0$ (Kestner, 1966).

TABLE 4.9. *Coefficients of R^{-n} occurring in pair functions of (72) (in atomic units)*

	n				
	6		8	10	
	$b_1 =$ -0.39418 $\beta = 0.0$ $\gamma = 0.0$	$b_1 =$ -0.14539 $\beta = 0.485$ $\gamma =$ -0.015	$b_2 =$ -0.11146 $\beta = 0.445$ $\gamma = 0.0$ (dipole–quadrupole)	$b_3 =$ -0.32587 $\beta = 0.0$ $\gamma = 0.0$ (quadru-pole–quadru-pole)	$b_3 =$ -0.34404 $\beta = 0.0$ $\gamma = 0.0$ (dipole–octopole)
	(dipole–dipole)				
$4\tilde{\varepsilon}_{13}^{(a)}$	−1.4680	−1.6443	−14.2768	−53.759	−105.73
16 ◸	0.1636	0.1668	0.4736	0.6332	7.150
8 ▽	0.1268	0.1246	0.5532	0.6720	4.126
Total	−1.1776	−1.3529	−13.25	−52.454	−94.15

Table 4.9 shows that the coupling terms are very large, and one is inclined to question whether the pair energies can be minimized independently. Secondly, it is seen that the R^{-8} and R^{-10} terms can contribute to the potential significantly at distances around 5–6a_0. Notice the magnitude of the dipole–octupole term which is so often neglected.

To test the validity of minimizing each pair function independently, we consider a model equation involving two types of pairs. Let us write in place of $E - E_{HF}$ in (64) the simple expression

$$E_{corr} = Ab + Bb^2 + Dc + Kc^2 + b^2F + bcG. \tag{73}$$

We shall assume that $K = B$, $D = \beta A$, $F = \alpha B$, and $G = \omega B$. The approximate equivalence with (64) can be made apparent if we neglect all of the small normalization corrections in (64) and identify $Dc + Kc^2$ with the *intra*-atomic pair correlation energy and $Ab + Bb^2$ with the *inter*atomic correlation energy. Note that β^2 is roughly the ratio of the intra to interorbital correlation energies. F represents three-electron effects involving two weak interorbital pairs, while G involves an inter- and a large intra-atomic pair function. Equation (73) can be minimized exactly and the large intra-atomic correlation energy, $-(A^2/4B)\beta^2$, can be substracted to find ΔE_{corr}, the contribution of the correlation energy to the intermolecular potential energy. The approximate results discussed previously arise from minimizing $\beta Ac + Bc^2$ and $Ab = Bb^2$ separately, and then using the values of b and c to obtain

$$\Delta E_{\text{corr}}^{\text{approx}} = -\frac{A^2}{4B}[1-\alpha-\varkappa], \tag{74}$$

where $\varkappa = \beta\omega$ (a constant). In the limit of large β the exact result is

$$\Delta E_{\text{corr}}^{\text{exact}} = -\frac{A^2}{4B}\frac{1-\varkappa+\varkappa^2/4}{(1+\alpha)} \tag{75}$$

and the ratio of the error to the value of the exact correlation energy becomes

$$\frac{\text{difference}}{\Delta E_{\text{corr}}^{\text{exact}}} = \frac{1-\alpha-\varkappa}{(1-\varkappa/2)^2}(1+\alpha). \tag{76}$$

As seen from Table 4.2, $\alpha = 0.076$ and $\varkappa = 0.101$; hence the relative error is only 1.8% in spite of the large three-electron coupling terms represented by F and G in our model.

The exact result in the limit of very large β can also be obtained by substituting the value of the intra-atomic pair correlation parameter, $c = -A\beta/2B$, into (73) and minimizing the whole of (73) minus the large intra-atomic correlation energy. This procedure was employed in the calculations leading to Table 4.4, which shows all minimization and normalization effects. It is somewhat astonishing that the final result is but a fraction of a per cent lower than the approximate result despite the large values of α and $\varkappa$, a fact which is very encouraging in our continuing study to find simple methods of calculation.

In Table 4.10 we compare some of the results obtained for the R^{-6} coefficient at large R. The latest sum rule result (Bell, 1965) and the elaborate calculations of Davison (1966) yield a value of -1.47 in atomic units. This should be accurate to within about 1%. The Slater–Kirkwood value is low by about 10% because of the neglect of the intra-atomic correlation. London's result

TABLE 4.10. *Coefficients of Various parts of the R^{-6} Interaction between Two Non-overlapping Atoms (in atomic units)*

	Restricted calculation (independent pairs) $b = -0.14539$ $\beta = 0.485$ $\gamma = -0.015$	Unrestricted-complete calculation (dependent inter-atomic pairs) $b = -0.06688$ $\beta = 1.050$ $\gamma = -0.090$
$4\,\tilde{\varepsilon}_{13}$	-1.6443	
$8\,\triangledown$	0.1246 (9.5%)	0.0929 (6.8%)
$16\,\ulcorner$	0.1668 (12.7%)	0.1458 (10.8%)
$-8\,\tilde{\varepsilon}_{13}\,\langle \rho_{12}^2 \rangle$	0.0211	0.0208
$-8\,\tilde{\varepsilon}_{12}\,\langle \rho_{13}^2 \rangle$	0.0209	0.0082
$-16\,\triangledown\,\langle \rho_{12}^2 \rangle$	-0.0019	-0.0014
$-32\,\ulcorner\,\langle \rho_{12}^2 \rangle$	-0.0025	-0.0022
	-1.31	-1.362

neglects the continuum effects (see Chapter 2) and should be high. Karplus and Kolker neglect all coupling terms (which contribute about 20%). The experimental result undoubtedly includes some of the R^{-8} effect and is therefore low (Table 4.11).

The procedure described here is correct through the first order in the intra-atomic electron correlation. Corrections to the value of -1.362 must arise from higher order effects. Such contributions involve coupling of the inter- and intra-atomic electron correlations, i.e. three- and four-electron correlations in the state function [(48) and (49)]. Estimates of their contributions range from 8% to 13% of the -1.362 value. Such a correction would bring all calculations into agreement (Kestner, 1966).

For comparison we note that Davison's (1966) result for the R^{-8} coefficient is -14.2 in atomic units. The earlier result of Page (1938) was -11.19. The errors in the R^{-8} and R^{-10} coefficients due to three- and four-electron correlations are likely to be smaller than for the R^{-6} term since configurations containing *d* and *f* atomic orbitals do not mix strongly in the helium *atom*. It is such mixing that accounts for these contributions, for they interfere with induced quadrupole and induced octupole,

TABLE 4.11. *Comparison of R^{-6} Coefficients for the Helium–Helium Interactions*

	Coefficient of R^{-6} (in atomic units)	Method
Theoretical		
Slater–Kirkwood (1931)	−1.557	Variational (not an upper bound)
London (1930)	−1.33	Perturbational
Salem (1960)[a]	−1.561	Using accurate polarizability and matrix elements over exact ground state of atom
Karplus–Kolker (1964)	−1.655	HF variational perturbation (not upper limit)
Davison (1966)	−1.47	Elaborate trial functions
Margenau (1939)	−1.45	Sum rules (1939)
Kingston (1964)	−1.456	Sum rules (1964)
Bell (1965)	−1.47	Sum rules (1965)
Kestner (1966)	−1.353	First approximation to many-electron theory
	−1.362	Final calculation (Table 4.10)
Experimental		
de Boer, Michels, Lunbeck[b]	−1.57	Virial data with quantum corrections

[a] The value quoted here differs from that reference due to the use of more accurate data from Perkeris (1959).

[b] de Boer and Michels (1938); de Boer and Lunbeck (1948), discussed in Hirschfelder, Curtiss and Bird (1954) p. 1068.

R^{-8} and R^{-10}, interactions which also arise from *d* and *f* orbital mixing (see, for example, eqn. (72) and Chapter 2). Because of the close relation between intermolecular forces and polarizability it is not surprising that, for example, the effect of electron correlation on polarizability is also largest for those cases in which *p*-orbital mixing is very large, e.g. in the beryllium atom (Dalgarno, 1962).

Let us now return to the general results and compare them with the experimental minima of depth from 7.6°K to 10°K. Table 4.9 shows that some of the error in our potential between $4a_0$ and $9a_0$ stems from the lack of sufficient variational parameters; we used $\beta = \gamma = 0$ for the general results. Recalling the approximate calculation for hydrogen, it seems reasonable to add to the calculated potential the corrections coming from further improvements in the long-range interactions. In Fig. 4 we illustrate how the potential changes upon adding $-0.29\,R^{-6}$. This represents the difference between the correct R^{-6} interaction and our $\beta = \gamma = 0$ result in Table 4.9. Further, we estimate the *d*-orbital contributions to a general superposition of configurations by adding the R^{-8} and R^{-10} interactions and allowing small corrections for second-order exchange results.† Finally, we correct our potential for the use of an approximate HF orbital by adding $\Delta E_{\mathrm{HF}} - \Delta E_{\mathrm{LCAO\text{-}MO}}$. While this procedure is of questionable accuracy, it does suggest that addition of all *p* orbitals would yield a potential minimum no deeper than 5.68°K. The major contribution is made by the *pd* configurations, which introduce terms such as $x_1 x_4^2$ in our pair functions $\hat{\mu}_{14}$. These could lead to a potential of depth 8.12°K at $5.8a_0$. The R^{-10} contributions are estimated to lower the potential by only about 0.3°K further. Again, it is of interest that even such relatively simple estimates lead to reasonable agreement with experimental results, especially the latest exp–6 result (MR–5) in Table 4.7.

The preceding analysis, while it has not been carried far enough to yield an intermolecular force law of highest precision, is thought to be very useful because it exposes to view the various detailed physical effects which collaborate to produce the total interaction. As to its limitations, the potential appears to exaggerate repulsion at small separations. Calculations (unpublished) show that it yields poor values for second virial coefficients, as might be expected for the reason stated.

Summarizing the experience gained from the foregoing review, we note that the following points must be heeded in a reasonably accurate calculation:

(1) At least all *p* and *d* orbitals must be included in the configurations composing the total state function. Hirschfelder and Linnett's two-electron function, (71), must be extended to include these additional terms.

† If these corrections are very small one can obtain a potential as low as 12 °K. See footnote on p. 124.

(2) All three-electron coupling terms must be included. Intra-atomic correlation can not be neglected.
(3) The like and unlike spin pairs must be allowed to vary independently.
(4) While we found that the molecular orbital calculation using a linear combination of HF atomic orbitals LCAO–MO was a good approximation to the SCF results, for increased accuracy the difference between the two results must be included.
(5) To attain high accuracy at distances near the minimum of the potential and beyond, three-electron excitations should be considered.
(6) In most cases localized orbitals provide the most efficient method of calculation. At very short distances ($R < 3a_0$ for helium), however, the delocalized or molecular orbitals are preferable since distortion of the atomic orbitals is severe.

In an effort to understand why configuration interaction methods, as applied to molecular orbital theory, converge so slowly to the correct potential, we now consider the pair correlation functions in the molecular orbital and the localized orbital description. We deal with moderately large separations where overlap effects are only minor corrections. Several possible two-electron substitutions occur in a function for the helium interaction when it is constructed by superposition of configurations: we can replace the two $1\sigma_g$ orbitals, the two $1\sigma_u$ orbitals, or one of each. In the latter case two distinct types of pair functions result and these depend on the spin function of the orbitals selected. When properly antisymmetrized these replacements involve the following functions:

$$\begin{aligned}
&\mathcal{A}[1\sigma_g(1)\alpha(1)1\sigma_g(2)\beta(2)],\\
&\mathcal{A}[1\sigma_u(3)\alpha(3)1\sigma_u(4)\beta(4)],\\
&\mathcal{A}[1\sigma_u(3)\alpha(3)1\sigma_g(2)\beta(2)],\\
&\mathcal{A}[1\sigma_u(3)\alpha(3)1\sigma_g(1)\alpha(1)].
\end{aligned} \tag{77}$$

When overlap is small they can be expanded in terms of atomic orbitals about the two centers, A and B (see Chapter 3) as follows:

$$\tfrac{1}{2}\mathcal{A}[(A_1A_2+B_1A_2+A_1B_2+B_1B_2)\alpha(1)\beta(2)],$$
$$\tfrac{1}{2}\mathcal{A}_2[(A_3A_4-B_3A_4-A_3B_4+B_3B_4)\alpha(3)\beta(4)],$$
$$\tfrac{1}{2}\mathcal{A}[(A_3A_2-B_3A_2+A_3B_2-B_3B_2)\alpha(3)\beta(2)],$$
$$\tfrac{1}{2}\mathcal{A}[(A_3A_1+B_3A_1-A_3B_1-B_3B_1)\alpha(3)\alpha(1)]. \qquad (78)$$

The last expression, because of antisymmetrization, reduces to

$$\tfrac{1}{2}\mathcal{A}[(B_3A_1-A_3B_1)\alpha(1)\alpha(3)]. \qquad (79)$$

In contrast, two-electron substitutions in the localized representation which would, in fact, be the Heitler–London function, generate the forms

$$\mathcal{A}(A_1A_2\alpha(1)\beta(2))\} \quad \text{leading to intra-atomic correlation,} \qquad (80)$$

$$\left.\begin{matrix}\mathcal{A}(A_1B_4\alpha(1)\beta(4))\\ \mathcal{A}(A_1B_3\alpha(1)\alpha(3))\end{matrix}\right\} \quad \text{leading to inter-atomic correlation.} \qquad (81)$$

Upon comparing (78) with (80) and (81) we see that only for the like-spin, interorbital pair do separate inter- and intra-atomic contributions occur explicitly in the molecular orbital treatment. For all unlike-spin substitutions a combination of inter- and intra-atomic correlations results.

At large separations the correlation energies for these various two-electron, molecular orbital substitutions are related to the term ΔE_{corr} of (68), which in turn is related to certain terms arising from localized orbitals, in the following fashion:

$$\begin{pmatrix}\text{molecular orbital}\\ \text{correlation energy}\\ \text{for unlike-spin}\\ \text{pairs}\end{pmatrix} = \begin{pmatrix}\text{one-half of}\\ \text{atomic}\\ \text{correlation}\\ \text{energy}\end{pmatrix} + \begin{pmatrix}\text{interatomic}\\ \text{interaction}\\ \text{energy}\end{pmatrix} \qquad (82)$$

$$\begin{pmatrix}\text{molecular orbital}\\ \text{correlation energy}\\ \text{for like-spin}\\ \text{pairs}\end{pmatrix} = \begin{pmatrix}\text{interatomic}\\ \text{interaction}\\ \text{energy}\end{pmatrix} \qquad (83)$$

The molecular orbitals distribute electron density equally among all centers unless this is forbidden by the Pauli principle. Thus for unlike-spin pairs, regardless of whether they inhabit the same orbital or not, each electron spends half of its time on each center. Hence each *pair* is attached to one particular center one fourth of the time. Since there are two centers, this leads to the factor one-half in (82).

In the case of like-spin correlation functions the exclusion principle dominates via the $\mathcal{A}$ operator and a simple result, (79), is obtained.

In the localized description, the electrons are at large separations assigned to a particular nucleus, and the individual two-electron substitutions lead to a specific effect, inter- or intra-atomic. Thus in a Heitler–London calculation, if only substitutions of the form of (81) are introduced, the total ΔE calculated will be reasonably accurate. One thereby neglects completely the intra-atomic correlation pair functions at all separations. Since this error is relatively independent of the separation, the potential curve calculated in this manner should be very close to those resulting from calculations which include all intra-atomic correlation effects. This accounts for the results of Kim and Moore. Their only major error was in neglecting coupling between inter- and intra-atomic effects. In calculations based on configuration interaction via the molecular orbital description, it is impossible to introduce *only* interatomic effects since all unlike spin, two-electron substitutions make contributions to both the intra- and the interatomic effects. Since intra-correlation energies are 10^3 to 10^4 larger than intereffects, most of the added configurations correct these energy components first and therefore lead to a slow convergence of this expansion toward the correct potential. At very small separations, where the distinction between inter- and intra-atomic effects is meaningless, the molecular orbital approach is ideal since it introduces all effects simultaneously.

4.3. Several General Methods Applicable to Larger Systems

Thus far attention has been restricted to hydrogen and helium interactions.

One of the advantages of the many-electron theory discussed in the last section is that it can be generalized and made applicable to larger systems. Since only two-electron variational equations need to be solved the complexity of the problem grows rather slowly as the number of electrons N increases. Then, too, many pair functions are identical. In helium we encountered only three types. In neon, of the 36 interatomic pair functions, only 8 are independent. The theory can be used at all distances either in the MO or localized form. Furthermore, care in determining each part accurately and evaluating all cross terms allows one to obtain bounds on the interaction, a feature usually missing in other theories.

However, there are two difficulties connected with this approach. The first, not only affecting this particular many-electron theory but most other methods as well, is the need to know atomic correlation functions.† While the theory under consideration provides ways and means for finding them, they have not been determined in detail for atoms with more than four electrons. The most serious limitation of the method is the need for molecular HF orbitals. Their construction requires much computer time and money. However, if interest is confined to a calculation of intermolecular energies, one may use functions very similar to the LCAO–MO orbitals, which are more easily obtained.‡ We now turn to these.

Murrell, Randic, and Williams (1965) systematized a perturbation approach starting from a Heitler–London function. The development was in terms of two parameters: powers of the potential V as in the ordinary perturbation theory of intermolecular forces *and* powers of the overlap integral between the two atoms $\langle A \mid B \rangle$ in the notation of (78). Their zero-order function was constructed from exact atomic functions. The total state function had several parts in addition to the Heitler–London term and was written in the form

$$\Psi = \psi_{\mathrm{HL}} + \psi_{\mathrm{ind}} + \psi_{vdW} + \psi_{CT}. \tag{84}$$

ψ_{ind} reflects distortions induced by the other atom. It consists of one- and two-electron substitutions of the covalent type which we discussed in section 4.1. ψ_{vdW} is the collection of all two-electron substitutions (*pp*, *pd*, *dd*, *df*, etc.) which at large separations yield the dispersion interaction. It forms the equivalent of the interatomic pair-function terms. ψ_{CT} is the sum of all charge transfer determinants and comprises all one- and two-

† This is not true for the C_6, C_8, and C_{10} coefficients (Deal and Kestner, 1966).

‡ See further discussion in Chapter 3 and in papers by P. Cade, A. C. Wahl, and W. Huo.

electron substitution terms of the ionic form discussed in connection with hydrogen interactions in section 4.1. They are called charge transfer terms since they impart more electrons to one atom than to another, as if a charge were actually transferred from one atom to the other.

The expression for the interaction through second-order perturbation theory is†

$$\Delta E = E_1 + E_2 + \Delta E_2 + E_{\text{ind}} + E_{CT} \tag{85}$$

where E_1 and E_2 are defined as previously and ΔE_2 is the second-order exchange term and is called the exchange polarization term by Murrell *et al.* (1965). E_1 is calculated with φ_0 alone. E_{ind} and E_{CT} are energies which correspond to their respective terms in (84). The authors also present methods of obtaining E_1 which do not involve explicit differences between small terms.

From our review of the hydrogen and helium calculations it appears that this procedure may be very useful. We saw that the effects of ψ_{ind} are minor for separations near the potential minimum and beyond. The ψ_{CT} terms for the hydrogen interaction (in the triplet state) are important. Donath and Pitzer, (1956) in fact, found in their approximate SCF calculation that

$$E_1 + E_{CT} = \Delta E_{\text{SCF}}.$$

This suggests that SCF or HF contributions to the intermolecular potential can be calculated rather easily.

The principal defect of the method employed by Murrell *et al.* (1965) is its need to use exact atomic wave functions. This is tantamount to assuming all intra-atomic correlation functions to be known and already included. The other shortcoming of the procedure is its reliance on perturbation theory. Ground state solutions of the atoms are not well known and the excited states needed in perturbation theory are even less certain.

It is also possible to combine the best features of the methods of Murrell *et al.*, those of Donath and Pitzer, and of Sinanoglu into a simple approximate method of calculation. Such possibilities are currently being investigated and preliminary results are encouraging. The zero-order function is a linear combination of HF atomic orbitals to which are added one and two electron corrections which can be determined variationally.

† In the paper by Murrell *et al.* (1965) these terms are each developed in powers of atomic overlap also.

CHAPTER 5

Nonadditivity of Intermolecular Forces†

5.1. Long-range Forces

THE elementary Coulomb interactions from which molecular forces are compounded have two important properties: they are central and they are additive. The first implies the absence of torques; the second means this: when many particles interact, the total force on one is the vector sum of the forces exerted upon it by all other particles, considered one at a time. A consequence of this characteristic with respect to the potential energy V of a system is its *pairwise* additivity,

$$V = \sum_{i>j} V_{ij}, \tag{1}$$

where V_{ij} is the potential energy of particle i relative to j. It follows from these facts that every classical Hamiltonian from which intermolecular forces are derived has the property of (pairwise) additivity.

When two *sets* of elementary charges interact it is in general not possible to find a point within each set about which the other will not produce a torque; hence the central property is lost for a collection of charges, except for sets of spherical symmetry. Multipoles, for instance, do not exhibit central forces.

But the additive property persists so long as all individual charges are rigidly held in position. Thus the potential energy of three rigid dipoles, fixed in their orientations, can be computed as the sum of intersystem pair potentials. However, if only the centers of the dipoles are fixed and the orientations allowed to adjust them-

† An earlier form of this chapter has been published in *Advances in Quantum Chemistry*, Vol. 3, p. 129, by Margenau and Stamper.

selves for minimum potential energy, additivity is lost. The relative orientation of two will be altered when a third is introduced, and the potential energy of the first two will therefore depend on the position of the third.

In quantum mechanics, the change from additivity of the classical Hamiltonian to nonadditivity of intermolecular potentials occurs in the same way: the charges within each molecule readjust their motions in response to every new molecule entering its force range. The question of additivity therefore arises naturally; yet it cannot be dealt with in a uniform way since its aspects are different for the different types of intermolecular forces.

Some features are apparent from the manner in which these forces are calculated. Interactions which appear in first-order perturbation theory are clearly not additive, for they are mean values over functions having complicated symmetries with respect to intermolecular exchanges, or else the roots of secular equations. In the second order of perturbation theory, however, certain notable regularities occur, and these will first be summarized.

The long-range forces between unexcited molecules, i.e. the dispersion forces and those arising from all higher multiples, are additive in second order of perturbation theory (Margenau, 1939). This statement has been questioned by Wojtala (1964), and we shall show below in what sense his assertions are correct.

Let there be n molecules, their separations being such that it is proper to write their state as a product of individual-molecule states:

$$|k\rangle = |k_1\rangle |k_2\rangle \ldots |k_n\rangle.$$

They are subject to the perturbation $V = \sum_{i>j} V(ij)$, (ij) denoting all coordinates pertaining to the ith and jth molecules. We now assume for $V(ij)$ its multipole expansion (see section 2.1). This obeys the relation

$$\langle k_i k_j | V(ij) | \varkappa_i \varkappa_j \rangle = f(R)[1-\delta(k_i\varkappa_i)][1-\delta(k_j\varkappa_j)]. \tag{2}$$

The complete second-order perturbation energy is

$$\Delta_2 E = \sum_{\{\varkappa\}} \frac{|\langle k | V | \varkappa \rangle|^2}{\sum_{i=1}^{n} [E(k_i) - E(\varkappa_i)]}, \tag{3}$$

where $\{\varkappa\}$ denotes the set of quantum numbers $\varkappa_1, \varkappa_2, \ldots, \varkappa_n$ and the E's are unperturbed energies. Now

$$\langle k|\,V\,|\varkappa\rangle = \sum_{i>j} \langle k_i k_j\,|\,V(ij)\,|\,\varkappa_i\varkappa_j\rangle \prod_{s\neq ij} \delta(k_s, \varkappa_s). \tag{4}$$

When this is squared *and use is made of eqn.* (2), only the squares of individual matrix elements remain and we find

$$\Delta_2 E = \sum_{i>j} \left[\sum_{\varkappa_i\varkappa_j}{}' \frac{|\langle k_i k_j\,|\,V(ij)\,|\,\varkappa_i\varkappa_j\rangle|^2}{E(k_i)-E(\varkappa_i)+E(k_j)-E(\varkappa_j)} \right]. \tag{5}$$

The bracketed expression, however, represents the second-order interaction between molecules i and j; hence additivity is established.

Wojtala (1964) did not expand V in a multipole series; hence his V does not satisfy eqn. (2). As a result he obtains cross terms which are not present in the asymptotic expansion. Their meaning will be discussed later in connection with Fig. 5.

Nonadditive contributions from the multipole series first occur in the third order of perturbation theory. In the literature attention has been confined for the most part to the dipole–dipole term in a configuration of three molecules. A qualitative formula for the triple dipole interaction was first given by Axilrod and Teller (1943); it was established in detail by Muto (1943); Axilrod (1951) and Midzuno as well as Kihara (1956) later published fuller derivations. In developing it here we employ a useful technique devised by Sinanoglu and applied to this problem by Kestner and Sinanoglu (1963).

Three atoms, a, b, and c, are represented by state functions $\varphi_a(1)$, $\varphi_b(2)$, and $\varphi_c(3)$. They are far enough apart so that their charge clouds do not overlap. Interatomic exchanges will therefore be neglected; each set of electrons (denoted by a numeral) is fixed to its atom, and summations over 1, 2, 3 are equivalent to summations over a, b, c. The complete function is the product

$$\phi_0 = \varphi_a\varphi_b\varphi_c, \tag{6}$$

$$\begin{aligned} H_0\phi_0 &= [H_0(a)+H_0(b)+H_0(c)]\phi_0 = E_0\phi_0 \\ &= [E_0(a)+E_0(b)+E_0(c)]\phi_0, \end{aligned} \tag{7}$$

and we define $H_0(a)-E_0(a) \equiv e_a$, so that

$$H_0 - E_0 = e_a + e_b + e_c. \tag{8}$$

We note that each e acts only on the electron coordinates of the atom whose subscript it carries.

The first-order perturbation energy is

$$H_1 = V_{ab}+V_{ac}+V_{bc} \tag{9}$$

and we mean by V_{ab} the dipole–dipole energy which has the form

$$V_{ab} = \frac{e^2}{R^3}\left(\boldsymbol{r}_a\cdot\boldsymbol{r}_b-3\,\frac{\boldsymbol{r}_a\cdot\boldsymbol{R}\,\boldsymbol{r}_b\cdot\boldsymbol{R}}{R^2}\right) \tag{9a}$$

provided $e\boldsymbol{r}_a$ and $e\boldsymbol{r}_b$ are the dipole moments of the electrons in a and b, and $\boldsymbol{R}$ is the intermolecular vector between a and b. (See eqn. 9a of Chapter 2.)

The third-order perturbation energy (Dalgarno in Bates, 1961, vol. 1, p. 184) may be written

$$\Delta_3 E = \langle \phi_1 | H_1 - E_1 | \phi_1 \rangle. \tag{10}$$

Since the first-order energy $E_1 = \langle \phi_0 | H_1 | \phi_0 \rangle = 0$, E_3 is simply the perturbation H_1 averaged over the first-order state function ϕ_1. We compute the latter as follows.

It is well known that ϕ_1 satisfies in general the equation

$$(H_0-E_0)\phi_1 = -H_1\phi_0+E_1\phi_0. \tag{11}$$

In our case the last term is absent. Hence, in view of (6), (8), and (9),

$$\phi_1 = -(e_a+e_b+e_c)^{-1}(V_{ab}+V_{ac}+V_{bc})\varphi_a\varphi_b\varphi_c. \tag{12}$$

It is understood that the reciprocals of the e operators, undefined as they stand, take on meaning when inverted or expanded. We consider first the contribution to ϕ_1 which comes from V_{ab}, and we define the new operator

$$M_c = (e_a+e_b)^{-1}e_c. \tag{13}$$

In terms of it,

$$\begin{aligned}\phi_1^{ab} &= -(e_a+e_b+e_c)^{-1}V_{ab}\varphi_a\varphi_b\varphi_c\\ &= -(1+M_c)^{-1}(e_a+e_b)^{-1}V_{ab}\varphi_a\varphi_b\varphi_c\\ &= -(1-M_c+M_c^2-\cdots)(e_a+e_b)^{-1}V_{ab}\varphi_a\varphi_b\varphi_c.\end{aligned}$$

At this point we observe that, because $e_c\varphi_c = 0$,

$$M_c\phi_c = 0. \tag{14}$$

Furthermore, φ_c commutes with $(e_a+e_b)^{-1}V_{ab}\varphi_a\varphi_b$ and may therefore be written after the first parenthesis in the expression for ϕ_1^{ab}, where it causes every power of M_c to disappear. Hence

$$\phi_1^{ab} = \varphi_c u_{ab} \tag{15}$$

provided

$$-u_{ab} = (e_a+e_b)^{-1}V_{ab}\varphi_a\varphi_b. \tag{16}$$

The usefulness of (15) arises from the manner in which ϕ_1^{ab} factors into a function of the coordinates of atom c alone (φ_c) and a pair-function involving a and b.

In dealing with the contributions of V_{ac} and V_{bc} we introduce operators

$$M_b = (e_a+e_c)^{-1}e_b$$
$$M_a = (e_b+e_c)^{-1}e_a$$

and proceed similarly. Two other pair functions, u_{ac} and u_{bc}, appear, and they are constructed in analogy with (16). Thus one obtains

$$\phi_1 = u_{bc}\varphi_a+u_{ac}\varphi_b+u_{ab}\varphi_c \tag{17}$$

and the u-functions satisfy the inhomogeneous differential equations

$$(e_a+e_b)u_{ab} = -V_{ab}\varphi_a\varphi_b \tag{18}$$

etc.

Next, (17) is to be inserted in (10),

$$\Delta_3 E = \langle\phi_1|(V_{ab}+V_{bc}+V_{ac})\phi_1\rangle. \tag{19}$$

There are 27 terms when E_3 is expanded. Each contains, aside from atomic state functions φ, two u-factors and one V-factor. The first, for instance, is

$$\langle u_{bc}\varphi_a|V_{ab}u_{bc}\varphi_a\rangle. \tag{20}$$

It vanishes when the integration over atom a is carried out [cf. (2)]. If, following Sinanoglu, we represent individual brackets by graphs, using solid lines for u-links and dotted lines for V-links between the atoms, element (20) corresponds to

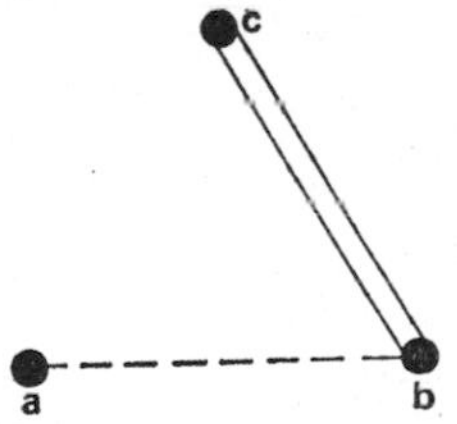

All elements of this type, i.e. with diagrams having two solid lines between one pair of atoms and a dotted line extending to the third atom, vanish for the reason just given.

Another bracket has the form $\langle u_{bc}\varphi_a | V_{bc} u_{bc}\varphi_a \rangle$. It corresponds to

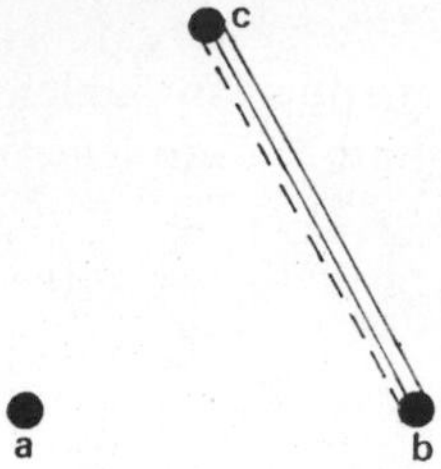

and vanishes upon integration over b or c. So do all elements of this type.

A third type is exemplified by $\langle u_{ab}\varphi_c | V_{ab} u_{ac}\varphi_b \rangle$. Its diagram is

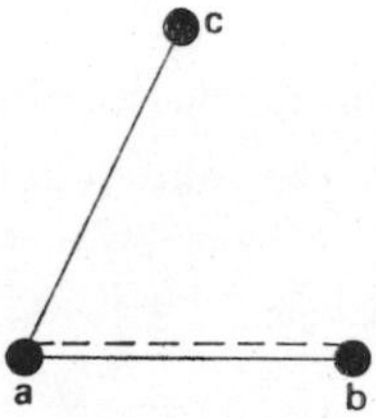

It is zero because φ_c and u_{ac} are orthogonal with respect to integration over the coordinates of c.

We are thus left with six matrix elements of the form

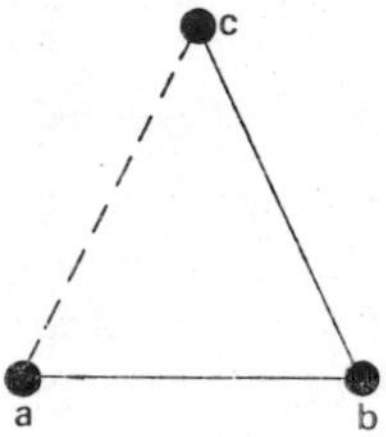

They are

$$\Delta_3 E = 2\langle u_{ab}\varphi_c | V_{ac} u_{bc}\varphi_a \rangle + 2\langle u_{ab}\varphi_c | V_{bc} u_{ac}\varphi_b \rangle + 2\langle u_{ac}\varphi_b | V_{ab} u_{bc}\varphi_a \rangle. \quad (21)$$

The u-functions, which must be obtained as solutions of (18), are most simply constructed as expansions involving φ_a, φ_b and their excited states.† If, writing φ_a^λ, φ_b^μ for the latter, we put

$$u_{ab} = \varphi_a\varphi_b + \sum_{\lambda\mu} a_{\lambda\mu}\varphi_a^\lambda\varphi_b^\mu$$

† In Chapter 4 equations similar to this were solved for many-electron atoms by variational methods.

and substitute in (18) we find after the usual integrations

$$a_{\lambda\mu} = -\langle \lambda\mu | V_{ab} | 00 \rangle / (\Delta E_a^{\lambda} + \Delta E_b^{\mu})$$

with the excited-state energies $\Delta E_a^{\lambda} = E_a^{\lambda} - E_a^0$, $\Delta E_b^{\mu} = E_b^{\mu} - E_b^0$.

The first integral in (21), which we now evaluate, involves the factor

$$\int u_{ab}^* u_{bc} d\tau_b = \varphi_a^* \varphi_c^* + \sum_{\lambda\mu\nu} \frac{\langle 00 | V_{ab} | \lambda\mu \rangle}{\Delta E_a^{\lambda} + \Delta E_b^{\mu}} \frac{\langle \mu\nu | V_{bc} | 00 \rangle}{\Delta E_b^{\mu} + \Delta E_c^{\nu}}$$

and

$$\langle u_{ac}\varphi_c | V_{ac} u_{bc} \varphi_a \rangle = \sum_{\lambda\mu\nu} \frac{\langle 00 | V_{ab} | \lambda\mu \rangle \langle \lambda 0 | V_{ac} | 0\nu \rangle \langle \mu\nu | V_{bc} | 00 \rangle}{(\Delta E_a^{\lambda} + \Delta E_b^{\mu})(\Delta E_b^{\mu} + \Delta E_c^{\nu})}. \tag{22}$$

With suitable permutations of subscripts we obtain the other terms in (21). In evaluating the summations we follow Kihara (1958). V_{ab} is expressed in the form

$$V_{ab} = \boldsymbol{p}_a \cdot \overleftrightarrow{\boldsymbol{T}}_{ab} \cdot \boldsymbol{p}_b \tag{23}$$

which is equivalent to (9a) provided p denotes the dipole moment operator of an atom and $\overleftrightarrow{\boldsymbol{T}}_{ab}$ the *dyadic* $(\overleftrightarrow{\boldsymbol{1}} - 3\boldsymbol{e}_{ab}\boldsymbol{e}_{ab})/R_{ab}^3$. The dots signify ordinary dot products between vectors, $\overleftrightarrow{\boldsymbol{1}}$ is the unit dyadic and $\boldsymbol{e}_{ab}$ the unit vector pointing from a to b. Note the following rules for dyadics, $\overleftrightarrow{\boldsymbol{A}}$, $\overleftrightarrow{\boldsymbol{B}}$. In writing them we employ dyadic notation on the left, tensor notation on the right of the equations.

$$(\overleftrightarrow{\boldsymbol{A}} \cdot \overleftrightarrow{\boldsymbol{B}})_{ij} = \sum_{\lambda} A_{i\lambda} B_{\lambda j}$$

$$\overleftrightarrow{\boldsymbol{A}} : \overleftrightarrow{\boldsymbol{B}} = \sum_{\lambda\mu} A_{\lambda\mu} B_{\lambda\mu}$$

$$\overleftrightarrow{\boldsymbol{1}} \cdot \overleftrightarrow{\boldsymbol{A}} = \overleftrightarrow{\boldsymbol{A}}$$

$$\overleftrightarrow{\boldsymbol{1}} : \overleftrightarrow{\boldsymbol{A}} = \mathrm{Tr}\, A.$$

If x and y are vectors, we have $\boldsymbol{x} \cdot \overleftrightarrow{\boldsymbol{A}} \cdot \boldsymbol{y} = \overleftrightarrow{\boldsymbol{xy}} : \overleftrightarrow{\boldsymbol{A}}$.

Now the summations in (22) include those over "magnetic" quantum numbers, which are equivalent to rotations in space (correspond to an irreducible representation of the rotation group). Hence a dyadic like

$$\sum \langle 0 | \boldsymbol{p}_a | \lambda \rangle \langle \lambda | \boldsymbol{p}_a | 0 \rangle,$$

the sum extending over the space quantum numbers, can at once be replaced by $\sum \langle 0 | p_{ax} | \lambda \rangle \langle \lambda | p_{ax} | 0 \rangle \overleftrightarrow{\boldsymbol{1}}$.

Thus we find

$$\langle u_{ac}\varphi_c | V_{ac} u_{bc}\varphi_a \rangle =$$

$$\sum_{\lambda\mu\nu} \frac{|\langle 0|p_{ax}|\lambda\rangle|^2 |\langle 0|p_{bx}|\mu\rangle|^2 |\langle 0|p_{cx}|\nu\rangle|^2}{(\Delta E_a^\lambda + \Delta E_b^\mu)(\Delta E_b^\mu + \Delta E_c^\nu)} \text{Trace}(\overleftrightarrow{T}_{ab}\cdot\overleftrightarrow{T}_{bc}\cdot\overleftrightarrow{T}_{ca}). \tag{24}$$

For the trace one obtains

$$Tr(\overleftrightarrow{T}_{ab}\cdot\overleftrightarrow{T}_{bc}\cdot\overleftrightarrow{T}_{ca}) = 3(R_{ab}R_{bc}R_{ac})^{-3}\{1-3+3[(\boldsymbol{e}_{ab}\cdot\boldsymbol{e}_{ac})^2+(\boldsymbol{e}_{ab}\cdot\boldsymbol{e}_{bc})^2 + (\boldsymbol{e}_{bc}\cdot\boldsymbol{e}_{ac})^2] - 9(\boldsymbol{e}_{ab}\cdot\boldsymbol{e}_{ac})(\boldsymbol{e}_{ab}\cdot\boldsymbol{e}_{bc})(\boldsymbol{e}_{bc}\cdot\boldsymbol{e}_{ac})\}.$$

The dot products can be expressed in terms of the inner angles of the triangle *abc*:

$$\cos\theta_a = -\boldsymbol{e}_{ab}\cdot\boldsymbol{e}_{ac}, \quad \text{etc.,}$$

and

$$\cos^2\theta_a + \cos^2\theta_b + \cos^2\theta_c = 1 - 2\cos\theta_a\cos\theta_b\cos\theta_c.$$

On using these relations we find

$$\text{Tr}(T_{ab}\cdot T_{bc}\cdot T_{ca}) = 3(R_{ab}R_{bc}R_{ac})^{-3}(3\cos\theta_a\cos\theta_b\cos\theta_c + 1).$$

There now remains the calculation of the two last terms of (21). This merely duplicates the foregoing procedure and leads again to (24), but with properly modified denominators. The final result is therefore

$$\Delta_3 E = \frac{6(3\cos\theta_a\cos\theta_b\cos\theta_c+1)}{R_{ab}^3 R_{ac}^3 R_{bc}^3} \times \sum_{\lambda\mu\nu} |\langle 0|p_{ax}|\lambda\rangle|^2 |\langle 0|p_{bx}|\mu\rangle|^2 |\langle 0|p_{cx}|\nu\rangle|^2 \times \{[(\Delta E_a^\lambda+\Delta E_b^\mu)(\Delta E_b^\mu+\Delta E_c^\nu)]^{-1} + [(\Delta E_a^\lambda+\Delta E_b^\mu)(\Delta E_a^\lambda+\Delta E_c^\nu)]^{-1} + [(\Delta E_a^\lambda+\Delta E_c^\nu)(\Delta E_b^\mu+\Delta E_c^\nu)]^{-1}\}. \tag{25}$$

The first factor, containing the θ's, determines the sign of the nonadditive contribution to the dipole–dipole effect; it is positive when every $\theta < 117°$, negative if one $\theta > 126°$.

A simple result holds for isotropic oscillators, where $\Delta E_a^\lambda = h\nu_a\delta_{\lambda 1}$ (since $\boldsymbol{p}$ connects the ground state with the first excited state only) and the polarizability $P_a = (2/3)\langle 0|p_a^2|0\rangle / h\nu_a$. In this case, (25) reduces to

$$\Delta_3 E = \frac{3}{2} h \frac{(\nu_a+\nu_b+\nu_c)\nu_a\nu_b\nu_c}{(\nu_a+\nu_b)(\nu_a+\nu_c)(\nu_b+\nu_c)} \times P_a P_b P_c \frac{3\cos\theta_a\cos\theta_b\cos\theta_c+1}{R_{ab}^3 R_{ac}^3 R_{bc}^3}. \tag{26}$$

For like molecules we can write (26) in the form

$$\Delta_3 E = A\,\frac{3\cos\theta_a\cos\theta_b\cos\theta_c + 1}{R_{ab}^3 R_{bc}^3 R_{ac}^3} \tag{27}$$

with

$$A = 9h\nu P^3/16.$$

In the same approximation the two-body interaction reads

$$E_2 = -\frac{C_1}{R^6} \tag{II-29}$$

with

$$C_1 = 3h\nu P^2/4. \tag{28}$$

From (27) and (28) we find (Kihara, 1958)

$$A = 3C_1P/4. \tag{29}$$

Using experimentally determined polarizabilities, Davison and Dalgarno (1966) have shown that (29) is accurate to within a few per cent for hydrogen and rare gas interactions. This accuracy can hardly be justified by the derivation presented here. Equation (29) provides a very useful method for estimating long-range three-body interactions.

If the triple dipole effect is computed with the use of complex polarizabilities, the result is a formula like eqn. (60) of Chapter 2; specifically (McLachlan, 1963; Linder and Hoernschemeyer, 1964)

$$\Delta_3 E = \frac{1}{\pi}\,\mathrm{Tr}(T_{ab}\cdot T_{bc}\cdot T_{ca})\int_0^\infty P_a(u)P_b(u)P_b(u)\,du.$$

Writing once more

$$P_a(u) = 2e^2 \sum_{\lambda} \frac{\Delta E_a^{\lambda}\,|(Z_a)_{0\lambda}|^2}{(\Delta E_a^{\lambda})^2 + u^2}, \quad \text{etc.,}$$

and carrying through the analysis as before (there are now three poles instead of two) we find (25). If we use the simpler form,

$$P_a(u) = \frac{A}{a^2+u^2}, \quad a = \left(\frac{A}{P}\right)^{1/2}, \quad P \equiv P(0),$$

the result is similar to the Slater–Kirkwood formula [(55) of Chapter 2]:

$$\Delta E = \frac{3}{16}\,\mathrm{Tr}(T_{ab}\cdot T_{bc}\cdot T_{ca})P^3\left(\frac{e\hbar^3}{m^{1/2}}\right)\left(\frac{N}{P}\right)^{1/2}.$$

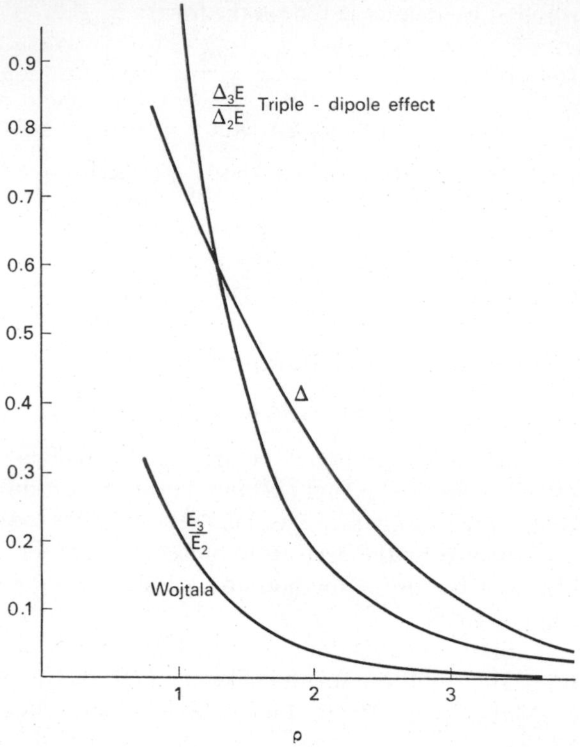

FIG. 5.

In Fig. 5 we have plotted the two second-order effects, ΔE_3 and the complicated result of Wojtala, for three hydrogen atoms in an equilateral triangle configuration. Ordinates represent the ratio of third-order to second-order energies, abscissae are lengths of the triangle sides. The third curve is the overlap integral, $\Delta^2 = e^{-2\varrho}(1+\varrho+(\varrho^2/3))^2$, $\varrho = R/a_0$. It was included to show that the nonadditive second-order effect is confined to the region where Δ is appreciable. Here, however, neglect of intermolecular electron exchange, i.e. the exclusion principle, is no longer legitimate. Hence the results of Wojtala are physically not very meaningful, and their lesson is to remind us that large nonadditive effects occur at small distances of separation where states of the separable form employed so far are inadequate.

The triple dipole effect, too, is of no interest at intermolecular distances where Δ is large; it may be seen, however, that its range is longer. At intermediate distances, therefore, it is worth consideration. Indeed, Graben and Present (1962)† have applied Axilrod's formulas in a calculation of the third virial coefficient for neon, argon, krypton, and xenon, obtaining better agreement with experimental values than is afforded by additive van der Waals' forces alone. In another study, Kestner and Sinanoglu (1963) have calculated "effective" dispersion forces between two molecules embedded in a nonpolar medium. They show, using the triple-dipole interaction between two molecules and their nearer solvent neighbors, that pair potentials are reduced by three-body effects as much as 32% (in carbon tetrachloride). Dispersion forces between (lateral base) pairs of atoms in a DNA double helix decrease by about 28% when the four adjacent bases of each pair, which have large polarizabilities, are included as third bodies in the calculation. How these forces are further modified by solvents is duscussed by Sinanoglu, Abdulnur, and Kestner (1964, p. 301). Higher orders of perturbation theory beyond the third, based on the dipole–dipole interaction alone, have been computed with the use of the oscillator model by Bade (1957) and applied to a linear lattice by Bade and Kirkwood (1957) and by Zwanzig (1963). An extension of this work to include the effects of distant perturbers was made by Doniach (1963). Finally, we record a publication by Ayres and Tredgold (1956) in which the three-body asymptotic interaction up to the dipole–dipole–quadrupole term is examined.

Let us now study briefly the role played by three-body forces in a solvent medium. Consider two atoms in free space. Atom a acts on atom b, polarizing it instantaneously. Atom b then interacts with atom a, and this reciprocal effect completes the interaction. The first phase is represented by the dyadic T_{ab} [Eqns. (23) et seq.] and the second by T_{ba}. The product of T_{ab} and T_{ba} thus determines the energy of interaction. When the two atoms are placed in a solvent there are new effects. The interaction represented by either T_{ab} or T_{ba} is modified by the solvent molecules. This change, which produces induced dipoles in the solvent molecules, gives rise to the effects which are related to the dielectric constant.

† See also Sherwood and Prausnitz (1964).

In the presence of the solvent the field acting on atom a due to a dipole moment in atom b is changed as follows (Brown, 1956, vol. 17, pp. 63–67):

$$\begin{aligned}\overleftrightarrow{T}_{ab}\cdot p_b \rightarrow \overleftrightarrow{T}_{ab}\cdot p_b &- \sum_j \overleftrightarrow{T}_{aj}\cdot\overleftrightarrow{P}_j\cdot\overleftrightarrow{T}_{jb}\cdot p_b \\ &+\sum_{jk}\overleftrightarrow{T}_{aj}\cdot\overleftrightarrow{P}_j\cdot\overleftrightarrow{T}_{jk}\cdot\overleftrightarrow{P}_k\cdot\overleftrightarrow{T}_{kb}\cdot p_b+\ldots \\ &\equiv \overleftrightarrow{T}_{ab}\cdot p_b/\varkappa.\end{aligned}$$

The last expression defines the dielectric constant $\varkappa$.

The summations extend over all molecules of the solvent. The first term is the field effect of b on a, unaffected by the solvent. The next term is the first solvent correction in which the interaction between a and b is mediated by only one solvent molecule. The field produced by atom b ($\overleftrightarrow{T}_{jb}\cdot p_b$) induces a dipole moment of strength $-\overleftrightarrow{P}_j\cdot\overleftrightarrow{T}_{jb}\cdot p_b$ in solvent molecule j (here $\overleftrightarrow{P}_j$ is the polarizability tensor of the solvent molecule). The sum of all such induced dipole fields yields the second term. The third term includes all effects in which the interaction of a and b is mediated by *two* solvent molecules. The field due to b creates a dipole in solvent molecule k, its field induces another dipole in solvent molecule j. The total contribution to the field at atom a is therefore represented by the double summation. This argument can be carried to higher order processes. It is equally valid for the effect of the solvent on the field due to atom a, $\overleftrightarrow{T}_{ba}\cdot p_a$. The effective field in the solvent is the free-space value divided by the dielectric constant ε.

Thus $T_{ab}T_{ba}$ in the two-body interaction becomes $T_{ab}T_{ba}/\varkappa^2$. Since each atom "oscillates" at many frequencies, we can not easily specify the value of $\varkappa(u)$. The simplest way to express our results is in analogy to eqn. (60) of Chapter 2, that is, by way of an integral over polarizability at imaginary frequencies,†

$$\begin{aligned}E &= \frac{1}{2\pi}\int_0^\infty P_a(u)P_b(u)\mathrm{Tr}(T_{ab}\cdot T_{ba}/\varkappa^2(u))\,du \\ &= -\frac{3}{\pi R^6}\int_0^\infty \frac{P_a(u)P_b(u)}{\varkappa^2(u)}\,du.\end{aligned}$$

† For a more thorough discussion of solvent effects, see McLachlan (1965). Other comments in the same volume are also of interest in this context.

If $1/\varkappa^2$ were expanded in accordance with the foregoing transformations, the first additional term would be the three-body contribution. For one solvent molecule we find

$$E_3 = \frac{1}{2\pi}\int_0^\infty P_a(u)P_b(u)P_c(u)\mathrm{Tr}(T_{ac}\cdot T_{cb}\cdot T_{ba}+T_{bc}\cdot T_{ca}\cdot T_{ab})\,du$$

$$= \frac{1}{\pi}\mathrm{Tr}\,(T_{ac}\cdot T_{cb}\cdot T_{ba})\int_0^\infty P_a(u)P_b(u)P_c(u)\,du,$$

which is the previous result. In general we would need to know $\varkappa(u)$ not only for all frequencies but for all molecular separations since the normal asymptotic value usually quoted applies only when the atoms are very far apart. While the above equations provide in principle the connection between three-body forces and $\varkappa$, there are no instances in which they prove numerically useful.

A more serviceable formula can be derived if only long-range three-body effects contribute. Consider molecules a and b in a solvent c. Each molecule of the solvent acts as a third body. Thus, to obtain the entire contribution of the solvent to the interaction of molecules a and b we must sum (26) over the entire liquid. This requires detailed knowledge of the structure of the liquid, information not easily available. To estimate the effect, however, we can consider an extreme representation of the liquid, a continuum model. In this case, the effective interaction between molecules a and b turns out to be

$$\Delta E_{ab,\,\text{liquid}}/\Delta E_{ab,\,\text{vacuum}} = 1-\frac{\Delta E_c(\Delta E_a+\Delta E_b+\Delta E_c)}{(\Delta E_a+\Delta E_c)(\Delta E_b+\Delta E_c)}\,P_c^\omega K_{\text{cont}}, \tag{30}$$

where P_c^ω is the polarizability of the solvent per unit volume,

$$P_c^\omega = \frac{P_c\varrho N_0}{M_c}, \tag{31}$$

P_c the polarizability of the solvent molecule, ϱ the solvent density (in g/ml), N_0 Avogadro's number, and M_c the molecular weight of the solvent molecules. K_{cont} is a dimensionless number, a constant equal to 9.4 for large separations. At smaller separations, the value of K_{cont} decreases. Representative values are listed in Table 5.1.

TABLE 5.1. *Factors K_{cont} for Calculating Effective Three-body Contributions to the Pair Potential*

Continuum result[a]	
R/R_0	K_{cont}
1.0	6.88
1.5	8.34
2.0	8.86
3.0	8.97
4.0	9.06
5.0	9.11
∞	9.4

[a] R_0 is the effective diameter of the molecule.

At very small separations in regular liquids the continuum model is not trustworthy. Calculations based on a lattice model for the liquid, another extreme, are also available (Kestner and Sinanoglu, 1963). If overlap can be neglected in the complete many-body expansion, effects of higher order than the three-body effects contribute less than 20% to eqn. (30) for most liquids.

Similarly, a different constant can be derived for the interaction of two polymer molecules by taking into account the solvent molecules excluded by the solvent. The appropriate value of K is then between 4.5 and 5.5 depending on the separation of the polymer molecules. For typical polymers this value leads to a 17–22% reduction in the dispersion interactions between chains in water.

To evaluate the contribution of three-body effects to the energy of the entire liquid we need the sum

$$\sum_{a>b} \Delta E'_{ab,\text{ liquid}} = E_{\text{total}}. \tag{32}$$

However, for a pure liquid, (30) counts all three-body effects three times and therefore one needs to use in $E'_{ab,\text{ liquid}}$ a value of K_{cont} one-third of that presented in Table 5.1.

Reductions in the total energy of carbon tetrachloride, which is a typical example, are about 11%. The reductions are proportional to the polarizability of the solvent per unit volume.

Dispersion forces between lateral base pairs in the DNA double helix decrease by about 28% when the four adjacent bases of the selected pair are included as third bodies in the calculation.

These three-body effects contribute strongly to physical adsorption, where the third body is the solid. Sinanoglu and Pitzer (1960), McLachlan (1964), and Mavroyannis (1963) have considered various aspects of this problem. Experimental evidence here is in accord with a large three-body contribution.† A more thorough discussion is contained in Chapter 9. The application of all these results involves intermolecular distances at which overlap can not be neglected with confidence. Thus, while quantitative conclusions represent important indications of the nonadditivity of even long-range effects, they do not tell the entire story and must not be trusted in detail. Hence we turn now to the calculation of first-order effects at small separations, using properly antisymmetrized electron wave functions.

5.2. Short range Three-body Forces

In dealing with multiple interactions it becomes essential to use a convenient and nonredundant notation. No confusion will arise if we use the letters *a*, *b* and *c* in two senses: as the points at which the nuclei of the three atoms are situated and as atomic orbitals constructed about these points. When signifying the latter they will usually carry arguments specifying the electron occupying the orbital, e.g. $a(1)$. Furthermore, to simplify writing the Hamiltonian, we abbreviate

$$\frac{e^2}{r_{ai}} \equiv \alpha_i, \quad \frac{e^2}{r_{bj}} \equiv \beta_j, \quad \text{etc., and} \quad \frac{e^2}{r_{ij}} \equiv \varrho_{ij}. \tag{33}$$

Here r_{ai} is the distance between a and electron i, r_{ij} that between electrons i and j.

In the following we shall encounter a variety of elementary exchange integrals, which can be classified into two-center and

† See Johnson and Klein (1964); Barker and Everett (1962); Sams, Jr., Constabaris, and Halsey, Jr. (1962); Steele and Kebbekus (1965); and Everett (1965). The latter reference points out the discrepancies with theoretical predictions.

three-center integrals, with a possible subdivision into one-electron and two-electron classes. These can be labeled perspicuously in a manner useful for more complicated later problems (see section 7.2), viz.:

$$(a\beta c) \equiv \int a^*(1) \frac{e^2}{r_{b1}} c(1)\, d\tau_1$$

$$(ab\varrho bc) \equiv \int a^*(1) b^*(2) \frac{e^2}{r_{12}} b(1) c(2)\, d\tau_1\, d\tau_2, \quad \text{etc.}$$

As usual, the overlap integral is denoted by

$$\Delta_{ab} \equiv (ab) = \int a^*(1) b(1)\, d\tau_1 .$$

As to symmetry, we note that

$$(a\beta c) = (c\beta a)$$
$$(a\alpha b) = (a\beta b)$$
$$(ab\varrho bc) = (bb\varrho ac) = (ac\varrho bb) = (bc\varrho ab) = (ba\varrho cb)$$

provided the orbitals are real.

All one-center integrals, like $(a\alpha a)$, $(b\beta b)$ are, of course, equal and independent of the position of the nuclei. All two-center integrals,

$$(a\beta a),\ (a\beta b),\ (ab\varrho ab), \quad \text{and} \quad (ab\varrho ba)$$

occur in the pair interaction; they are well known from the Heitler–London theory of H_2 (see Chapter 3). Three-center integrals fall into three classes, characterized by

$$(a\beta c),\ (ac\varrho bc),\ (ac\varrho cb).$$

Our knowledge concerning these integrals is still incomplete. They are discussed by Slater (1963) and by various authors in Alder, Fernbach, and Rotenberg (1962).

The Hamiltonian for three interacting, *one-electron atoms* is

$$H = H_0 + V.$$

We suppose that $H_0 = H_a + H_b + H_c$, $H_a a(1) = E_a a(1)$, etc. Then, as before,

$$V = V_{ab} + V_{ac} + V_{bc}$$

$$\left.\begin{aligned} V_{ab} &= e^2 R_{ab}^{-1} + \varrho_{12} - \alpha_2 - \beta_1 \\ V_{ac} &= e^2 R_{ac}^{-1} + \varrho_{13} - \alpha_3 - \gamma_1 \\ V_{bc} &= e^2 R_{bc}^{-1} + \varrho_{23} - \beta_3 - \gamma_2 \end{aligned}\right\} \tag{34}$$

The normalized, unperturbed state of the triplet is

$$\Psi_0 = \sqrt{(\tfrac{1}{6})}(1-\delta_{abc})^{-1/2}\mathcal{A}a(1)b(2)c(3), \tag{35}$$

where $\mathcal{A}$ is the antisymmetrizing operator and

$$\delta_{abc} = \Delta_{ab}^2+\Delta_{ac}^2+\Delta_{bc}^2-2\Delta_{ab}\Delta_{ac}\Delta_{bc}. \tag{36}$$

Straightforward substitution of (34) and (35) and use of our symmetry rules leads to

$$\begin{aligned}\langle 0|H|0\rangle-3E_H = {} & e^2R_{ab}^{-1}+(1-\delta_{abc})^{-1}\{(\Delta_{ac}^2+\Delta_{bc}^2-2)(a\beta a)\\ & +2(\Delta_{ab}-\Delta_{ac}\Delta_{bc})(a\beta b)\\ & +(\Delta_{ac}-\Delta_{ab}\Delta_{bc})(a\beta c)+(\Delta_{bc}-\Delta_{ab}\Delta_{ac})(b\alpha c)\\ & +(ab\varrho ab)-(aa\varrho bb)+\Delta_{ac}[(ab\varrho bc)-(ab\varrho cb)]\\ & +\Delta_{bc}[(aa\varrho bc)-(ab\varrho ac)]\}+\ldots+\ldots\end{aligned} \tag{37}$$

The two sets of terms not written are constructed from those displayed by interchanging a with c and α with γ, then b with c and β with γ. Here $E_H = E_a = E_b = E_c$ represents the energy of a hydrogen atom. If atom c is removed to infinity, every term involving c in this expression drops out and we find

$$E_{ab} = \frac{e^2}{R_{ab}}+(1-\Delta_{ab}^2)^{-1}\{-2(a\beta a)+2\Delta_{ab}(a\beta b)+(ab\varrho ab)-(aa\varrho bb)\} \tag{38}$$

which is the repulsive Heitler–London energy between the two atoms [see eqn. (37) of Chapter 3].

The specific three-body contribution over and above the energy of pairs is

$$\langle 0|H|0\rangle-E_a-E_b-E_c-E_{ab}-E_{ac}-E_{bc} \equiv E_{abc}. \tag{39}$$

Thus the total energy

$$\langle 0|H|0\rangle \equiv E = E_a+E_b+E_c+E_{ab}+E_{ac}+E_{bc}+E_{abc}. \tag{40}$$

It is clearly of interest to know whether this series can be extended and, if so, whether it converges. For instance, one might suppose that the total energy of n interacting atoms can be written as a series

$$E = \sum_r E_r+\sum_{r<s} E_{rs}+\sum_{r<s<t} E_{rst}+\ldots+E_{rst\ldots n}. \tag{41}$$

in which the sums have progressively smaller values. Very little is known about this at the present time, but we shall return to this problem below.

The evaluation of (37) for specific configurations of hydrogen atoms, though not difficult, has not been carried out to our knowledge.† While it represents a possible mode of interaction of three hydrogen atoms, namely that in which all electron spins are parallel, experimental data are lacking. The effect is interesting for helium, however, and Rosen was the first to perform a calculation analogous to the above for six electrons attached to three helium nuclei (Rosen, 1963). His result can be written in fairly simple form, as follows.

Let ψ_a, ψ_b and ψ_c be antisymmetric functions for atomic helium, each composed of two atomic orbitals. Write $\mathcal{A}^{ab}$ for the interatomic antisymmetrizer, $\sum_{\lambda}(-1)^{\lambda_{ab}}P_{\lambda_{ab}}$, between atoms a and b. (The $P_{\lambda_{ab}}$ are all permutations of electrons *between a* and b.) Similarly, $\mathcal{A}^{abc}$ is the triatomic antisymmetrizing operator. Further, introduce

$$S_{ab} = \langle \psi_a\psi_b | \mathcal{A}^{ab}\psi_a\psi_b\rangle; \qquad \psi_a\psi_b\psi_c \equiv \Psi$$

$$S_{abc} = \langle \Psi | \mathcal{A}^{abc}\Psi\rangle, \qquad S = 1 + S_{ab} + S_{ac} + S_{bc} + S_{abc}$$

$$\bar{H} = \langle \Psi | H\mathcal{A}^{abc}\Psi\rangle$$

$$Q_a = \langle \Psi | (V_{ab} + V_{ac})\Psi\rangle$$

$$Q_b = \langle \Psi | (V_{ab} + V_{bc})\Psi\rangle$$

$$Q_c = \langle \Psi | (V_{ac} + V_{bc})\Psi\rangle .$$

Then

$$\left.\begin{aligned} E_{abc}S = \bar{H} + Q_a + Q_b + Q_c - S_{abc}(E_a + E_b + E_c + E_{ab} + E_{ac} + E_{bc}) \\ -(S_{ac} + S_{bc})E_{ab} - (S_{ab} + S_{bc})E_{ac} - (S_{ab} + S_{ac})E_{bc} . \end{aligned}\right\} \tag{42}$$

The quantities $\bar{H}$ and Q break up into elementary exchange integrals of the type discussed, most but not all of which have been tabulated.

Rosen evaluated equation (42) for two configurations, an equilateral triangle and three atoms equally spaced on a line.

† McGinnies and Jansen (1956) examine a simplified version of (7) in which atoms a and b are close together, but c is far enough away to permit a multipole expansion.

For the ratio of triple to pairwise interaction he finds (for $ZR > 3a_0$)

$$\frac{E_{abc}}{E_{ab}+E_{ac}+E_{bc}} = -1.15e^{-0.33(R_{ab}+R_{ac}+R_{bc})/a_0} \qquad (43)$$

(equilateral triangle),

$$= +9.8e^{-0.66(R_{ab}+R_{ac}+R_{bc})/a_0} \qquad (44)$$

(linear array).

The triangular configuration has maximum overlap, and it is therefore not surprising that (43) is larger than (44) for interatomic distances where nonadditivity matters. Unexpected is the change of sign. It looks as if in the triangular disposition the electrons enjoyed enough freedom to weaken the demands of the Pauli principle, which causes repulsion in the forces between pairs. The linear arrangement, on the contrary, offers no such freedom, in fact increases the repulsion to a small extent. It is also noteworthy that the sign of the nonadditive component coming from the present, first-order interactions agrees with that of the triple–dipole effect: it, too, is negative for $\theta = 60°$, positive for $\theta = 180°$.

The problem of three helium atoms was also investigated by Shostak (1955) who employed molecular orbitals (linear combination of atomic orbitals) in his calculation. He evaluated results only for Rosen's linear configuration, where molecular orbitals are likely to yield better precision. While in general agreement with Rosen's conclusions, his method gives larger nonadditive effects—at the shortest distances considered—by almost a factor 2.

Three-body interactions were studied in connection with crystal properties in a well-known publication by Löwdin (1948),† who found in them the cause of departures from Cauchy's relations (between the elastic constants of crystals). Starting from a complete many-body Hamiltonian he arrives at terms corresponding to nonadditive, three-body interactions which account for his interesting findings. His results, though obtained by an approach that is somewhat foreign to the methods em-

† Later work suggested by these investigations, but using a macroscopic model (polarizabilities) may be found in Linderberg (1964), and Linderberg and Bystrand (1964).

ployed in this article, are analogous to the three-body interactions discussed in the sequel. A careful comparison, which would be illuminating, has apparently not been made.

More recently Jansen and McGinnies (1956) seized upon many-body interactions to explain a renowned paradox in crystal structure theory. An illuminating account of it is given in a recent review (Jansen, 1965a, b). The facts to be explained are these.†

Among the rare gas elements only helium crystallizes in a hexagonal lattice, all others form face-centered cubic crystals. Pair potentials, computed in the usual way or derived semi-empirically, always favor the hexagonal lattice. That is to say, the hexagonal configuration invariably produces the lowest value for the Gibbs free energy at 0°K. Thus the behavior of helium seems regular, but that of the heavier rare gases anomalous. The effect computed by Rosen for helium *in*creases the stability of the hexagonal lattice and therefore provides little hope of accounting for the crystal forms of neon, argon, krypton, and xenon. To certify this expectation Jansen (1965a, b) and Zimering (1965) performed a first-order calculation of the kind outlined above for three hydrogen atoms, but with the following modifications for the sake of feasibility and application to heavier (spherical) atoms. Instead of the full shell of electrons in each atom they assume one to be effective in bringing about the interaction. This leads to (37) and (38). In evaluating the integrals which appear there, Jansen uses not hydrogen functions but Gaussian exponentials, i.e. simple harmonic oscillator functions of the form

$$a(1) = B^{3/2}\pi^{-3/4}e^{-(B^2r_1{}^2)/2}, \tag{45}$$

for they permit all integrals to be expressed in a simple form. The parameter B is obtained by fitting a formula for the dispersion forces, computed with function (45), to the empirical van der Waals potential. Since this model is useful in this and other respects, we list Jansen's expressions for the integrals that appear.

$$\Delta_{ab} = e^{-(B^2R_{ab}^2)/4}.$$

† The actual situation may be more complex than is suggested below. See an important note by Meyer, Barrett, and Haasen (1965).

Introducing $F(x^2) \equiv \dfrac{2e^2}{\sqrt{(\pi)}x}\displaystyle\int_0^x e^{-u^2}\,du \equiv \dfrac{e^2 erf(x)}{x}$, one finds for the other integrals

$$
\begin{aligned}
(a\beta a) &= \mathrm{B}\ \mathrm{F}(B^2R^2_{ab})\\
(a\alpha b) &= \Delta_{ab}B\ F(B^2R^2_{ab/4})\\
(b\alpha c) &= \Delta_{bc}B\ F(B^2R^2_{\overline{abc}})\\
(a\beta c) &= \Delta_{ac}B\ F(B^2R^2_{\overline{bac}})\\
\sqrt{(2)}(ab\varrho ab) &= B\ F(B^2R^2_{ab/2})\\
\sqrt{(2)}(aa\varrho bb) &= \frac{2}{\pi}\,Be^{-B^2R^2_{ab}/2}\\
\sqrt{(2)}(ab\varrho ac) &= \Delta_{bc}B\ F(B^2R^2_{\overline{abc}/2})\\
\sqrt{(2)}(ab\varrho cb) &= \Delta_{ac}B\ F(B^2R^2_{\overline{bac}/2})\\
\sqrt{(2)}(aa\varrho bc) &= \Delta_{ab}\Delta_{ac}B\ F(B^2R^2_{bc/8})\\
\sqrt{(2)}(ab\varrho bc) &= \Delta_{ab}\Delta_{ac}B\ F(B^2R^2_{ac/8}).
\end{aligned}
\qquad (46)
$$

Here R_{abc} denotes the distance from a to the midpoint of the line joining b and c.

Our account of Jansen's calculation differs from his version in these respects. Instead of computing $\langle 0|H|0\rangle - 3E_H$ he claims to calculate $\langle 0|V|0\rangle$, which he obtains by adding $\langle 0|V_{ab}|0\rangle$ to the two permuted expressions. The form of V_{ab} is given in (34). Now, as shown in Chapter 3, $\langle 0|V_{ab}|0\rangle$ when properly computed is without meaning, for it does not go to zero when R_{ab} becomes large. Neither do $\langle 0|V_{ac}|0\rangle$ and $\langle 0|V_{bc}|0\rangle$. Evidently, therefore, the meaning of these terms must have been changed somewhere in the calculation (e.g., use of $\mathcal{A}V$ instead of V in computing $\langle 0|V|0\rangle$ would have removed the finite part at $R \to \infty$). At any rate, the result given by Jansen is indeed $\langle 0|H|0\rangle - 3E_H$, and is correct. The use of oscillator functions, whose eigenvalues are not E_H, introduces further complications from the point of view of perturbation theory. But the procedure discussed seems very difficult to improve.

We have also noted elsewhere (Margenau and Rosen, 1953) that questions as to the meaning of intermolecular potentials arise when $H_0\varphi_0 \neq E_0\varphi_0$ (see also section 3.2). They emerge here. The quantity calculated with the use of oscillator functions is the one denoted by $\tilde{V}$ in eqn. (16) of Chapter 3.

Figure 6 illustrates magnitude and angular dependence of the three-body effect under study. The curves are drawn for the isosceles triangle, the distances being those appropriate for solid argon and xenon, and fractional energies are plotted as functions

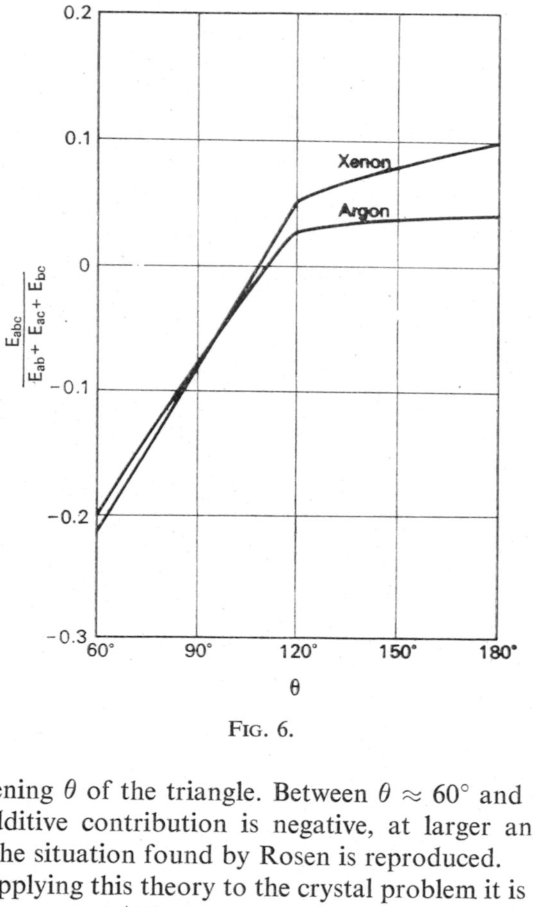

FIG. 6.

of the opening θ of the triangle. Between $\theta \approx 60°$ and $\theta \approx 110°$ the nonadditive contribution is negative, at larger angles it is positive; the situation found by Rosen is reproduced.

When applying this theory to the crystal problem it is evidently necessary to accept (41), supposing that terms beyond the third on the right can be neglected. Thus, in examining the stability of different lattices, Jansen selects a central atom and computes, first all pairwise interactions with its nearest neighbors $\left(\sum_{r<s} E_{rs}\right)$; then he constructs all possible triangles involving the central atom and two nearest neighbors and calculates $\sum E_{rst}$. When this

is done it become quite clear that the hexagonal close-packed configuration is the stable one.†

Hence the paradox remains. Let us recall, however, that the main forces which hold a rare-gas crystal together are those calculated in second order. Indeed, at the interatomic distances prevailing in a lattice the attractive second-order energies are greater, in absolute value, than the first-order exchange forces. Jansen and Zimering (1963) and Jansen (1963, 1964) therefore investigate the second-order exchange effect arising in the interaction of three model atoms whose charges are distributed in accordance with (45).

They apply perturbation theory in the usual way, writing

$$E_{2,abc} = \sum_{\lambda}{}' \frac{\langle 0 | V_{abc} | \lambda \rangle \langle \lambda | V_{abc} | 0 \rangle}{E_0 - E_\lambda} = -(\bar{E})^{-1}[\langle 0 | V^2_{abc} | 0 \rangle - \langle 0 | V_{abc} | 0 \rangle^2]$$

replacing $(E_0 - E_\lambda)^{-1}$ by some undefined average $\bar{E}^{-1}$. This suffers from lack of meaning, as before; it appears that what Jansen and Zimering actually calculate is

$$-(\bar{E})^{-1}[\langle 0 | (H - E_\infty)^2 | 0 \rangle - \langle 0 | H - E_\infty | 0 \rangle^2], \tag{47}$$

where $E_\infty = \lim_{R \to \infty} E = E_a + E_b + E_c$. This formula can be justified as follows.

Perturbation theory depends on the availability of a ϕ_0 which is an eigenfunction of some Hamiltonian H_0, preferably, in this case, of $H_a + H_b + H_c$. An antisymmetrized set of atomic orbitals like (35) is far from satisfying these requirements, nor does it yield $\langle 0 | H_0 + H_b + H_c | 0 \rangle = E_a + E_b + E_c$. Hence the best one can do is to employ the method of linear variation functions, assum-

† The reasoning is this. The Gibbs free energy $G = U - TS + PV$, U being the crystal energy (see, for example, Yourgrau, van der Merwe, and Raw (1966). The comparison involves no volume changes, hence PV drops out. The TS term can also be neglected at the low temperatures of interest (but possibly not at high temperatures and pressures. See Stillinger, Salsburg, and Kornegay (1965)). At least indications are that this is true. Thus G becomes U. But U consists of two parts: the static crystal energy and the zero-point energy of its constituents. On this last point, careful calculations mentioned in Jansen (1965a, b) exist suggesting that it can not cause the difference between the two structures. Hence Jansen holds that part of U which comes from all static interactions responsible for the lattice type.

ing the trial function to be

$$\phi = \phi_0 + c\phi_1. \tag{48}$$

Concerning ϕ_1 we suppose that (a) it is orthogonal to ϕ_0, and (b) that ϕ_0 and ϕ_1 together form a complete set. This, of course, is never true, but is equivalent to the Unsöld approximation, i.e. to replacing every $(E_0 - E_\lambda)^{-1}$ by $\bar{E}^{-1}$.

Choice of (48) leads to the condition for E:

$$\begin{vmatrix} H_{00}-E & H_{01} \\ H_{10} & H_{11}-E \end{vmatrix} = 0.$$

We may neglect the dependence of $H_{11}-E$ upon the R and write $H_{11}-E \equiv \bar{E}$. Then $E = H_{00} - \bar{E}^{-1}|H_{01}|^2 = H_{00} - \bar{E}^{-1}[(H^2)_{00} - H^2_{00}]$, since $|H_{00}|^2 + |H_{10}|^2 = (H^2)_{00}$. Therefore the van der Waals' energy

$$E - E_\infty = \langle 0 | H - E_\infty | 0 \rangle - \frac{(H^2)_{00} - H^2{}_{00}}{\bar{E}}.$$

We recognize the first expression on the right as the first-order energy, eqn. (37). Hence the "second-order" term is

$$E_{2,abc} = -\frac{(H^2)_{00} - H^2{}_{00}}{\bar{E}},$$

and this is identical with (47).†

The calculation of $(H^2)_{00}$ is lengthy and requires electronic computation. Details are found in the work of Jansen (1964), Lombardi and Jansen (1964), and Zimering (1965). They remove the uncertainty with respect to the value of $\bar{E}$ by forming the ratio $E_{2,abc}/E^0_{2,abc}$, the denominator being the long-range additive van der Waals energy, $E_{2,ab}+E_{2,ac}+E_{2,bc}$, as given by London's formula. The latter also contains an unknown energy difference; it is taken to be equal to the present $\bar{E}$ and therefore both drop out of the calculation. The result for $E_{2,abc}/E^0_{2,abc}$ is amazingly similar to the curves of Fig. 6. In Fig. 7 we plot this ratio together with $E_{1,abc}/E^0_{2,abc}$ (which is the ordinate of Fig. 6) for argon. The triangle is again an isosceles, and the value of BR has been chosen as 2.4, which corresponds to the lattice spacing of solid argon.

† This form was used by Dalgarno and Lynn (1956) for two atom interactions. It was mentioned in section 4.3.

Jansen and his collaborators perform extensive calculations of this type for all the triangles involving nearest neighbors which are encountered in rare-gas crystals and in alkali–halide crystals, both for the hexagonal and for the face-centered cubic configuration. Upon their results, which are typified by Fig. 7, they base the following remarkable conclusions.

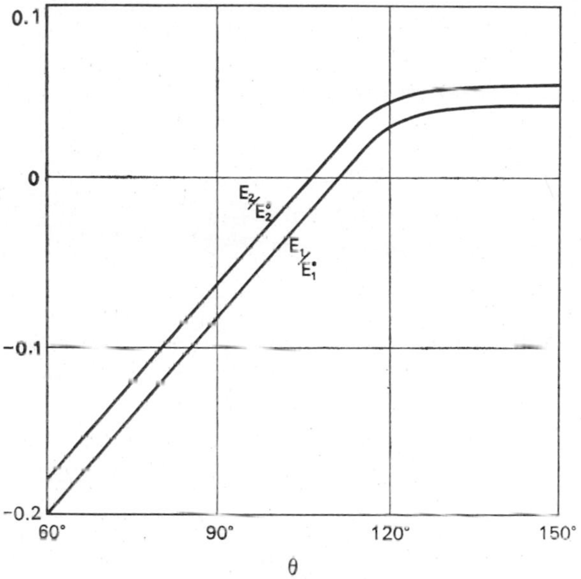

FIG. 7.

First-order and second order ratios (Fig. 7) are practically the same when plotted as functions of θ. Since $E^0_{2,abc}$ is negative while E^0_1 is positive, first-order nonadditive effects increase repulsion, second-order effects reduce it where both ratios are positive. While the precise manner of compounding the two effects is subject to ambiguities, it is nevertheless clear that in the region of the van der Waals minimum E^0_2 is numerically about twice as large as E^0_1; hence the second-order correction, which introduces predominantly attraction, carries twice the weight of the first. The consequences flowing from the analysis of $E_{1,abc}$ must therefore be reversed, and the net result is that, because of the second-order three-body interactions, rare-gas crystals favor the face-centered cubic configuration.

This leaves helium as the sole offender. In this case, however, one can see from Rosen's and Shostak's calculations that third-order effects are minimal, for the spacing of atoms is wide and the atoms are small. Jansen shows that for large BR the ratio of $(E_{1,abc}+E_{2,abc})/E^0_{abc}$ becomes small. Hence the three-body correction seems insufficient to make the f.c.c. configuration the stabler one for helium.

Having thus regularized the anomalies† of the rare-gas crystals he turns his attention to the alkali–halides and indicates how similar considerations produce gratifying agreement with observations. Their detailed presentation is beyond the scope of this book.

In view of these successes, a closer scrutiny of the basis on which the many-body theory rests becomes essential. In the next section, therefore, we present some new calculations designed to throw light on the convergence of the expansion of E as a series of multiparticle terms, eqn. (41). We shall extend previous treatments to include four-atom interactions. In a manner similar to Jansen's, a determination of the nonadditive four-body component will be made to see if, perchance, it remains smaller than the three-body component. For obvious practical reasons the calculation of four-body forces will be restricted to one order of perturbation theory, the first. The fact that the "second-order contribution" to three-body forces is, as we have seen, even greater than the first at distances near the lattice parameter, may seem disturbing. Nevertheless, it is interesting to know what happens within one consistent scheme of computation, and our results do become physically meaningful at smaller values of R, where first-order exchange forces always predominate.

We conclude this section by citing some recent evidence which suggests that the procedure of Jansen is limited to small separations and becomes inaccurate for larger values of R than occur in the solid. In comparing the electron density of Jansen's Gaussian function for argon (using the B determined by the method described) with the HF atomic electron density, it was found (Wenzel and Kestner, unpublished) that the Gaussian orbitals used by Jansen are much more extended than the HF results. This means that overlap and exchange effects are greatly exaggerated

† It might be noted, however, that the observations on which the foregoing reasoning is based require further scrutiny in view of the findings of Meyer, Barrett, and Haasen (1965).

in Jansen's calculations for separations over about 1.5 Å. Graben, Present, and McCulloch (1966)† attempting to use Jansen's results in calculating third virial coefficients, found that a 30% increase in B was necessary to yield reasonable results. They selected B by fitting the two-body repulsion calculated with Gaussian functions to the exponential repulsion in the region near the van der Waals' minimum. Their third virial coefficients were very sensitive to the value of B. Sherwood, DeRocco, and Mason (1966) have improved upon these results by first deriving a relation between nonadditive and additive contributions and then using a Lennard–Jones potential for the two-body interactions. In all cases investigated thus far the nonadditive short-range contributions to the third virial coefficients are much smaller than those Jansen would predict but not negligible compared with contributions from the long-range three-body effects discussed in section 5.1.

5.3. Short-range Four-body Forces‡

To the three atoms considered in the previous section we add a fourth, locate it at d and described it by the orbital $d(4)$. The total Hamiltonian is

$$H = H_a(1)+H_b(2)+H_c(3)+H_d(4)+V_{ab}(2,1)+V_{ac}(3,1)$$
$$+V_{ad}(4,1)+V_{bc}(3,2)+V_{bd}(4,2)+V_{cd}(4,3), \qquad (49)$$

where the V are defined in a manner analogous to (34) and $H_d(i)d(i) = E_d d(i)$. The normalized state function is, in the notation of (35),

$$\Psi_0 = \sqrt{\left(\frac{1}{4!}\right)}(1-\delta_{abcd})^{-1/2}\mathcal{A}a(1)b(2)c(3)d(4), \qquad (50)$$

where now

$$\delta_{abcd} = 1-\int a(1)b(2)c(3)d(4)\mathcal{A}a(1)b(2)c(3)d(4)\,d\tau, \qquad (51)$$
$$= (\Delta_{ab}^2+\Delta_{ac}^2+\Delta_{ad}^2+\Delta_{bc}^2+\Delta_{bd}^2+\Delta_{cd}^2)-(\Delta_{ab}^2\Delta_{cd}^3+\Delta_{ac}^2\Delta_{bd}^2+\Delta_{ad}^2\Delta_{bc}^2)$$
$$-2(\Delta_{ab}\Delta_{ac}\Delta_{bc}+\Delta_{ab}\Delta_{ad}\Delta_{bd}+\Delta_{ac}\Delta_{ad}\Delta_{bd}+\Delta_{bc}\Delta_{bd}\Delta_{cd})$$
$$+2(\Delta_{ab}\Delta_{ac}\Delta_{bd}\Delta_{cd}+\Delta_{ab}\Delta_{ad}\Delta_{bc}\Delta_{cd}+\Delta_{ac}\Delta_{ad}\Delta_{bc}\Delta_{bd}). \qquad (52)$$

† Later work (Graben and Present, 1966) suggests an increase of 35% for xenon and krypton.

‡ Calculations in this section were performed by Stamper (1965).

A clear understanding of the meaning of the terms in (41) is necessary; hence we insert a few further comments. Let the total energy of n interacting bodies of the same kind be $E(n)$; we shall denote by $E^{(\alpha)}(n)$ the α-body component of $E(n)$. The $E^{(\alpha)}(n)$ derives its meaning only from $E(n)$ and cannot be separately defined. We label the bodies $a, b, c, \ldots, n$ and understand that indices like i, j, k take on values from a to n. One may then write

$$E^{(1)}(1) = E_a = E_b = \ldots E_n,$$

where E_a, etc., are the energies of the isolated bodies.

$$E^{(1)}(n) = nE^{(1)}(1) = nE_a$$

$$E^{(2)}(2) = E_{ab}, \quad E^{(2)}(n) = \frac{1}{2!}\sum_{i \neq j} E_{ij}$$

$$E^{(3)}(3) = E_{abc}, \quad E^{(3)}(n) = \frac{1}{3!}\sum_{i \neq j \neq k} E_{ijk}, \text{ etc.}$$

Then
$$E(n) = \sum_{\alpha=1}^{n} E^{(\alpha)}(n). \tag{53}$$

In the four-body case,

$$E(4) = 4E_a + \frac{1}{2}\sum_{i \neq j} E_{ij} + \frac{1}{6}\sum_{i \neq j \neq k} E_{ijk} + E_{abcd}. \tag{54}$$

The total interaction energy may be written

$$E(4) - 4E_a = \langle 0 | H | 0 \rangle - 4E_a.$$

Use of (49) leads to

$$\begin{aligned} E(4) - 4E_a = (1-\delta_{abcd})^{-1} \int a(1)b(2)c(3)d(4)\mathcal{A}\{[V_{ab}(2, 1) \\ + V_{ac}(3, 1) + V_{ad}(4, 1) + V_{bc}(3, 2) + V_{bd}(4, 2) \\ + V_{cd}(4, 3)]a(1)b(2)c(3)d(4)\}\, d\tau. \end{aligned} \tag{55}$$

We now abbreviate

$$(1-\delta_{abcd})^{-1} \int a(1)b(2)c(3)d(4)\mathcal{A}[V_{ab}(2, 1)a(1)b(2)c(3)d(4)]\, d\tau \equiv v_{ab,\,cd}. \tag{56}$$

This expression remains unchanged by an interchange of the variables of integration 3 and 4, which has the same effect as an interchange of c and d. Hence $v_{ab,\,cd}$ is independent of the order of c and d and is sufficiently characterized as v_{ab}. Nevertheless, to indicate that $v_{ab,\,cd}$ is the interaction between a and b in the presence of two other atoms we will retain the full notation. From $v_{ab,\,cd}$ the other five terms in (55) are obtained by permutations of a, b, c, and d, together with α, β, γ, and δ. We write

$$E(4)-4E_a = (v_{ab,\,cd}+v_{ac,\,bd}+v_{ad,bc}+v_{bc,\,ad}+v_{bd,\,ac}+v_{cd,\,ab}). \tag{57}$$

Direct evaluation yields

$$\begin{aligned}
v_{ab,\,cd} = \frac{e^2}{R_{ab}} + (1-\delta_{abcd})^{-1}\Big\{ & -2\Big[1-\frac{1}{2}(\Delta_{ac}^2+\Delta_{ad}^2+\Delta_{bc}^2+\Delta_{bd}^2 \\
& +2\Delta_{cd}[\Delta_{cd}-\Delta_{bc}\Delta_{bd}-\Delta_{ac}\Delta_{ad}])\Big](\alpha\beta a) \\
& +2[\Delta_{ab}-\Delta_{bc}\Delta_{ac}-\Delta_{bd}\Delta_{ad}+\Delta_{cd}(\Delta_{ac}\Delta_{bd}+\Delta_{bc}\Delta_{ad}-\Delta_{cd}\Delta_{ab})](a\alpha b) \\
& +[\Delta_{ac}-\Delta_{ab}\Delta_{bc}-\Delta_{ad}\Delta_{cd}+\Delta_{bd}(\Delta_{ab}\Delta_{cd}+\Delta_{ad}\Delta_{bc}-\Delta_{ac}\Delta_{bd})](\alpha\beta c) \\
& +[\Delta_{bc}-\Delta_{ab}\Delta_{ac}-\Delta_{bd}\Delta_{cd}+\Delta_{ad}(\Delta_{ab}\Delta_{cd}+\Delta_{ac}\Delta_{bd}-\Delta_{ad}\Delta_{bc})](b\alpha c) \\
& +[\Delta_{bd}-\Delta_{ab}\Delta_{ad}-\Delta_{bc}\Delta_{cd}+\Delta_{ac}(\Delta_{ab}\Delta_{cd}+\Delta_{ad}\Delta_{bc}-\Delta_{ac}\Delta_{bd})](b\alpha d) \\
& +[\Delta_{ad}-\Delta_{ab}\Delta_{bd}-\Delta_{ac}\Delta_{cd}+\Delta_{bc}(\Delta_{ab}\Delta_{cd}+\Delta_{ac}\Delta_{bd}-\Delta_{ad}\Delta_{bc})](a\beta d) \\
& +(1-\Delta_{cd}^2)[(ab\varrho ab)-(aa\varrho bb)]+(\Delta_{ac}-\Delta_{ad}\Delta_{cd})[(ab\varrho bc) \\
& \quad -(ab\varrho cb)] \\
& +(\Delta_{bc}-\Delta_{bd}\Delta_{cd})[(aa\varrho bc)-(ab\varrho ac)]+(\Delta_{bd}-\Delta_{bc}\Delta_{cd})[(aa\varrho bd) \\
& \quad -(ab\varrho ad)] \\
& +(\Delta_{ad}-\Delta_{ac}\Delta_{cd})[(ab\varrho bd)-(ab\varrho db)]+(\Delta_{ac}\Delta_{bd}-\Delta_{ad}\Delta_{bc})[(ab\varrho cd) \\
& \quad -(ab\varrho dc)]\Big\}.
\end{aligned} \tag{58}$$

Three-body effects have already been computed; they are given by (37). The one term written there on the right (with δ_{abc} given by (36)), is $v_{ab,\,c}$. The two-body interaction is represented by (38), which in the present notation would read v_{ab}.

Clearly, $E^{(2)}(4)$ is

$$E^{(2)}(4) = E_{ab}+E_{ac}+E_{ad}+E_{bc}+E_{bd}+E_{cd}. \tag{59}$$

Since $E_{ab} = v_{ab}$, etc., we find from (57) and (59) that

$$\begin{aligned}
E^{(3)}(4)+E^{(4)}(4) &= E(4)-4E_a-E^{(2)}(4) \\
&= (v_{ab,\,cd}-v_{ab})+(v_{ac,\,bd}-v_{ac})+(v_{ad,\,bc}-v_{ad}) \\
&\quad +(v_{bc,\,ad}-v_{bc})+(v_{bd,\,ac}-v_{bd})+(v_{cd,\,ab}-v_{cd}),
\end{aligned} \tag{60}$$

while

$$E^{(3)}(4) = E_{abc}+E_{abd}+E_{acd}+E_{bcd}, \tag{61}$$

$$E_{abc} = (v_{ab,\,c}-v_{ab})+(v_{ac,\,b}-v_{ac})+(v_{bc,\,a}-v_{bc}), \quad \text{etc.}$$

We consider two specific configurations, (a) the regular tetrahedron and (b) the square.

(a) From symmetry, all parenthetical expressions of (60) are equal and have the value

$$v_{ab,\,cd}-v_{ab} = \frac{4\Delta}{(1-\Delta^2)(1+3\Delta)}\{-\Delta(a\beta a)-\Delta(a\alpha b)+(1+\Delta)(a\beta c)$$

$$+\frac{\Delta}{1-\Delta}[(ab\varrho ab)-(aa\varrho bb)]+\frac{1+\Delta}{1-\Delta}[(ab\varrho bc)-(ab\varrho cb)]\}. \tag{62}$$

In writing (62) use has been made of the following equalities, each deriving from symmetry considerations.

$$\Delta_{ab} = \Delta_{ac} = \Delta_{ad} = \Delta_{bc} = \Delta_{bd} = \Delta_{cd} \equiv \Delta$$

$$(a\beta c) = (b\alpha c) = (b\alpha d) = (a\beta d)$$

$$(ab\varrho bc) = (aa\varrho bc) = (aa\varrho bd) = (ab\varrho bd)$$

$$(ab\varrho cb) = (ab\varrho ac) = (ab\varrho ad) = (ab\varrho db).$$

Similarly, all E_{ijk} are equal, as are the three parentheses in each. In particular

$$v_{ab,\,c}-v_{ab} = \frac{2\Delta}{(1-\Delta^2)(1+2\Delta)}\{-\Delta(a\beta a)-\Delta(a\alpha b)+(1+\Delta)(a\beta c)$$

$$+\frac{\Delta}{1-\Delta}[(ab\varrho ab)-(aa\varrho bb)]+\frac{1+\Delta}{1-\Delta}[(ab\varrho bc)-(ab\varrho cb)]\}. \tag{63}$$

Comparison shows that the contents of the curly brackets in (62) and (63) are the same. If we call them A we arrive at the simple formulas

$$E^{(3)}(4)+E^{(4)}(4) = \frac{24\Delta A}{(1-\Delta^2)(1+3\Delta)} \tag{64}$$

and

$$E^{(3)}(4) = \frac{24\Delta A}{(1-\Delta^2)(1+2\Delta)}\,. \tag{65}$$

Hence

$$\frac{E^{(4)}(4)}{E^{(3)}(4)} = -\frac{\Delta}{1+3\Delta}. \tag{66}$$

Several conclusions can be drawn from (66).

(1) Since $E^{(3)}$ is negative, $E^{(4)}$ must be positive, like $E^{(2)}$, and increase the repulsive exchange forces.

(2) The magnitude of $E^{(4)}/E^{(3)}$ is 1/4 at $R = 0$ and decreases monotonically as the atoms are separated. This is in accord with physical intuition: the importance on n-body effects relative to $(n-1)$-body effects should diminish as each atom becomes less effective in influencing the $(n-1)$-atom interaction.

(3) The ratio $E^{(4)}/E^{(3)}$ depends only on the overlap of the atomic wave function. This, of course, is a fortunate peculiarity of the regular tetrahedron and can not be expected for other configurations.

(4) We have calculated four-body interactions using antisymmetrized state functions; Wojtala (1964) has shown that they also exist when exclusion is not respected. He calculated $E^{(3)}(3)$ and $E^{(2)}(3)$, as already noted, but also $[E^{(3)}(4)+E^{(4)}(4)]/E^{(2)}(4)$ for a regular tetrahedron, and has presented these in the form of graphs. This permits us to make the following comparisons. Let

$$\frac{E^{(3)}(3)}{E^{(2)}(3)} = C(R), \qquad \frac{E^{(3)}(4)+E^{(4)}(4)}{E^{(2)}(4)} = D(R).$$

Since $E^{(3)}(4) = 4E^{(3)}(3)$ and $E^{(2)}(4) = 2E^{(2)}(3)$ we find

$$D(R) = 2C(R)\left[1+\frac{E^{(4)}(4)}{E^{(3)}(4)}\right],$$

whence

$$\frac{E^{(4)}(4)}{E^{(3)}(4)} = \frac{D(R)}{2C(R)} - 1.$$

Now, study of Wojtala's curves shows that, while D and C reach appreciable magnitudes (in regions, to be sure, where exchange can not be neglected), D is always near $2C$, so that in view of our last equation the ratio of four-body to three-body interactions in the case of a regular tetrahedron *without exchange* is practically zero. Hence we conclude that exchange forces are responsible for four-body effects.

It is tempting to examine what bearing our results have upon Jansen's work. Clearly, they do not affect it directly because his crucial conclusions are based upon second-order effects, which we shrink from attacking. But it is easy to make comparisons within the first order provided we adopt Gaussian wave functions. This leads to the integrals tabulated as equations (46) plus one four-center integral,

$$(ab\varrho cd) = B\Delta_{ac}\Delta_{bd}\, F\left(\frac{B^2R^2_{\overline{ac}\,\overline{bd}}}{2}\right), \tag{67}$$

$R_{\overline{ac}\,\overline{bd}}$ meaning the distance between the midpoints of $\overline{ac}$ and $\overline{bd}$. For our tetrahedron, of course, we do not need this result and rely on formula (66) with $\Delta = \exp(-\frac{1}{4}B^2R^2)$. Choosing the proper values of BR for argon, krypton, and xenon we obtain Table 5.2.

TABLE 5.2

	BR	$\left.\frac{E^{(3)}(3)}{E^{(2)}(3)}\right\vert_{\theta=60°}$	$\frac{E^{(4)}(4)}{E^{(3)}(4)}$
Argon	2.40	−0.200	−0.1385
Krypton	2.10	−0.210	−0.1663
Xenon	1.99	−0.214	−0.1757

The second column contains values read from the curves given by Jansen and Zimering (1963). When the total interaction energy of the regular tetrahedron is computed, and written in the form (54),

$$E(4)-4E_a = E^{(2)}(4)+E^{(3)}(4)+E^{(4)}(4) = E^2(4)\,[1+\varepsilon_3+\varepsilon_4],$$

we obtain for the three crystals the values given in Table 5.3.

The appearance of rapid convergence here may be deceptive. Four-body forces are, in this instance, computed for a single tetrahedron, which contains four triangles (and therefore four terms contributing to ε_3) but only one E_{abcd}. In a crystal, where a domain of interacting atoms comprises something like one atom plus its nearest neighbors, ε_3 has $(n/3)$ and ε_4, $(n/4)$ terms, and these numbers are comparable for $n \approx 12$. It is very likely, therefore, that Table 5.3 overestimates the rate of convergence for a large aggregate of

TABLE 5.3.

	ε_3	ε_4
Argon	-0.40	$+0.055$
Krypton	-0.42	$+0.070$
Xenon	-0.43	$+0.075$

atoms. It will also be seen that the story is quite different for the square, to which we now turn.

(b) The atoms comprising the square are labeled as shown in the diagram below:

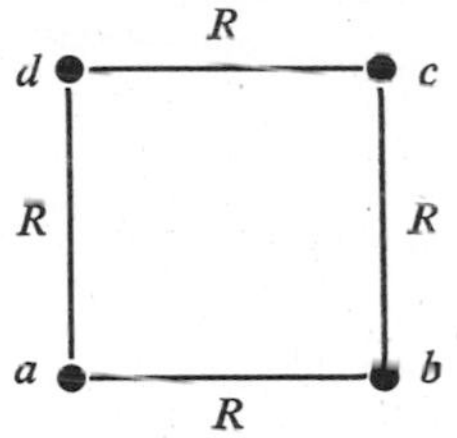

Symbols have the same meaning as before but the distances are different.

The departure from pairwise additivity is once again given by (60), which, in view of the symmetry properties inherent in the diagram, takes the form

$$E^{(3)}(4)+E^{(4)}(4) = 4(v_{ab,\,cd}-v_{ab})+2(v_{ac,\,bd}-v_{ac}). \tag{68}$$

In the evaluation of $(v_{ab,\,cd}-v_{ab})$ use is made of the following equalities:

$$\begin{aligned} \Delta_{ab} = \Delta_{bc} = \Delta_{cd} = \Delta_{ad} &\equiv \Delta \\ \Delta_{ac} = \Delta_{bd} &= \Delta^2 \\ (a\beta c) &= (b\alpha d) \\ (b\alpha c) &= (a\beta d) \\ (ab\varrho bc) &= (ab\varrho cb) \\ (aa\varrho bc)-(ab\varrho ac) &= (ab\varrho bd)-(ab\varrho db) \\ (ab\varrho cd)-(ab\varrho dc) &= -\Delta^2[(ab\varrho ab)-(aa\varrho bb)]. \end{aligned} \tag{69}$$

Here, specific appeal is made to the Gaussian character of the atomic orbitals. Thus, our treatment of the square represents a straightforward extension of the three-atom formalism, but lacks exact applicability to hydrogen. From (69) we find that

$$v_{ab,\,cd}-v_{ab}=\frac{2\Delta}{(1-\Delta^2)^2}\Big\{-\Delta(a\beta a)+\Delta^2(a\alpha b)-\Delta(a\beta c)+(b\alpha c)$$

$$+\frac{\Delta}{1-\Delta^2}[(ab\varrho ab)-(aa\varrho bb)]+\frac{1}{1-\Delta^2}[(aa\varrho bc)-(ab\varrho ac)]\Big\}. \tag{70}$$

Obtaining the proper expression for $v_{ac,\,bd}$ requires an interchange of b and β with c and γ respectively. Specifically,

$$v_{ac,\,bd}=\frac{e^2}{R_{ac}}+(1-\delta_{abcd})^{-1}\{-2[1-\tfrac{1}{2}(\Delta_{ab}^2+\Delta_{ad}^2+\Delta_{bc}^2+\Delta_{cd}^2)$$
$$+\Delta_{bd}(\Delta_{bd}-\Delta_{bc}\Delta_{cd}-\Delta_{ab}\Delta_{ad})](a\gamma a)$$
$$+2[\Delta_{ac}-\Delta_{bc}\Delta_{ab}-\Delta_{cd}\Delta_{ad}+\Delta_{bd}(\Delta_{ab}\Delta_{cd}+\Delta_{bc}\Delta_{ad}-\Delta_{bd}\Delta_{ac})](a\alpha c)$$
$$+[\Delta_{ab}-\Delta_{ac}\Delta_{bc}-\Delta_{ad}\Delta_{bd}+\Delta_{cd}(\Delta_{ac}\Delta_{bd}+\Delta_{ad}\Delta_{bc}-\Delta_{ab}\Delta_{cd})](a\gamma b)$$
$$+[\Delta_{bc}-\Delta_{ac}\Delta_{ab}-\Delta_{cd}\Delta_{bd}+\Delta_{ad}(\Delta_{ac}\Delta_{bd}+\Delta_{ab}\Delta_{cd}-\Delta_{ad}\Delta_{bc})](b\alpha c)$$
$$+[\Delta_{cd}-\Delta_{ac}\Delta_{ad}-\Delta_{bc}\Delta_{bd}+\Delta_{ab}(\Delta_{ac}\Delta_{bd}+\Delta_{ad}\Delta_{bc}-\Delta_{ab}\Delta_{cd})](c\alpha d)$$
$$+[\Delta_{ad}-\Delta_{ac}\Delta_{cd}-\Delta_{ab}\Delta_{bd}+\Delta_{bc}(\Delta_{ac}\Delta_{bd}+\Delta_{ab}\Delta_{cd}-\Delta_{ad}\Delta_{bc})](a\gamma d)$$
$$+(1-\Delta_{bd}^2)[(ac\varrho ac)-(aa\varrho cc)]+(\Delta_{ab}-\Delta_{ad}\Delta_{bd})[(ac\varrho cb)-(ac\varrho bc)]$$
$$+(\Delta_{bc}-\Delta_{cd}\Delta_{bd})[(aa\varrho cb)-(ac\varrho ab)]+(\Delta_{cd}-\Delta_{bc}\Delta_{bd})[(aa\varrho cd)$$
$$-(ac\varrho ad)]$$
$$+(\Delta_{ad}-\Delta_{ab}\Delta_{bd})[(ac\varrho cd)-(ac\varrho dc)]+(\Delta_{ab}\Delta_{cd}-\Delta_{ad}\Delta_{bc})[(ac\varrho bd)$$
$$-(ac\varrho db)]\}. \tag{71}$$

Similarly,

$$v_{ac,\,b}=\frac{e^2}{R_{ac}}+(1-\delta_{abc})^{-1}\{[-2+\Delta_{ab}^2+\Delta_{bc}^2](a\gamma a)+$$
$$2[\Delta_{ac}-\Delta_{ab}\Delta_{bc}](a\alpha c)$$
$$+[\Delta_{ab}-\Delta_{ac}\Delta_{bc}](a\gamma b)+[\Delta_{bc}-\Delta_{ac}\Delta_{ab}](b\alpha c)+[(ac\varrho ac)-(aa\varrho cc)]$$
$$+\Delta_{ab}[(ac\varrho cb)-(ac\varrho bc)]+\Delta_{bc}[(aa\varrho cb)-ac\varrho ab)]\} \tag{72}$$

and

$$v_{ac}=\frac{e^2}{R_{ac}}+(1-\Delta_{ac}^2)^{-1}\{-2(a\gamma a)+2\Delta_{ac}(a\alpha c)+(ac\varrho ac)-(aa\varrho cc)\}. \tag{73}$$

From the relations

$$(a\gamma b) = (b\alpha c) = (c\alpha d) = (a\gamma d)$$
$$(ac\varrho cb)-(ac\varrho bc) = (aa\varrho cd)-(ac\varrho ad) = (ac\varrho cd)-(ac\varrho dc)$$
$$= (aa\varrho cb)-(ac\varrho ab) \equiv (aa\varrho bc)-(ab\varrho ac) \tag{74}$$

one finds that

$$v_{ac,\,bd}-v_{ac} = \frac{4\Delta}{(1-\Delta^2)^2}\left\{-\frac{\Delta}{1+\Delta^2}(a\gamma a)-\frac{\Delta}{1+\Delta^2}(a\alpha c)+(b\alpha c)\right.$$
$$\left.+\frac{\Delta}{1-\Delta^4}[(ac\varrho ac)-(aa\varrho cc)]+\frac{1}{1-\Delta^2}[(aa\varrho bc)-(ab\varrho ac)]\right\}. \tag{75}$$

Hence, in view of (70) and (75), (68) reads

$$E^{(3)}(4)+E^{(4)}(4) = \frac{8\Delta}{(1-\Delta^2)^2}\left\{-\Delta(a\beta a)-\frac{\Delta}{1+\Delta^2}(a\gamma a)+\Delta^2(a\alpha b)\right.$$
$$-\frac{\Delta}{1+\Delta^2}(a\alpha c)-\Delta(a\beta c)+2(b\alpha c)+\frac{\Delta}{1-\Delta^2}[(ab\varrho ab)-(aa\varrho bb)]$$
$$\left.+\frac{2}{1-\Delta^2}[(aa\varrho bc)-(ab\varrho ac)]+\frac{\Delta}{1-\Delta^4}[(ac\varrho ac)-(aa\varrho cc)]\right\}. \tag{76}$$

Clearly, all E_{ijk} are equal, so that

$$E^{(3)}(4) = 4E_{abc} \tag{77}$$
$$= 8(v_{ab,\,c}-v_{ab})+4(v_{ac,\,b}-v_{ac}). \tag{78}$$

Equations (37) and (38), together with equations (69), provide that

$$v_{ab,\,c}-v_{ab} = \frac{\Delta}{1-\Delta^2}\left\{-\Delta(a\beta a)+(b\alpha c)\right.$$
$$\left.+\frac{\Delta}{1-\Delta^2}[(ab\varrho ab)-(aa\varrho bb)]+\frac{1}{1-\Delta^2}[(aa\varrho bc)-(ab\varrho ac)]\right\} \tag{79}$$

while eqns. (72–74) yield

$$v_{ac,b}-v_{ac} = \frac{2\Delta}{1-\Delta^2}\left\{-\frac{\Delta}{1+\Delta^2}(a\gamma a)-\frac{\Delta}{1+\Delta^2}(a\alpha c)+(b\alpha c)\right.$$
$$\left.+\frac{\Delta}{1-\Delta^4}[(ac\varrho ac)-(aa\varrho cc)]+\frac{1}{1-\Delta^2}[(aa\varrho bc)-(ab\varrho ac)]\right\}. \tag{80}$$

Hence, (78) becomes

$$E^{(3)}(4) = \frac{8\Delta}{1-\Delta^2}\Bigg\{-\Delta(a\beta a) - \frac{\Delta}{1+\Delta^2}(a\gamma a) - \frac{\Delta}{1+\Delta^2}(a\alpha c) + 2(b\alpha c)$$

$$+\frac{\Delta}{1-\Delta^2}[(ab\varrho ab)-(aa\varrho bb)] + \frac{2}{1-\Delta^2}[(aa\varrho bc)-(ab\varrho ac)]$$

$$+\frac{\Delta}{1-\Delta^4}[(ac\varrho ac)-(aa\varrho cc)]\Bigg\}. \qquad (81)$$

In view of (81), it is apparent that (76) can be written as follows;

$$E^{(3)}(4)+E^{(4)}(4) = \frac{1}{1-\Delta^2}E^{(3)}(4) + \frac{8\Delta}{(1-\Delta^2)^2}[\Delta^2(a\alpha b)-\Delta(a\beta c)]$$

or

$$E^{(4)}(4) = \frac{\Delta^2}{1-\Delta^2}E^{(3)}(4) + \frac{8\Delta^2}{(1-\Delta^2)^2}[\Delta(a\alpha b)-(a\beta c)]. \qquad (82)$$

We have

$$(a\alpha b) = \Delta BF\left(\frac{B^2R^2}{4}\right) = \frac{2\Delta e^2}{R}\operatorname{erf}\left(\frac{BR}{2}\right)$$

$$(a\beta c) = \Delta^2 BF\left(\frac{B^2R^2}{2}\right) = \frac{\sqrt{2}\Delta^2 e^2}{R}\operatorname{erf}\left(\frac{BR}{\sqrt{2}}\right).$$

Thus,

$$\frac{E^{(4)}(4)}{E^{(3)}(4)} = \frac{\Delta^2}{1-\Delta^2}\left\{1+\frac{\dfrac{8e^2}{R}\dfrac{\Delta^2}{1-\Delta^2}\left[\operatorname{erf}\left(\dfrac{BR}{2}\right)-\dfrac{1}{\sqrt{2}}\operatorname{erf}\left(\dfrac{BR}{\sqrt{2}}\right)\right]}{\left[\dfrac{E^{(3)}(3)}{E^{(2)}(3)}\right]E^{(2)}(4)}\right\}, \qquad (83)$$

where on the right we have replaced $E^{(3)}(4)$ by $2[E^{(3)}(3)/E^{(2)}(3)] \times E^{(2)}(4)$.

It is not difficult to calculate $E^{(2)}(4)$ for each of the heavy rare gases and, once again, Jansen's results for $E^{(3)}(3)/E^{(2)}(3)$ may be used. The results are presented in Table 5.4.

TABLE 5.4

	BR	$\left.\frac{E^{(3)}(3)}{E^{(2)}(3)}\right\|_{\theta=90°}$	$\frac{E^{(4)}(4)}{E^{(3)}(4)}$
Argon	2.40	−0.075	−0.9438
Krypton	2.10	−0.075	−1.031
Xenon	1.99	−0.075	−1.198

If, as before, we put $E(4)-4E_a = E^{(2)}(4)(1+\varepsilon_3+\varepsilon_4)$, there results Table 5.5.

TABLE 5.5

	ε_3	ε_4
Argon	−0.150	+0.142
Krypton	−0.150	+0.155
Xenon	−0.150	+0.180

The outcome is somewhat unexpected and disturbing, for it shows that for a square configuration four-body forces may outweigh three-body forces. How this affects applications, such as those reviewed in this article, is at present difficult to foresee, chiefly because our calculation was performed only in first-order of perturbation theory. Extension in two directions is needed to resolve these issues: other configurations must be included, and second-order effects require investigation. Unfortunately, both of these tasks are time consuming. But it is worth knowing that thoughtless use of the multibody series, equation (53), is fraught with risks.

Our first-order results are not wholly without physical meaning. For as R decreases, first-order effects predominate. Hence the failure of the multibody series for very closely packed atoms is already clear.

The total departure from additivity, $(E^{(3)}+E^{(4)})/E^{(2)}$, is about 2% for the square configuration, but 35% for the regular tetrahedron, where the arrangement is more compact. The sign of $E^{(4)}$ remains positive for both configurations, opposing the three-body effects.

In conclusion the following points may well be recalled concerning the role of many-body forces.

(1) For dense states of matter they are quite important, may even be crucial in determining structure.
(2) There is no conclusive evidence to show that the multibody series, eqn. (41), has convenient features of rapid convergence.
(3) Relevant calculations are at present based for the most part on simple state functions, i.e. on one-electron oscillator functions. This introduces errors which have not been estimated. However, the use of more correct atomic functions may well improve the convergence of the multibody series even at distances of interest in crystal structures.

The work in this section was based on first-order perturbation theory, which limits our inferences to small distances of separation. In section 5.2, second-order effects were also considered. The method of introducing them, however, i.e. of adding them to the first-order effects, which seems dictated by practical necessity, is open to grave objections in view of our findings in Chapter 4.

Failure of the multibody expansion would ultimately invalidate the calculation of crystal energies via intermolecular forces and necessitate an approach like Löwdin's (1948). We would then be facing the complexities of the many-body problem even here.

The recent literature contains critical comments upon Jansen's procedures (Swenberg, 1967; Present, 1967) and a reply (Jansen and Lombardi, 1967).

CHAPTER 6

Retarded Dispersion Forces; Relativistic Effects

6.1. The Effect of Radiative Coupling on Intermolecular Forces between Ground State Atoms

Thus far in this book we have neglected all effects of radiation fields. When fully written our Hamiltonian actually had the form

$$H = H_0 + V + H_{\text{rad}}, \tag{1}$$

where H_0 is the Hamiltonian of the noninteracting molecules, V is the Coulomb interaction between molecules, and H_{rad} is the Hamiltonian for the radiation field, i.e. the photon Hamiltonian. Since no terms in (1) couple H_0 or V with H_{rad}, we previously did not even consider H_{rad} and simply used H_0+V as the effective Hamiltonian. In this chapter we include a coupling term in H but treat only the interaction of nonoverlapping molecules since it is only for large separations that radiation fields are important; and therefore the potential V may be expanded at once in a multipole series, the first term of which is the dipole–dipole interaction discussed in Chapter 2.

The complete Hamiltonian reads

$$H_{\text{total}} = H_0 + V + H_{\text{rad}} + H_{\text{int}}. \tag{2}$$

The last term can be neglected in calculating intermolecular forces at small separations. At large separations it contributes to dispersion interactions since radiation effects are of long range. As will be shown, at very large separations the interaction energy of two neutral molecules in their ground states changes from R^{-6} to R^{-7}. An elementary explanation for this effect is this. Dispersion forces arise from the interaction of one instantaneous dipole inducing another dipole in a neighboring molecule. This new induced dipole then interacts with the original dipole. Because of the finite velocity of electromagnetic radiation the total time elapsed is $2R/c$ (c being the velocity of light). During this time the original dipole has changed its orientation. The returning field is then "retarded" with respect to the initial field, hence the name "retarded dispersion forces". Since the maximum interaction occurs for no retardation or no change in orientation, the corrected interaction must be weaker. It is weaker by a factor depending on R at large separations. Such a reduction was first postulated to explain the apparent small value of dispersion interactions between colloidal particles at large separations (Verwey, 1947).

The first detailed theoretical treatment of the problem was carried out by Casimir and Polder (1948). It was followed by simpler treatments by Aub, Power, and Zienau (1957), Dzialoshinskii (1957), Mavroyannis and Stephen (1962), and McLachlan (1963).

In this discussion we shall follow the work of Casimir and Polder (1948) and of Power and Zienau (1957a), concentrating our attention chiefly on physical concepts.

Casimir and Polder (1948) begin by writing the total Hamiltonian in the Coulomb gauge (see Appendix C for details):†

$$H_{\text{total}} = H_0 + H_{\text{rad}} + V - \frac{e}{mc}\,\mathfrak{p}_a \cdot \boldsymbol{A}_a + \frac{e^2}{2mc^2}\,A_a^2 - \frac{e}{mc}\,\mathfrak{p}_b \cdot \boldsymbol{A}_b + \frac{e^2}{2mc^2}\,A_b^2. \tag{3}$$

Here $\boldsymbol{A}_i$ is the vector potential of the radiation field at atom i, $\mathfrak{p}_a$ is the total momentum of the electrons of atom $\boldsymbol{a}$

$$\mathfrak{p}_a = \sum_i (\mathfrak{p}_a)_i$$

where the sum extends over all electrons in the atom. It is related to the electric dipole moment $\boldsymbol{p}_a$ by

$$e\mathfrak{p}_a = m(d/dt)\boldsymbol{p}_a. \tag{4}$$

The interaction energy is found by perturbation theory, using the last five terms in (3) as the perturbation. Since dispersion for-

† We also refer the reader to standard treatments of electromagnetic theory. See, in particular, Heitler (1954, Chapter 1); or Jackson (1963, sections 6.3–6.6).

ces are proportional to the fourth power of the electric charge, contributions from the electrostatic perturbation V occur in the second-order energy, as do important contributions from the A^2 terms. The $\mathfrak{p} \cdot A$ terms, however, must be evaluated in the fourth order.

The zero-order state function of $H_0 + H_{\text{rad}}$ is the product of a state function of the electronic system (itself a product of functions centered about the two atoms), $|nl\rangle$, and a function for the radiation field, $|k)$:

$$\Psi_0 = |00\rangle\,|0) \tag{5}$$

We shall distinguish between the two functions by using round bra and ket notations for the photons and standard Dirac notation for the electronic states. In the ground state, (5), the first two zeros imply that both atoms are in their lowest energy states, while the photon function indicates that no photons are present, i.e. the radiation field is in its ground state.

Consideration of the various contributions in each order of perturbation theory is greatly simplified by introduction of diagrams to represent the various terms. A vertical line traced upward represents a succession of states of an atom in time. In our case, one line is drawn for atom a, one for atom b. Dots (sometimes called vertices) denote changes in state and are therefore equivalent to matrix elements which couple atoms a and b. The nature of these matrix elements is suggested by the type of line which joins a and b at the vertices. For instance, a dotted horizontal line designates an element of V, which acts instantaneously; a wavy line indicates a photon exchange via H_{int}. Since this takes time, all wavy lines are slanted. A few examples will illustrate this technique.

The dispersion energy from second-order perturbation theory (see Chapter 2) is

$$E_2 = \sum_{(\lambda_1, \lambda_2) \neq 0} \frac{|V_{00,\lambda_1\lambda_2}|^2}{E_a^0 + E_b^0 - E_a^{\lambda_1} - E_b^{\lambda_2}}, \tag{6}$$

where $\lambda_1\lambda_2$ is an excited state of the two-atom system and $E_a^{\lambda_1}$ is the energy of atom a in state λ_1. The diagram symbolizing one term of (6) is shown in Fig. 8:

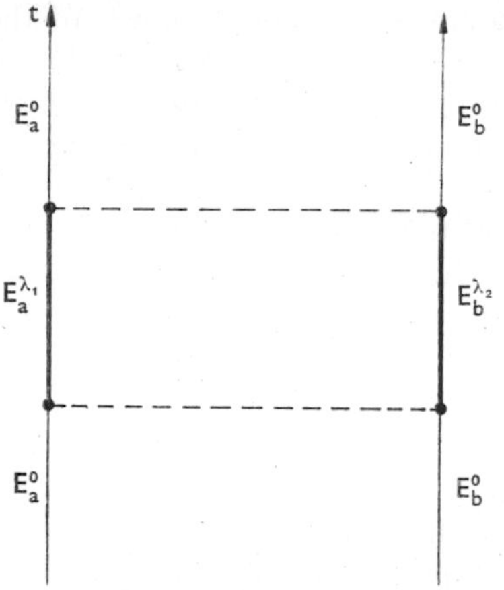

FIG. 8.

where we have explicitly written the energies of the states involved. The energy denominators are obtained by subtracting the intermediate states (heavy portions of the lines) from the zero-order energy. In our diagram we have an instantaneous interaction, the atoms are both excited to virtual states, and then another instantaneous interaction returns the atoms to their ground states.

Consider now the interaction of an atom with light of wave vector $\boldsymbol{k}$ and polarization σ. From time-dependent second-order perturbation theory (see, for example, Eyring, Walter, and Kimball, 1944, chap. 8) the energy of the system† is found to be

$$\sum_{\mathrm{I}} \left\langle 0 \,|\, \left(\boldsymbol{k}\sigma \,|\, -\frac{e}{mc}\, \mathfrak{p}\cdot \boldsymbol{A} \,|\, 0\right) |\, \mathrm{I} \right\rangle \frac{1}{E_a^0 + \hbar kc - E_a^{\mathrm{I}}} \left\langle \mathrm{I} \,|\, (0\,|\, -\frac{e}{mc}\,\mathfrak{p}\cdot\boldsymbol{A}\,|\,\boldsymbol{k}\sigma)\,|_0 \right\rangle$$

$$+\sum_{\mathrm{II}} \left\langle 0 \,|\, \left(\boldsymbol{k}\sigma \,|\, -\frac{e}{mc}\, \mathfrak{p}\cdot \boldsymbol{A} \,|\, \boldsymbol{k}\sigma + \boldsymbol{k}\sigma\right) |\, \mathrm{II} \right\rangle \frac{1}{(E_a^0 + \hbar kc) - (E_a^{\mathrm{II}} + 2\hbar kc)} \left\langle \mathrm{II} \,|\, \left(\boldsymbol{k}\sigma + \boldsymbol{k}\sigma \,|\, -\frac{e}{mc}\,\mathfrak{p}\cdot\boldsymbol{A}\,|\,\boldsymbol{k}\sigma\right) |\, 0 \right\rangle, \quad (7)$$

† This is related to the polarizability P, as $-(1/2)\, P\,|\boldsymbol{A}|^2$.

where in a new notation I and II now denote the excited states of the atom. The first term corresponds to absorption of a photon of energy $\hbar kc$ and polarization σ followed by emission of a photon of the same energy and polarization. The second term belongs to the reverse process: emission followed by absorption. Correspondingly the diagrams (Fig. 9) for the two processes are

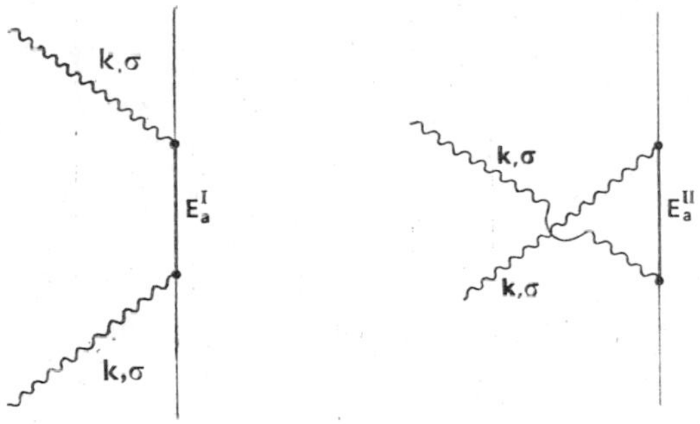

FIG. 9.

Similar diagrams for the inelastic processes (emission and absorption at different frequencies) occur in the theory of Raman scattering.

Using the diagrams we calculate the intermolecular interaction, employing the Hamiltonian in (3).

In second-order perturbation theory typical terms are given by Fig. 10(a)–(b)

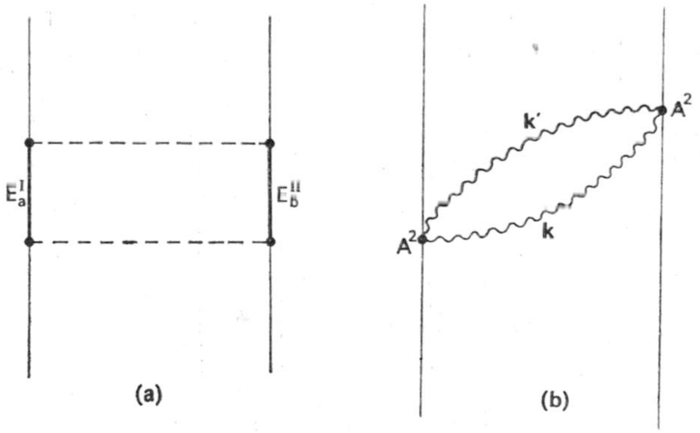

FIG. 10. (a – b)

Vertices which involve matrix elements of the A^2 term in (3) are labeled A^2 when confusion might result.

In the third-order energy, the important terms are given in Fig. 10(c)–(g) where an unlabeled vertex involves either a matrix element of V or of $\mathfrak{p}\cdot\boldsymbol{A}$, the distinction between them being easily made since a photon contributes only to the latter.

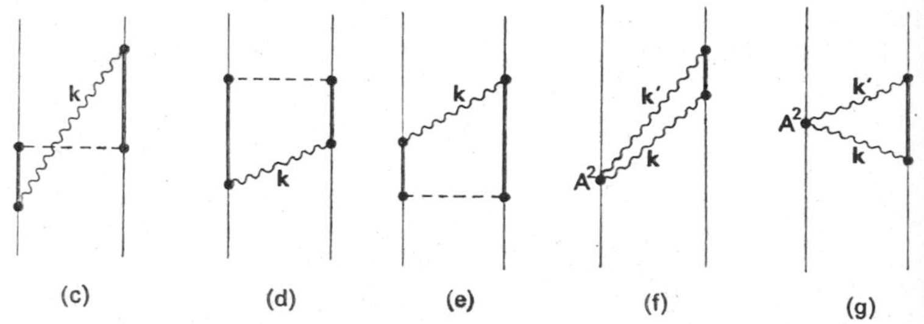

FIG. 10. (c – g)

The fourth-order energy terms only involve matrix elements of $\mathfrak{p}\cdot\boldsymbol{A}$. They have the form given in Fig. 10 (h)–(m).

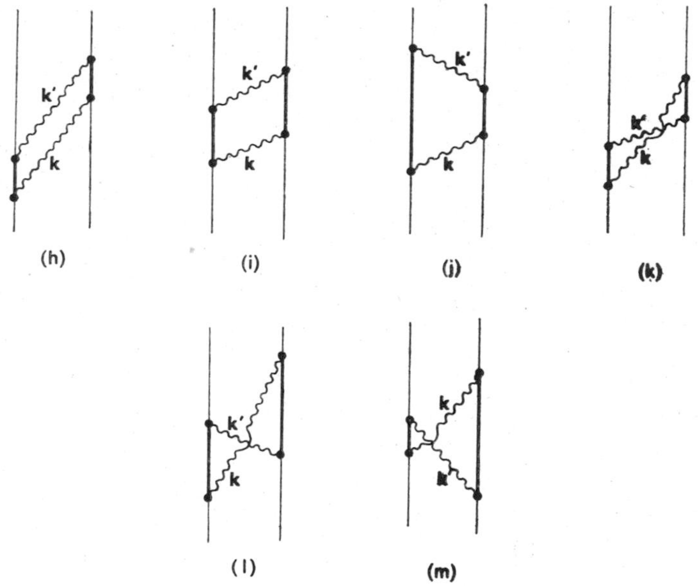

FIG. 10. (h – m)

Again, two intermediate photons are involved and a summation must be extended over them in calculating the total interaction. The above diagrams differ only in the composition of the intermediate state, for it now becomes important to decide whether atom ***b*** becomes excited before atom ***a*** is de-excited, or vice versa.

We shall not evaluate individual contributions here since the final result can be derived in a more straightforward manner. Details of the above analysis are given by Casimir and Polder (1948) and by Power (1965, chap. 7). The calculation involves many effects, the net result of which is to cancel out the long-range R^{-6} interaction and replace it by an R^{-7} potential. The London result, Fig. 10(a), is exactly canceled at large separations by parts of the fourth-order energy contribution, Fig. 10(h)–(m). This is because the A propagator has a static contribution as well as its radiation components (Power, 1965, appendix 4).

The complications caused by the many cancellations can be avoided if the atom is incorporated consistently in the Hamiltonian as the source of the quantized radiation field. In the electric dipole approximation we show in Appendix C that the effective interaction Hamiltonian, the perturbation, is simply

$$H' = -\mathbf{p}_a \cdot \mathbf{E}^{\perp}(a) - \mathbf{p}_b \cdot \mathbf{E}^{\perp}(b) \tag{8}$$

where $\mathbf{p}_i$ is the dipole moment of atom i and $\mathbf{E}^{\perp}$ is the transverse electric field at the position of atom i.

With this new Hamiltonian all contributions occur in the fourth-order energy which is (cf. Hirschfelder, Byers Brown, and Epstein, 1964)

$$E_4 = \sum_{\mathrm{I,\,II,\,III} \neq 0} \frac{\langle 0 | H' | \mathrm{III} \rangle \langle \mathrm{III} | H' | \mathrm{II} \rangle \langle \mathrm{II} | H' | \mathrm{I} \rangle \langle \mathrm{I} | H' | 0 \rangle}{(E_0 - E_{\mathrm{I}})(E_0 - E_{\mathrm{II}})(E_0 - E_{\mathrm{III}})} - \sum_{I \neq 0} \frac{\langle 0 | H' | I \rangle \langle I | H' | 0 \rangle}{E_0 - E_I} \sum_{I \neq 0} \left| \frac{\langle 0 | H' | I \rangle}{E_0 - E_I} \right|^2 . \tag{9}$$

The last term in (9) does not contribute to the interaction energy. It is a higher order renormalization correction of the state function.

Since H' is composed of two terms we find six distinct contributions from (9) which depend on p_a^2 and p_b^2 and thus depend on the separation between the two atoms. All other terms are similar to $p_a p_b^3$ which vanishes for neutral atoms or p_a^4 which does not contribute to the interaction energy, that is, it has no R dependence.

The diagrams of the six terms are given in Fig. 11 below (All vertices here are $\boldsymbol{p}\cdot\boldsymbol{E}^{\perp}$ matrix elements).

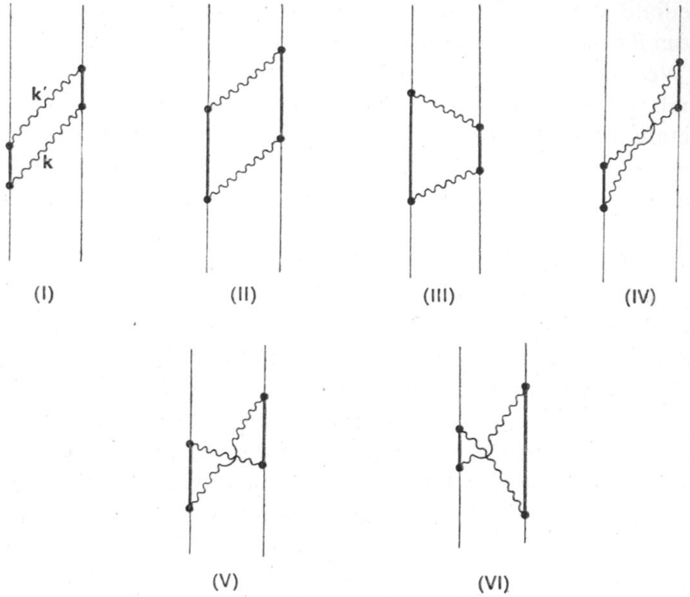

FIG. 11.

We notice immediately that Fig. 11 (II), (III), (V), and (VI) involve virtual states with energy denominators in which both atoms are excited, while in Fig. 11 (I) and (IV) only one atom is excited at one time.

In terms of virtual states Fig. 11 (I) implies that atom a is excited from its ground state and creates a virtual photon of energy $\hbar kc$. This state of energy $E_a^l + \hbar ck$ then decays to the ground electronic state of the system but in so doing creates another photon of energy $\hbar ck'$. This state of energy $c\hbar(k+k')$ persists until one photon is absorbed (destroyed) and atom b is excited. The resulting state of energy $E_b^n + c\hbar k$ lasts until both the atom and the radiation field decay to the ground state again. The terminology used here should not imply that real excitations take place, rather that virtual states are involved in perturbation theory similar to the superposition of configurations discussed in the previous chapters. They are the terms necessary to achieve an accurate representation of the true state function. None the less, the language used here aids in understanding what excited states will enter most strongly in the perturbation formalism.

Two matrix elements occur in the evaluation of the energy contributions; that of the electric dipole moment between two states of the atom, and the matrix element of $E^{\perp}$ between two states of the radiation field. The first was encountered in Chapter 2 when we calculated dispersion forces, but the second requires knowledge of the quantized $E^{\perp}$ field. Quantization of $E^{\perp}$ is necessary in order that spontaneous absorption and emission may enter as virtual states in perturbation theory.

If the entire system of two atoms or molecules is placed in a box of unit volume, the quantized field is

$$E^{\perp} = \sum_{k,\lambda} \hat{e}(k\sigma)\, i\sqrt{(2\pi\hbar ck)}\,\{-a_{k\sigma}^{+}e^{-ik\cdot r} + a_{k\sigma}e^{ik\cdot r}\}, \tag{10}$$

where $\hat{e}(\boldsymbol{k}, \sigma)$ is a unit vector specifying the polarization direction of a photon of wave vector $\boldsymbol{k}$ and polarization σ ($\sigma = 1, 2$ since $E^{\perp}$ is transverse). $a_{k\sigma}^{+}$ and $a_{k\sigma}$ are creation and annihilation operators satisfying†

$$a_{k\sigma}^{+}|0) = |\boldsymbol{k}\sigma)$$

and

$$a_{k\sigma}|k\sigma) = |0). \tag{11}$$

Upon using (10) the matrix element involving the quantized radiation field becomes

$$(k\sigma|E^{\perp}|0) = -i\sqrt{(2\pi\hbar ck)}\,\hat{e}(k\sigma)\,e^{-ik\cdot r} \tag{12}$$

These matrix elements allow us to evaluate the various fourth-order contributions from (9) represented by the diagrams in Fig. 11. Specifically, Fig. 11 (I) represents (in double Dirac notation)

$$I = \sum_{\lambda,\lambda'} \sum_{l,n,kk'} \langle 00|(0|p_a\cdot E^{\perp}|k\sigma)|l0\rangle \frac{1}{E_a^0 - E_a^l - \hbar ck}$$
$$\langle l0|(k\sigma|p_a\cdot E^{\perp}|k\sigma + k'\sigma')|00\rangle \frac{1}{-\hbar c(k+k')}$$
$$\langle 00|(k\sigma + k'\sigma'|p_b\cdot E^{\perp}|k'\sigma')|0n\rangle \frac{1}{E_b^0 - E_b^n - k'c\hbar}$$
$$\langle 0n|(k'\sigma'|p_b\cdot E^{\perp}|0)|00\rangle, \tag{13}$$

† For additional information on quantized fields see Power (1965), Heitler (1954), or Louisell (1964).

(where l and n are quantum numbers of the excited states of atoms a and b respectively). Each energy denominator is equal to the ground state energy minus the energy of the intermediate state. This accounts for the negative sign in the two-photon propagator, the second energy denominator. The two-photon state functions are represented as the sum of two $\boldsymbol{k}\sigma$ designations. Upon evaluating the matrix elements we find in easy steps, using photon functions given by (10) and a simpler notation for matrix elements,

$$I = -\sum_{\lambda,\lambda'}\sum_{l,n,k,k'} \frac{\langle \boldsymbol{p}_a\rangle_{l0}\cdot\hat{e}(k\sigma)ie^{ik\cdot r_a}}{\Delta E_a^l+\hbar ck}$$

$$\langle \boldsymbol{p}_a\rangle_{0l}\cdot\hat{e}(\boldsymbol{k}'\sigma')ie^{ik'\cdot r_a}\,\frac{1}{\hbar c(k+k')}$$

$$\langle \boldsymbol{p}_b\rangle_{n0}\cdot\hat{e}(\boldsymbol{k}\sigma)\,(-i)e^{-ik\cdot r_b}\,\frac{1}{\Delta E_b^n+\hbar ck'}$$

$$\langle \boldsymbol{p}_b\rangle_{0n}\cdot\hat{e}(k'\sigma')\,(-i)e^{-ik'\cdot r_b}(4\pi^2\hbar^2c^2kk'), \tag{14}$$

$$= \sum_{\lambda\lambda'}\sum_{l,n,k,k'}\sum_{ij\sigma\tau}(4\pi^2\hbar^2c^2)\,\langle p_a^i\rangle_{0l}\langle p_a^j\rangle_{l0}\langle p_b^j\rangle_{0n}\,\langle p_b^\tau\rangle_{n0}$$

$$\times\hat{e}_i(k\sigma)\hat{e}_j(\boldsymbol{k}'\sigma')\hat{e}_\sigma(k\sigma)\hat{e}_\tau(k'\sigma')e^{-i(k+k')\cdot R}$$

$$\times\frac{kk'}{\Delta E_a^l+\hbar ck}\,\frac{1}{\hbar c(k+k')}\,\frac{1}{\Delta E_b^n+c\hbar k}, \tag{15}$$

where $\boldsymbol{R} = \boldsymbol{r_b}-\boldsymbol{r_a}$, $\Delta E_a^l = -E_a^0+E_a^l$, $\Delta E_b^n = -E_b^0+E_b^n$, while the superscripts on the dipole operators and the subscripts on the unit vectors denote components over which one must sum. $\langle p_a^i\rangle_{0l}$ denotes the matrix element of the ith component of the dipole moment of atom $\boldsymbol{a}$ between atomic states 0 and l.

The summation over all polarizations leads to (Power, 1965, pp. 68–69)

$$\sum_\sigma \hat{e}_i(k\sigma)\hat{e}_j(k\sigma) = \delta_{ij}-\hat{k}_i\hat{k}_j. \tag{16}$$

Thus we obtain

$$I = (2\pi)^2\hbar^2c^2\sum_{l,n,k',k}\sum_{ij\sigma\tau}e^{-i(k+k')\cdot R}\,kk'(\delta_{ij}-\hat{k}_i\hat{k}_j)(\delta_{\sigma\tau}-\hat{k}_\sigma\hat{k}_\tau)$$

$$\langle p_a^i\rangle_{0l}\langle p_a^\sigma\rangle_{l0}\langle p_b^j\rangle_{0n}\langle p_b^\tau\rangle_{n0}\,\frac{1}{\Delta E_a^l+\hbar ck}\,\frac{1}{\hbar c(k+k')}\,\frac{1}{\Delta E_b^n+\hbar ck'}. \tag{17}$$

If the volume containing the radiation field is large we can replace the summation over $\boldsymbol{k}$ and $\boldsymbol{k}'$ by an integration with the weight factor $k^2\,dk\,d\Omega_k(k')^2\,dk'\,d\Omega_{k'}/(2\pi)^6$. The integrations over angles yield the following component of a matrix $\mathcal{T}$ (McLone and Power, 1965):

$$\frac{1}{4\pi}\int e^{-i(k+k')\cdot R}\,(\delta_{ij}-\hat{k}_i\hat{k}_j)\,d\Omega_k \equiv \mathcal{T}_{ij}(kR)$$
$$=\left\{\alpha_{ij}\frac{\sin kR}{kR}-\beta_{ij}\left[\frac{\cos kR}{k^2R^2}-\frac{\sin kR}{k^3R^3}\right]\right\}, \tag{18}$$

where

$$\alpha_{ij} = (\delta_{ij}-|\hat{R}_i||\hat{R}_j|)$$
$$\beta_{ij} = (\delta_{ij}-3|\hat{R}_i||\hat{R}_j|).$$

Finally, therefore,

$$(\mathrm{I}) = -\frac{\hbar^2c^2}{\pi^2}\sum_{l,n}\sum_{i,j,s,t}\langle p_a^i\rangle_{0l}\langle p_a^s\rangle_{l0}\langle p_b^j\rangle_{0n}\langle p_b^t\rangle_{n0}$$
$$\int_0^\infty\int_0^\infty k^3(k')^3\mathcal{T}_{ij}(kR)\mathcal{T}_{st}(k'R)D(I)dk\,dk', \tag{19}$$

where

$$D(I) = \frac{1}{\hbar c(\Delta E_a^l+\hbar ck)(k+k')(\Delta E_b^n+\hbar ck')}. \tag{20}$$

In a similar way the other six diagrams are seen to yield an identical result except for a different factor D, since only the order and form of the intermediate states differ from diagram to diagram. From the other diagrams in Fig. 11 we can write by inspection

$$D(\mathrm{II}) = \frac{1}{(\Delta E_a^l+\hbar ck)(\Delta E_a^l+\Delta E_b^n)(\Delta E_b^n+\hbar ck')}, \tag{21}$$

$$D(\mathrm{III}) = \frac{1}{(\Delta E_a^l+\hbar ck)(\Delta E_a^l+\Delta E_b^n)(\Delta E_a^l+\hbar ck')}, \tag{22}$$

$$D(\mathrm{IV}) = \frac{1}{\hbar c(\Delta E_a^l+\hbar ck)(k+k')(\Delta E_b^n+\hbar ck)}, \tag{23}$$

$$D(\mathrm{V}) = \frac{1}{(\Delta E_a^l+\hbar ck)(\Delta E_a^l+\Delta E_b^n+\hbar ck+\hbar ck')(\Delta E_b^n+\hbar ck)}, \tag{24}$$

$$D(\mathrm{VI}) = \frac{1}{(\Delta E_b^n+\hbar ck)(\Delta E_a^l+\Delta E_b^n+\hbar ck+\hbar ck')(\Delta E_b^n+\hbar ck)}. \tag{25}$$

From this point on we shall assume that the two atoms are identical and that only one term in the l and n summations is important. Furthermore we assume that $\Delta E_a^l = \Delta E_b^n = E$, thereby greatly simplifying the mathematics.

In the interaction of ground state atoms only those terms survive in which $s = i$ and $t = j$. This, coupled with our previous assumption that only one excited state of each atom needs to be considered, permits us to write in somewhat simplified notation

$$\langle p^i \rangle_{0l} \langle p^s \rangle_{l0} = |\langle p^i \rangle_{0l}|^2 \, \delta_{is} . \tag{26}$$

The total interaction is the sum of the six terms calculated above, multiplied by a factor of two since we also have the diagrams corresponding to a reflection of each one listed in Fig. 11 ($a \to b, b \to a$). Thus

$$\begin{aligned}\Delta E = &-\frac{2\hbar^2 c^2}{\pi} \sum_{ij} |\langle p_a^i \rangle_{0l}|^2 |\langle p_b^j \rangle_{0n}|^2 \int_0^\infty \int_0^\infty k^3 (k')^3 \, dk \, dk' \\ &\times \frac{\mathcal{T}_{ij}(kR) \mathcal{T}_{ij}(k'R)}{(E+\hbar ck)} \left[\frac{1}{\hbar c(k+k')(E+c\hbar k')} + \frac{2}{2E(E+\hbar ck')} \right. \\ &+ \frac{1}{\hbar c(k+k')(E+\hbar ck)} + \frac{1}{[2E+\hbar c(k+k')][E+\hbar ck]} \\ &\left. + \frac{1}{[2E+\hbar c(k+k')]\,[E+\hbar ck]} \right]\end{aligned} \tag{27}$$

$$\begin{aligned}= &-\frac{2\hbar^2 c^2}{\pi^2} \sum_{ij} |\langle p_a^i \rangle_{0l}|^2 |\langle p_b^j \rangle_{0n}|^2 \int_0^\infty \int_0^\infty \frac{k^3 (k')^3 \, dk \, dk'}{E+\hbar ck} \mathcal{T}_{ij}(kR) \mathcal{T}_{ij}(k'R) \\ &\left[\frac{1}{\hbar c(k+k')(E+\hbar ck')} + \frac{1}{E(E+\hbar ck')} + \frac{1}{(E+\hbar ck)(E+\hbar ck')} \right. \\ &\left. + \frac{1}{\hbar c(k+k')(E+\hbar ck)} \right].\end{aligned} \tag{28}$$

To simplify the remaining integrations we convert one of the integrations from 0 to ∞ to $-\infty$ to ∞. For example:

$$\begin{aligned}&\frac{1}{(E+k\hbar c)} \left\{ \frac{1}{(E+\hbar ck')\hbar c(k+k')} + \frac{1}{2E(E+\hbar ck')} \right\} \\ &= \frac{2E+\hbar c(k+k')}{(E+\hbar ck')(E+\hbar ck)\hbar c(k+k')(2E)} \\ &= \frac{1}{(E+\hbar ck)\hbar c(k+k')2E} + \frac{1}{(E+\hbar ck')\hbar c(k+k')(2E)}\end{aligned} \tag{29}$$

We also write

$$\frac{1}{(E+\hbar ck')(2E)(E+\hbar ck)} = \frac{1}{2E(E+k)}\left[\frac{1}{\hbar c(k'-k)} - \frac{E+\hbar ck}{\hbar c(E+\hbar ck')(k'-k)}\right]. \tag{30}$$

Combining (29) and (30) we find

$$\frac{1}{2E(E+\hbar ck)\hbar c}\left[\frac{1}{k+k'}+\frac{1}{k'-k}\right]+\frac{1}{2E(E+\hbar ck)\hbar c}\left[\frac{1}{k+k'}-\frac{1}{k'-k}\right].$$

Since the integration in (27) is invariant to the interchange of k and k', we can write this as

$$\frac{1}{E(E+\hbar ck)\hbar c}\left[\frac{1}{k+k'}-\frac{1}{k-k'}\right]. \tag{31}$$

Likewise

$$\frac{1}{(E+\hbar ck)^2(E+\hbar ck')} = \frac{1}{(E+\hbar ck)^2\hbar c}\left[\frac{1}{k'+k}-\frac{(E+\hbar ck)}{(E+\hbar ck)(k'-k)}\right]$$

$$= \frac{1}{\hbar c(E+\hbar ck)^2(k'+k)} - \frac{1}{\hbar c(E+\hbar ck)(E+\hbar ck')(k'-k)}. \tag{32}$$

Again, because of the invariance of the result under the substitution of k for k' and vice versa, the last term in (32) vanishes. Combining the first term with the remaining term in (28) we find

$$\frac{1}{(E+\hbar ck)^2\hbar c}\left[\frac{1}{k+k'}-\frac{1}{k-k'}\right]. \tag{33}$$

Therefore, using (33) and (31) we are able to reduce (28) to the form

$$\Delta E = \frac{2\hbar c}{\pi^2}\sum_{ij}|\langle p_a^i\rangle_{0l}|^2\int_0^\infty dk'\int_0^\infty dk k^3(k')^3$$

$$\frac{\mathcal{T}_{ij}(kR)\mathcal{T}_{ij}(k'R)}{E(E+\hbar ck)^2}(2E+\hbar ck)\left[\frac{1}{k'+k}-\frac{1}{k'-k}\right]|\langle p_b^j\rangle_{0n}|^2 \tag{34}$$

$$= -\frac{2\hbar c}{\pi^2}\sum_{ij}|\langle p_a^i\rangle_{0l}|^2|\langle p_b^j\rangle_{0n}|^2\int_{-\infty}^\infty dk'\int_0^\infty dk(kk')^3$$

$$\frac{\mathcal{T}_{ij}(kR)\mathcal{T}_{ij}(k'R)}{E(E+\hbar ck)^2}(2E+\hbar ck)\left[\frac{1}{k'+k}\right]. \tag{35}$$

The k' integration yields†

$$\int_{-\infty}^{\infty} \frac{(k')^3 \mathcal{T}_{ij}(k'R)}{k'+k}\, dk' = \frac{\alpha_{ij}}{R} \int_{-\infty}^{\infty} \frac{(k')^2 \sin k'R}{k'+k}\, dk'$$

$$+\frac{\beta_{ij}}{R^2} \int_{-\infty}^{\infty} \frac{k' \cos k'R}{k'+k}\, dk' - \frac{\beta_{ij}}{R^3} \int_{-\infty}^{\infty} \frac{\sin k'R}{k'+k}\, dk' \tag{36}$$

$$= \alpha_{ij} \frac{\pi k^2}{R} \cos kR - \beta_{ij} \frac{\pi k}{R^2} \sin kR - \beta_{ij} \frac{\pi}{R^3} \cos kR. \tag{37}$$

Using (37) in (35) we obtain

$$\Delta E = -\frac{2\hbar c}{\pi^2} \sum_{ij} |\langle p_a^i \rangle_{0l}|^2 |\langle p_b^j \rangle_{0n}|^2 \pi \int_0^{\infty} \frac{(2E+\hbar ck)^2}{E(E+\hbar ck)}\, dk$$

$$\left\{ \frac{\alpha_{ij}\alpha_{ij}}{R^2} k^4 \sin kR \cos kR + \frac{\alpha_{ij}\beta_{ij}}{R^3} \left[k^3 \cos^2 kR - \frac{k^2}{R} \sin kR \cos kR \right] \right.$$

$$- \frac{\alpha_{ij}\beta_{ij}}{R^3} \left[k^3 \sin^3 kR + \frac{k^3}{R} \sin kR \cos kR \right]$$

$$+ \frac{\beta_{ij}\beta_{ij}}{R^4} \left[-k^2 \sin kR \cos kR - \frac{k}{R} (\cos^2 kR - \sin^2 kR) \right.$$

$$\left. \left. + \frac{1}{R^2} \sin kR \cos kR \right] \right\}. \tag{38}$$

Since the results are symmetric in i and j, we can reduce the last expression to

$$\Delta E = -\frac{\hbar c}{\pi} \sum_{ij} |\langle p_a^i \rangle_{0l}|^2 |\langle p_b^j \rangle_{0n}|^2 \int_0^{\infty} \frac{(2E+\hbar ck)}{E(E+\hbar ck)^2}\, dk$$

$$\left\{ \frac{\alpha_{ij}\alpha_{ij}}{R^2} k^4 \sin 2kR + \frac{2\alpha_{ij}\beta_{ij}}{R^3} \left[k^3 \cos 2kR - \frac{k^2}{R} \sin 2kR \right] \right.$$

$$\left. + \frac{\beta_{ij}\beta_{ij}}{R^4} \left[-k^2 \sin 2kR - \frac{2k}{R} \cos 2kR + \frac{1}{R^2} \sin 2kR \right] \right\}. \tag{39}$$

† The pole at $k = k'$ is to be treated as a principal value. There is no pole on the left-hand side of (30); individually the two terms on the right do have a pole at $k = k'$, but their sum has not. For a more thorough treatment of these poles and the imaginary components as they affect the lifetimes of the various states, see Philpott (1966).

From their definitions (18) we note that

$$\alpha_{ij} = \begin{pmatrix} 1 & 0 & 0 \\ 0 & 1 & 0 \\ 0 & 0 & 0 \end{pmatrix} \quad \text{and} \quad \beta_{ij} = \begin{pmatrix} 1 & 0 & 0 \\ 0 & 1 & 0 \\ 0 & 0 & -2 \end{pmatrix}$$

and obtain

$$\Delta E = -\frac{2\hbar c e^4}{\pi} |\langle z_a \rangle_{0l}|^2 |\langle z_b \rangle_{0n}|^2 \int_0^\infty \frac{(2E+\hbar ck)}{E(E+\hbar ck)^2} dk$$

$$\left\{\frac{k^4}{R^2} \sin 2kR + \frac{2}{R^3}\left(k^3 \cos 2kR - \frac{k^2}{R} \sin 2kR\right)\right.$$

$$\left. + \frac{3}{R^4}\left(-k^2 \sin 2kR - \frac{2k}{R} \cos 2kR + \frac{1}{R^2} \sin 2kR\right)\right\}. \tag{40}$$

All dipole integrals are now written in terms of one component, namely the z-component. This result is usually listed in another form (Power and Zienau, 1957a; Casimir and Polder, 1948; McLone and Power, 1965). Let $kR = x$, then

$$\Delta E = -\frac{2\hbar c e^4}{\pi ER} |\langle z_a \rangle_{0l}|^2 |\langle z_b \rangle_{0n}|^2 \frac{1}{2i} \int_0^\infty \frac{x^4[2(ER)+\hbar cx]}{(ER+\hbar cx)^2}$$

$$\times \left[e^{2ix}\left(1 - \frac{2}{ix} - \frac{5}{x^2} + \frac{6}{ix^3} + \frac{3}{x^4}\right) + \text{complex conjugate}\right] dx. \tag{41}$$

Introducing $u = \pm ix/R$, one value for each part of (41), and combining the terms we obtain the integral of Casimir and Polder, namely,

$$\Delta E = -\frac{4}{\pi} \frac{e^4 |\langle z_a \rangle_{0l}|^2 |\langle z_b \rangle_{0n}|^2 E^2}{\hbar^3 c^3 R^2} \int_0^\infty \frac{u^4 e^{-2uR}}{[(E/\hbar c)+u^2]^2}$$

$$\left(1 + \frac{2}{uR} + \frac{5}{u^2R^2} + \frac{6}{u^3R^3} + \frac{3}{u^4R^4}\right) du. \tag{42}$$

McLone and Power (1965) as well as Philpott (1966) and Meath and Hirschfelder (1966b) have expressed this result in terms of cosine and sine integrals.

Following Meath and Hirschfelder, we rewrite (42) as

$$\Delta E = -\frac{4}{\pi} \frac{e^4 |\langle z_a \rangle_{0l}|^2 |\langle z_b \rangle_{0n}|^2 m^2}{\hbar c R^3} X(m, m), \tag{43}$$

where

$$X(m, m) = I_4 + 2I_3 + 5I_2 + 6I_1 + 3I_0 \tag{44}$$

and

$$I_n = I_n(m, d) = \int_0^\infty \frac{e^{-2y}y^n\,dy}{(m^2+y^2)(d^2+y^2)}. \tag{45}$$

m and d refer to the two different atoms and are defined by

$$m = \Delta E_a^l R/\hbar c,$$
$$d = \Delta E_b^n R/\hbar c. \tag{46}$$

One may write

$$I_n = \frac{1}{d^2-m^2}\{J_n(m)-J_n(d)\}, \tag{47}$$

where

$$J_n(m) = \int_0^\infty \frac{y^n e^{-2y}\,dy}{(m^2+y^2)}. \tag{48}$$

If $m = d$, as we have assumed above, one has a similar relationship:

$$I_n(m, m) = -\frac{1}{2m}\frac{\partial}{\partial m} J_n(m). \tag{49}$$

The J_n can be evaluated in terms of cosine and sine integrals (C_i and S_i), or better, their auxiliary functions $f(y)$ and $g(y)$ (Gautschi and Cahill, 1964, chap. 5).

Using the definitions

$$f(x) = C_i(x)\sin x-(S_i(x)-\pi/2)\cos x, \tag{50}$$
$$g(x) = -C_i(x)\cos x-(S_i(x)-\pi/2)\sin x, \tag{51}$$

we obtain in the general case of two different atoms

$$X(m, d) = \frac{1}{d^2-m^2}\left[f(2m)\left(m^3-5m+\frac{3}{m}\right)+g(2m)\,(6-2m^2)-\frac{m^2}{2}\right.$$
$$\left.-f(2d)\left(d^2-5d+\frac{3}{d}\right)-g(2d)(6-2d^2)+\frac{d^2}{2}\right]. \tag{52}$$

Likewise

$$X(a, a) = f(2a)\left[\frac{a}{2}-\frac{7}{2a}+\frac{3}{2a^3}\right]+g(2a)\left[a^2-3+\frac{3}{a^2}\right]-1/2+3/a^2. \tag{53}$$

It is also possible to express (40) directly in terms of sine and cosine integrals† using formulas such as

$$\int_0^\infty \frac{\sin 2kR}{m+k}\,dk = \sin(2mR)C_i(2mR) - \cos(2mR)S_i(2mR). \tag{54}$$

With these various representations for the integrals it is easy to find the behavior of the interaction at large and small R.

CASE 1. $\quad m < 1,\ d < 1,\ R < \bar{\lambda} = \hbar c/E$

$$\Delta E(R \to 0) = U_6 + U_4 + U_3 + U_2. \tag{55}$$

In this case, since R is small, we can neglect the exponent in (42). Thus

$$U_6 = -\frac{4}{\pi}\frac{3e^4|\langle z_a\rangle_{0l}|^2|\langle z_b\rangle_{0n}|^2}{\hbar^3c^3R^6}E^2\int\frac{du}{((E/\hbar c)^2+u^2)^2}$$

$$= -\frac{3e^4|\langle z_a\rangle_{0l}|^2|\langle z_b\rangle_{0n}|^2}{ER^6} \tag{56}$$

or, if atoms a and b are different,

$$U_6 = -\frac{6e^4|\langle z_a\rangle_{0l}|^2|\langle z_b\rangle_{0n}|^2}{(\bar{E}_a+\bar{E}_b)R^6}. \tag{57}$$

Similarly we find

$$U_4 = \frac{2e^4}{R^4\hbar^2c^2}|\langle z_a\rangle_{0l}|^2|\langle z_b\rangle_{0n}|^2\frac{\bar{E}_a\bar{E}_b}{(\bar{E}_a+\bar{E}_b)} \tag{58}$$

and

$$U_3 = -\frac{14}{3\pi}\frac{e^4}{R^3\hbar^3c^3}|\langle z_a\rangle_{0l}|^2|\langle z_b\rangle_{0n}|^2\bar{E}_a\bar{E}_b. \tag{59}$$

If we use the corresponding approximation of energy denominators in calculating the polarizability, $P_a = 2e^2|\langle z_a\rangle_{0l}|^2/\bar{E}_a$, we obtain the London formula discussed in Chapter 2 for U_6 as well as

$$U_4 = \frac{1}{2R^4\hbar^2c^2}\frac{(\bar{E}_a\bar{E}_b)}{(\bar{E}_a+\bar{E}_b)}P_aP_b. \tag{60}$$

† McLone and Power (1965). $\bar{E}_a$ and $\bar{E}_b$ are average energy expressions for the two atoms.

and

$$U_3 = -\frac{7}{6\pi}\,\frac{1}{\hbar^3 c^3 R^3}\,(\bar{E}_a)^2(\bar{E}_b)^2 P_a P_b \tag{61}$$

$$= -\frac{7}{6\pi R^3}\,\frac{n_a n_b}{\hbar^3 c^3}, \tag{62}$$

where n_a and n_b are the effective numbers of electrons in atom a and b, best obtained from the sum rule discussed in Chapter 2.

U_4 and U_3 are relativistic corrections to the R^{-6} energy at small separations. However, U_4/U_6 is of the order of $\alpha^2 R^2$ or $(R/137)^2$, α being the fine structure constant. Thus this contribution is important only for large R and is uninteresting in the small separation limit which we are at present considering.

For example, for $m = 0.3$, ΔE as determined from (55) differs from U_6 by only 2%. In fact only for $m > 0.45$ does this difference exceed 5%. $U_6 + U_4$ depart from ΔE by only 5% at $m = 0.56$.

CASE 2. $\quad m > 1,\ d > 1,\ R > \lambda\!\!\!^{-} = \hbar c/E$.

In this case we can neglect u^2 against m^2 to obtain

$$\Delta E(R \to \infty) = U_7 + U_9 + U_{11} + \ldots . \tag{63}$$

The first term of the series is obtained from (42) if, in the integral appearing there, we neglect u^2 against $(E^2/\hbar c)$. Hence

$$U_7 = -\frac{4}{\pi}\, e^4 \frac{|\langle z_a\rangle_{0l}|^2 |\langle z_b\rangle_{0n}|^2}{E^2 \qquad R^2}\hbar c$$

$$\int_0^\infty u^4 e^{-2uR}\left(1 + \frac{2}{uR} + \frac{5}{u^2R^2} + \frac{6}{u^3R^3} + \frac{3}{u^4R^4}\right) du, \tag{64}$$

$$= -\frac{23\, e^4 |\langle z_a\rangle_{0l}|^2 |\langle z_b\rangle_{0n}|^2 \hbar c}{4\pi R^7 E^2}, \tag{65}$$

$$= -\frac{23 P_a P_b}{4\pi R^7}\,\hbar c. \tag{66}$$

To be sure, we have used an approximate form of the polarizability to obtain (66), but we will later see that the result is independent of this assumption.

Likewise we find

$$U_9 = \frac{129}{8\pi\hbar^3 c^3 R^9}\,\frac{P_a P_b[(\bar{E}_a)^2 + (\bar{E}_b)^2]}{(\bar{E}_a)^2(\bar{E}_b)^2} \tag{67}$$

The ratio of U_9 to U_7 behaves as $(\alpha R)^{-2}$, but in this case R must be very large indeed for this conclusion to be meaningful. Numerical results indicate that U_7 differs from ΔE [Eq. (63)] by less than 5% when $m > 10$. For most cases of interest U_9 can be neglected.

The U_7 term can be calculated easily by realizing that for large $R (R \gg \lambda\!\!\!^{-} = \hbar c/E)$ the large contributions come from the regions where k and $k' \lll E/\hbar c$, since in each energy term an exponential factor appears. Also at large separations simultaneity of interactions no longer holds. This suggests that only those diagrams in Fig. 11 in which *no* simultaneous two-atom excitations appear should be considered, namely Fig. 11 (I) and (IV). Their contribution is

$$-\frac{4\hbar c}{\pi^2}\sum_{ij}\frac{|\langle p_a^i\rangle_{0l}|^2|\langle p_b^j\rangle_{0n}|^2}{E^2}\int_0^\infty\int_0^\infty k^3(k')^3dkdk' \frac{\mathcal{T}_{ij}(kR)\,\mathcal{T}_{ij}(k'R)}{k+k'} \tag{68}$$

$$= -\frac{4\hbar c}{\pi^2}\sum_{ij}\frac{|\langle p_a^i\rangle_{0l}|^2|\langle p_b^j\rangle_{0n}|^2}{E^2}\int_0^\infty dk\left\{\frac{\alpha_{ij}\alpha_{ij}}{R^2}k^4\sin 2kR + \frac{2\alpha_{ij}\beta_{ij}}{R^3}\left[k^3\cos 2kR - \frac{k^2}{R}\sin 2kR\right] + \frac{\beta_{ij}\beta_{ij}}{R^4}\left[-k^2\sin 2kR - \frac{2k}{R}\cos 2kR + \frac{1}{R^2}\sin 2kR\right]\right\} \tag{69}$$

and this, when summed over the magnetic quantum numbers with use of the matrix elements of α_{ij} and β_{ij}, leads to the result

$$-\frac{4}{\pi}\frac{e^4|\langle z_a\rangle_{0l}|^2|\langle z_b\rangle_{0n}|^2}{E^2R^2}\hbar c\int_0^\infty u^4e^{-2uR}\left(1+\frac{2}{uR}+\frac{5}{u^2R^2}+\frac{6}{u^3R^3}+\frac{3}{u^4R^4}\right)du \tag{70}$$

which is identically U_7.†

In a similar way it may be shown that the other diagrams in Fig. 11 contribute terms of order $E^{-1}R^8$ for large R.

The general solution for the interaction of two isotropic molecules can be written in the form

$$\Delta E = -G(R)\left[\frac{3}{4}\frac{EP_a^2}{R^6}\right]. \tag{71}$$

† This was first shown by Power (1965, pp. 109–11).

In Fig. 12 we give a graph of G versus R in units of the wave vector of energy E, i.e. $R/\lambda\!\!\!^{-} = RE/\hbar c$. For large separations $G = 23\hbar c/3\pi ER$ or $23\lambda\!\!\!^{-}/(3\pi R)$.

For small R a series expansion of K can be obtained from the work of Meath and Hirschfelder (1966b):

$$G(R) = 1 - \frac{E^2R^2}{3\hbar^2c^2} + \dots \quad (R \ll \lambda\!\!\!^{-}) . \tag{72}$$

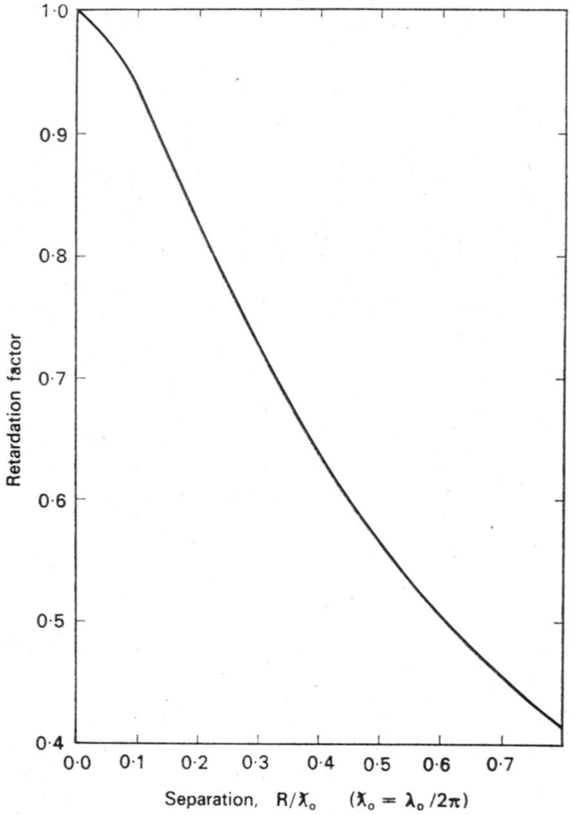

FIG. 12. Retardation factor for identical molecules. This factor should be used in eq. (71) to estimate retardation effects. The separation is measured in units of λ_0, the wavelength of the most important electronic transition.

To exhibit the magnitude of these effects we consider the interaction of two argon atoms where $E = 34.2$ ev (the wavelength, $2\pi\lambda\!\!\!^{-}$, being about 362 Å). From the graph at a value of R equal to about 0.6 $\lambda\!\!\!^{-}$ the interaction is reduced to one half. Therefore for argon we can expect retardation effects in sizeable amounts for distances of a hundred angstroms or more. For most organic molecules $E \sim 10$–20 ev, so that the retardation effects are significantly less. However, large molecules such as polymers can be expected to show these effects appreciably.

It is important also to see what assumptions led to the long and short distance approximations. One finds that the deciding feature is the magnitude of the energy E in comparison with the photon energy $\hbar ck$.

This is to be expected, for E represents the excitation of the atom while k and k' label photon energies. If the characteristic frequency of the atom ν_0 ($\nu_0 = c/\lambda_0$, $h\nu_0 = E$) is small compared with the photon frequencies, which are measures of the rapidity of variation in the fields between atoms one and two, then the electrostatic case where one neglects retardation is correct. In this limit we neglect E versus $\hbar ck$ or kR versus unity.

In the alternative approximation (large R) the wave vector of the atom's characteristic frequency (ν_0) is large (the frequency is high) compared with that of the photons. In this case retardation is important. We may here neglect $\hbar ck$ and $\hbar ck'$ versus E. This means that simultaneous excitation of both atoms is unlikely since this requires more energy, i.e. Eq. (12) has a bigger denominator. Notice that this also means that only Fig. 11 (I) and (IV) contribute to the retarded result. Physically this means that retardation removes the simultaneity.

There have been other deviations of the results. Dzialoshinskii (1956, 1957) used the Lorentz gauge and Feynman's perturbation techniques in conjunction with S-matrix theory. The results are, of course, identical. A more thorough presentation of his method which includes magnetic dipole terms is given by Mavroyannis and Stephen (1962).

6.2. Relativistic Dispersion Forces

Corrections to the R^{-6} interaction of two neutral molecules occur when a relativistic Hamiltonian is used. Meath and Hirschfelder (1966a) have calculated this correction, finding a term which is proportional to the square of the fine structure constant α. Their work will now be summarized.

The relativistic Hamiltonian correct to order α^2 is

$$H = H_0 + \alpha^2 H_{\text{rel}} + V, \tag{73}$$

where H_0 is the unperturbed nonrelativistic Hamiltonian of the molecules and V represents the classical electrostatic interaction between them. H_{rel} is given by the generalized Breit–Pauli Hamiltonian†

$$H_{\text{rel}} = H_{LL} + H_{SS} + H_{LS} + H_P + H_D, \tag{74}$$

$$H_{LL} = -\frac{e^2}{2m^2c^2} \sum_{j>k} \left\{ \frac{1}{r_{ij}^3} [r_{ij}^2 \mathfrak{p}_j \cdot \mathfrak{p}_k + \boldsymbol{r}_{jk} \cdot (\boldsymbol{r}_{jk} \cdot \mathfrak{p}_j) \cdot \mathfrak{p}_k] \right\}, \tag{75}$$

$$H_{SS} = \frac{e^2\hbar^2}{m^2c^2} \sum_{k>j} \left\{ -\frac{8\pi}{3} (\boldsymbol{S}_j \cdot \boldsymbol{S}_k)\, \delta^{(3)}(\boldsymbol{r}_{jk}) + \frac{1}{r_{jk}^5} [r_{jk}^2\, \boldsymbol{S}_j \cdot \boldsymbol{S}_k - 3(\boldsymbol{S}_j \cdot \boldsymbol{r}_{jk})(\boldsymbol{S}_k \cdot \boldsymbol{r}_{jk})] \right\}, \tag{76}$$

$$H_{LS} = \frac{e^2\hbar^2}{2m^2c^2} \left\{ \sum_{j\beta} \frac{Z_\beta}{r_{j\beta}^3} (\boldsymbol{r}_{j\beta} \times \mathfrak{p}_j) \cdot \boldsymbol{S}_j - \sum_{i,\, j\neq i} \frac{1}{r_{jk}^3} [(\boldsymbol{r}_{jk} \cdot \mathfrak{p}_j) \cdot \boldsymbol{S}_j - 2(\boldsymbol{r}_{jk} \cdot \mathfrak{p}_k) \cdot \boldsymbol{S}_j] \right\}, \tag{77}$$

$$H_P = -\frac{1}{8m^3c^3} \sum_j \mathfrak{p}_j^4, \tag{78}$$

$$H_0 = \frac{\pi e^2\hbar}{m^2c^2} \sum_j \left[\sum_\beta Z_\beta\, \delta^{(3)}(\boldsymbol{r}_{j\beta}) - \sum_{k\neq j} \delta^{(3)}(\boldsymbol{r}_{jk}) \right] + \frac{ie^2\hbar}{4m^2c^2} \sum_j \left[\sum_\beta \frac{Z_\beta}{r_{j\beta}^3} (\boldsymbol{r}_{j\beta} \cdot \mathfrak{p}_j) - \sum_{k\neq j} \frac{1}{r_{jk}^3} (\boldsymbol{r}_{jk} \cdot \mathfrak{p}_j) \right]. \tag{79}$$

The nuclei are labeled $\beta(\beta = a, b)$ and the electrons i, j, and k. $\delta^{(3)}$ is a three-dimensional delta function.

As to the meaning of the various terms, H_{LL} is the classical electromagnetic coupling of electrons through the interaction of the

† See, for example, Bethe and Salpeter (p. 170 ff.) or Hirschfelder, Curtiss and Bird (1954, p. 1044).

magnetic fields created by their motion.† It correlates the orbital magnetic moments of the electrons.

H_{SS} is the interaction between spin magnetic moments of the electrons. Its second term has a well-known form which must, however be corrected at $r_{jk} = 0$ by the first, the Fermi contact term. The delta function appearing in it arises from the fact that the magnetic field is assumed constant within the nucleus. Other semi-classical arguments for its existence have also been proposed.

H_{LS} couples spin and orbital magnetic moments of the electrons. The first term of this expression connects spin and orbital magnetic moments of the same electron, while the second term represents coupling between spin of one electron and the orbital magnetic moments of the others.

H_P introduces changes in mass with velocity. H_0 results from the Dirac treatment and has no simple physical interpretation. Since we shall later use only the first-order corrections of H_{rel} the second term of H_0 will be neglected.

The total interaction is calculated through the first order in α^2 and the second order in V. The relativistic Hamiltonian of (74) is only correct through order α^2. The relativistic corrections can be obtained by use of a state function accurate to first order in V, i.e.

$$\Psi = \phi_0 + \phi_1 , \tag{80}$$

where

$$(H_0 - E_0)\phi_1 = -V\phi_0 \tag{81}$$

and

$$(H_0 - E_0)\phi_0 = 0. \tag{82}$$

The interaction energy is given by those terms in the following expression which depend on R:

$$\begin{aligned} &\langle \phi_0 + \phi_1 \,|\, V + \alpha^2 H_{\text{rel}} \,|\, \phi_0 + \phi_1 \rangle \\ &\quad = E_2 + 2\alpha^2 \langle \phi_0 \,|\, H_{\text{rel}} \,|\, \phi_1 \rangle + \alpha^2 \langle \phi_1 \,|\, H_{\text{rel}} \,|\, \phi_1 \rangle \end{aligned} \tag{83}$$

provided E_2 is the long-range second-order energy discussed in Chapter 2. We assume here that the atoms do not overlap and the multipole expansion is valid. ϕ_1 then varies as R^{-3}. The term $\langle \phi_0 | H_{SS} + H_{LS} + H_{\text{LL}} \,|\, \phi_0 \rangle$ is not always zero, as will be seen in section 6.4 of this chapter.

† A very readable account of the derivation of all of these effects is found in Hameka (1965, chap. 3). Classical dipole interactions are also considered by Jackson (1962, p. 411).

Meath (Meath and Hirschfelder, 1966a) has succeeded in obtaining a multipole expansion of all terms in H_{rel} which we merely state (since the algebra is extensive):

$$H_{rel} = \sum_{\sigma} \sum_{m=0}^{\infty} H_{\sigma, m}/R^m. \tag{84}$$

Here

$$H_{\sigma, 0} = H_{\sigma, 0}(\text{molecule}_a) + H_{\sigma, 0}(\text{molecule}_b)$$

and σ stands for the following sets of indices:

$$\sigma = \{LL, SS, LS, P, D\}$$

Also

$$\left.\begin{aligned} H_{SL,1} &= H_{SS,1} = H_{SS,2} = 0 \\ H_{p,m} &= H_{D,m} = 0 \quad m > 0. \end{aligned}\right\} \tag{85}$$

The leading terms in H_{rel} and $H_{LL,1}$ and $H_{LS,2}$ correspond to (orbital current)–(orbital current) and (orbital current)–(spin-dipole) interactions, respectively. The expressions for each term are given by Meath and Hirschfelder (1966a) in the form of irreducible spherical tensors.

After a lengthy calculation, one finds that the dominant terms are

$$\Delta E = -\frac{C_1}{R^6} + \frac{\alpha^2 W_4}{R^4} + \frac{\alpha^2 W_6}{R^6}. \tag{86}$$

W_4 is an (orbital current)–(electrostatic dipole) dispersion energy and W_6 arises from $H_{LL,3}$ and $H_{SS,3}$. The constant C_1 is the London coefficient previously calculated.

The R^{-4} term is easily derived. It can only be due to the leading terms in ϕ_1 and H_{LL}, i.e.

$$2\alpha^2 \langle \phi_1 | H_{LL,1} | \phi_0 \rangle. \tag{87}$$

To write $H_{LL,1}$ we expand r_{ij} for one electron on center $\boldsymbol{a}$ and the other on center $\boldsymbol{b}$. Keeping only the leading terms, we simply substitute $\boldsymbol{R}$ for $\boldsymbol{r}_{ij}$ to obtain

$$H_{LL,1} = -\frac{e^2}{2m^2c^2} \frac{1}{R} [\mathfrak{p}_a \cdot \mathfrak{p}_b + (\mathfrak{p}_a)_z (\mathfrak{p}_b)_z]. \tag{88}$$

where $\mathfrak{p}_a$ is the total linear momentum (current operator) of atom a and $(\mathfrak{p}_a)_z$ is the total z-component (taken along the internuclear axis $\boldsymbol{R}$).

ϕ_1 is the sum of three terms

$$\phi_1 = -\sum_{l,n} \frac{\left\langle 00 \left| \frac{e^2(x_a x_b + y_a y_b - 2z_a z_b)}{R^3} \right| ln \right\rangle}{\Delta E_a^l + \Delta E_b^n} \langle ln|, \tag{89}$$

where the term in the matrix element is the leading one in the multipole expansion. X_a is the total x-coordinate of all the electrons in atom a, etc.

Using (88) and (89) we find

$$W_4 = +\frac{e^4}{m^2c^2} \sum_{l,n} \frac{\langle 00 | x_a x_b + y_a y_b - 2z_a z_b | ln \rangle \langle ln}{\Delta E_a^l + \Delta E_b^n} | \mathfrak{p}_a \mathfrak{p}_b + (\mathfrak{p}_a)_z (\mathfrak{p}_b)_z | 00 \rangle. \tag{90}$$

Because of symmetry this is equivalent to

$$W_4 = \frac{-2e^2}{m^2c^2} \sum_{l,n} \frac{\langle 0 | z_a | l \rangle \langle l | (\mathfrak{p}_a)_z | 0 \rangle \langle 0 | z_b | n \rangle \langle n | (\mathfrak{p}_b)_z | 0 \rangle}{\Delta E_a^l + \Delta E_b^n} \tag{91}$$

Matrix elements of z_a and $(\mathfrak{p}_a)_z$ are related by

$$H_0 z_a - z_a H_0 = \frac{\hbar}{m} (\mathfrak{p}_a)_z. \tag{92}$$

This relation leads immediately to (58) if one assumes that only one l and one n value contribute. If, in addition, we employ the definition of the oscillator strength from eqn. (24) of Chapter 2, (91) or (58) can be written in the form

$$W_4 = \frac{1}{2} \sum_l \sum_n \frac{f_{l,0} f_{n,0}}{\Delta E_a^l + \Delta E_b^n} \tag{93}$$

provided $f_{l,0}$ is the oscillator strength for the transition from the ground state to state l in molecule a while $f_{n,0}$ is the value for the $0 \to n$ transition in molecule b.

The primary contributions to W_6 come from H_{LL}. Estimates by Meath and Hirschfelder (1966a) indicate that W_6 for most atoms is comparable with C_1, and thus $\alpha^2 W_6$ is almost 5×10^{-5} times as strong as the dispersion interaction. More accurate calculations on helium suggest a value ten times as large, but still negligible compared with C_1.

The contributions from W_4 are not always negligible. Various estimates of (W_4/C_1) range from 0.2 to about 1.0. Assuming $W_4/C_6 = 1$, this effect would be

$$-\frac{C_1}{R^6}\left(1-\left(\frac{R}{137}\right)^2\right). \tag{94}$$

The actual correction term is probably smaller. In using (94) one must also remember that for large separations retardation effects are important. Equation (94) is valid only if R is less than the wavelength of radiation corresponding to the energy of ΔE_a^l for ΔE_b^n.

6.3. Semi-classical Calculations of Retardation Effects

Finally we review an approach to the calculation of the interaction of systems at large R which is suggested by classical physics. We have considered two such cases previously. The first is the common one where radiation fields never enter at all ($k = 0$) and the second deals with large distances where we neglected k and k' versus E. This, in effect, partially uncouples the radiation problem from the quantum mechanical problem of the atom itself. In this section we outline a method of computation which satisfies more closely the demands of physical intuition at all separations.

McLachlan (1963a) has shown that ΔE can be written as an integral over susceptibilities at imaginary frequencies. We have seen examples of this method in Chapters 2 and 5. Susceptibilities which contribute the important part of the long-range interactions are those which yield a dipole moment when the atom is acted upon by an external field. The simplest response to such an action is the tensor polarizability $\overleftrightarrow{\boldsymbol{P}}$ which yields the dipole moment

$$\boldsymbol{p} = \overleftrightarrow{\boldsymbol{P}}\cdot\boldsymbol{F}. \tag{95}$$

in an external field of strength $\boldsymbol{F}$.

According to McLachlan

$$\Delta E = \frac{\hbar}{2\pi}\int_0^\infty \alpha(u)\,\beta(u)\,du, \tag{96}$$

where α and β are the two polarizabilities evaluated at the complex frequency u. In our system of two molecules, one is acted upon not by an external field but by a field due to the other molecule. This field is induced by the first molecule. To avoid confusion we shall again label the two atoms a and b; $\boldsymbol{P}_a$ is the polarizability of atom a. The other polarizability which enters is the change in susceptibility of the system due to the presence of atom b:

$$\overleftrightarrow{\boldsymbol{B}} = \overleftrightarrow{\boldsymbol{T}}_{ab} \cdot \overleftrightarrow{\boldsymbol{P}}_b \cdot \overleftrightarrow{\boldsymbol{T}}_{ba} \tag{97}$$

since the field from a dipole is $\overleftrightarrow{\boldsymbol{T}}_{ba} \cdot \overleftrightarrow{\boldsymbol{P}}_a$ (see Chapters 2 and 5 for more details regarding this tensor). Equation (97) means that a field from a, represented by $\overleftrightarrow{\boldsymbol{T}}_{ab}$ creates a dipole moment $\overleftrightarrow{\boldsymbol{T}}_{ab} \cdot \overleftrightarrow{\boldsymbol{P}}_b$, at atom b, which then interacts with atom a via the field of the induced moment in b. As an indication that $\overleftrightarrow{\boldsymbol{B}}$ is a polarizability we notice that $\overleftrightarrow{\boldsymbol{B}} \cdot \boldsymbol{F}$ is the change in dipole moment of atom a in an external field due to the presence of atom b.

Substituting these susceptibilities into (96) we obtain the interaction

$$\Delta E = -\frac{\hbar}{2\pi} \int_0^\infty \overleftrightarrow{\boldsymbol{P}}_a(u) \cdot \overleftrightarrow{\boldsymbol{T}}_{ab}(u) \cdot \overleftrightarrow{\boldsymbol{T}}_{ab}(u) \cdot \overleftrightarrow{\boldsymbol{P}}_b(u)\, du. \tag{98}$$

This result can be derived from response functions by the method sketched here (see McLachlan for more details) or by time dependent Hartree theory (McLachlan, Gregory, and Ball, 1964). In the latter approach the interaction appears as the change in zero-point energy of the system of two oscillators as they are brought together. There are also higher terms arising from multiple scattering of the fields ($\overleftrightarrow{\boldsymbol{T}} \cdot \overleftrightarrow{\boldsymbol{P}} \cdot \overleftrightarrow{\boldsymbol{T}} \cdot \overleftrightarrow{\boldsymbol{P}} \cdot \overleftrightarrow{\boldsymbol{T}}$ etc) but these go to zero at large distances as R^{-12}.†

For spherical molecules (isotropic $\overleftrightarrow{\boldsymbol{P}}$), (98) becomes

$$\Delta E = -\frac{1}{2\pi} \int_0^\infty P_a(u) P_b(u)\, \mathrm{Tr}[\overleftrightarrow{\boldsymbol{T}}_{ab}(u) \cdot \overleftrightarrow{\boldsymbol{T}}_{ba}(u)]\, du, \tag{99}$$

where $P(u)$ is the isotropic polarizability used extensively in this book. It is evaluated at the imaginary frequency ($u = i\hbar ck$). The trace Tr must be taken over the tensor product.

The fields T_{ab} are those of an oscillating dipole and are given by

† When more than two atoms are considered the many-body effects discussed in Chapter 5 are easily found. In fact, an expression which includes the effects of all many-body dipole–dipole interactions has been derived by McLachlan. Three-body terms are discussed explicitly (McLachlan, 1963b); the general many-body results are treated by McLachlan, Gregory, and Ball (1963). See also Linder (1962) and Linder and Hoernschemeyer (1964).

radiation theory (see, for example, Panofsky and Phillips, 1955, chap. 13):

$$\overleftrightarrow{T} = [\nabla_b \nabla_a - \delta_{ab} \nabla^2] \frac{e^{-ik\cdot(r_a - r_b)}}{|r_b - r_a|}. \tag{100}$$

The components of this tensor are, if we write $\omega = ik$,

$$(\overleftrightarrow{T}_{ab})_{zz} = \left(\frac{2}{R^3}\right)(1+\omega R)e^{-\omega R}$$

$$(\overleftrightarrow{T}_{ab})_{xx} = (\overleftrightarrow{T}_{ab})_{yy} = -\left(\frac{1}{R^3}\right)(1+\omega R+\omega^2 R^2)e^{-\omega R}. \tag{101}$$

The zero frequency form used in Chapters 2 and 5 results from the fact that

$$\overleftrightarrow{T}_{ab}(0) = \frac{\overleftrightarrow{T} - 3\hat{R}\hat{R}}{R^3}, \tag{102}$$

where $\overleftrightarrow{T}$ is the unit tensor and $\hat{R}$ is a unit vector along the internuclear axis.

To relate this to results obtained earlier in this chapter, we note that

$$\overleftrightarrow{T}_{ab} = 2k^2 \mathcal{T}_{ab} \tag{103}$$

where $\mathcal{T}_{ab}$ is defined in (18).

By employing the form of the polarizability given in eqn. (60) of Chapter 2 for $P(u)$ and the present Eq. (101) it is a simple matter to prove that (99) is equivalent to (42), the result derived by considering the radiation field in detail.

In specific cases the connection between different formulas is easily seen. Using the static dipole field (102), we obtain the London result. For isotropic molecules

$$\Delta E = -\frac{1}{2\pi}\int_0^\infty P_a(u)P_b(u)\, du \operatorname{Tr}[\overleftrightarrow{T}_{ab}(0)\cdot\overleftrightarrow{T}_{ba}(0)]$$

$$= -\frac{3}{\pi R^6}\int_0^\infty P_a(u)P_b(u)\, du. \tag{104}$$

This is a formula derived in Chapter 2 by a different route.

In the long-range limit we neglect the energy of the photon relative to the excitation energy of the atom, that is, we consider only the low frequency part of the radiation field and obtain exactly

$$\Delta E = -\frac{\hbar c}{2\pi} P_a(0)P_b(0) \int_0^\infty \mathrm{Tr}[\boldsymbol{T}_{ab}(\omega)\cdot\boldsymbol{T}_{ba}(\omega)]\, du \tag{105}$$

$$= -\frac{\hbar c}{2\pi} P_a(0)P_b(0) \int_0^\infty [(T_{ba})_{zz}(T_{ba})_{zz}+2(T_{ab})_{xx}(T_{ab})_{xx})]\, d\omega$$

$$= -\frac{23P_a(0)P_b(0)}{4\pi R^7}\hbar c. \tag{106}$$

This is again the Casimir–Polder result, (66). The polarizability at zero frequency, the static polarizability, enters naturally without any need to adopt a mean excitation energy or some other approximation.

For nonisotropic molecules one must calculate each component separately. If the polarizability tensor is defined in terms of its components along the x, y, and z axes, P_{x_a} P_{y_a}, and P_{z_a}, the general interaction is

$$\Delta E = -\frac{1}{2\pi}\int_0^\infty [P_{x_a}(u)P_{x_b}(u)\,(T_{ab})_{xx}(T_{ba})_{xx}$$
$$+P_{y_a}(u)P_{y_b}(u)\,(T_{ab})_{xx}(T_{ba})_{xx}$$
$$+P_{z_a}(u)P_{z_b}(u)(T_{ab})_{zz}(T_{ab})_{zz}]\, du. \tag{107}$$

This leads to the following limiting expressions

$$\Delta E(R\to\infty) = -\frac{\hbar c}{2\pi}\frac{[13(P_{x_a}P_{x_b}+P_{y_a}P_{y_b})+20\,P_{z_a}P_{z_b}]}{R^7} \tag{108}$$

and

$$\Delta E = -\frac{1}{2\pi R^6}\int_0^\infty [P_{x_a}(u)P_{x_b}(u)+P_{y_a}(u)P_{y_b}(u)]\, du$$
$$-\frac{4}{2\pi R^6}\int_0^\infty P_{z_a}(u)P_{z_b}(u)\, du. \tag{109}$$

The last of these reduces to the simple result of Chapter 2 provided the mean excitation approximation is used and the corresponding energy is taken to be the same for all components of the polarizability:

$$\Delta E(R\to 0) = -\frac{(\bar{E}_a\bar{E}_b)}{(\bar{E}_a+\bar{E}_b)}\frac{P_{x_a}P_{x_b}+P_{y_a}P_{y_b}+4P_{z_a}P_{z_b}}{R^6}. \tag{110}$$

There is less dependence on relative orientations at large distances than at small, a consideration which could prove interesting in connection with radiation effects in crystals, i.e. excitons, metallic reflection, and so forth.

The approach just considered, for which we are indebted to McLachlan, is essentially classical, the only quantum mechanics occurring in the definition of the polarizability. Although we have only shown the equivalence of the results in the limiting cases, the present method can be used to derive the interaction at intermediate separations as well. The reason for the equivalence of our primarily classical argument with the quantum mechanical one is discussed by McLachlan (1963a). He also derives the polarizabilities quantum mechanically and establishes the identity with the classical approach given here.

So far we have studied only the interaction of two molecules and the effect of radiation upon the interaction. The problem of two interacting surfaces or other many-body systems is not simple. The multiple scattering terms neglected above must then be included since the electromagnetic waves scattered from the many molecules interfere and the effects of individual molecules are not additive. In Chapter 5 we discussed these interactions when radiation effects could be neglected. McLachlan (1966) has dealt with the problem extensively in an excellent review article. However, when treating the interaction of large bodies it is often simpler to consider fluctuations of the body as a whole, i.e. to introduce electrical susceptibilities of the entire body rather than a collection of smaller components. This has been done semi-classically by Lifshitz and his co-workers (Lifshitz, 1955; Dzialoshinskii, Lifshitz, and Pitaevskii, 1961; Abrikosov, Gorkov, and Dzialoshinskii, 1965, chap. 6; Landau and Lifshitz, 1960, chap. 13). Their approach has been further simplified by McLachlan (1963c).

If the effective polarizability of atoms a and b is P_a^* and P_b^* when in a solvent of dielectric constant $\varkappa$ the interaction energy becomes (McLachlan, 1966)

$$\Delta E(r) = -\frac{3}{\pi R^6}\int_0^\infty \frac{P_a^*(u)P_b^*(u)}{\varkappa^2(u)}\,du. \tag{111}$$

and

$$\Delta E(R\to\infty) = -\frac{23\hbar c}{4\pi R^7}\,\frac{P_a^*(0)P_b^*(0)}{\varkappa^{3/2}(0)}, \tag{112}$$

provided the liquid may be treated as a continuum. The effective polarizability can be defined experimentally by the change in dielectric constant $\varkappa$ with concentration, say N_A, of the atom in question,

$$P_a^* = \frac{1}{4\pi}\left(\frac{\partial\varkappa}{\partial N_A}\right). \tag{113}$$

In like manner the effect of temperature on the collective London interactions (McLachlan, 1963c; McLachan, Gregory, and Ball, 1963; Linder, 1962) has been calculated.

6.4. Spin and Magnetic Multipole Contributions

Two additional effects can contribute to intermolecular forces in special cases, both arising from magnetic interactions. They have appeared in various perturbation Hamiltonians, but we have thus far neglected them.

In discussing relativistic contributions to dispersion forces, it was said that

$$\alpha^2 \langle \phi_0 | H_{SS,3} + H_{SL,3} + H_{LL,3} | \phi_0 \rangle$$

is not always zero. Following Meath (1966) we now specify under what conditions these terms make appreciable contributions.

The atoms have magnetic moments:

$$M_a = -\frac{\alpha}{2}[L_a + 2S_a]$$

and

$$M_b = -\frac{\alpha}{2}[L_b + 2S_b],$$

where L_a, L_b are the total angular momenta, S_a, S_b the total spins of the two atoms. Semi-classically these should interact, yielding an energy

$$H'_{Sc} = \frac{1}{R^3}[M_a \cdot M_b - 3(M_a)_z(M_b)_z]. \tag{114}$$

From (75), (76), and (77) we see that this is also

$$H'_{Sc} = H_{LL,3} + H_{SS,3} + H_{SL,3} \tag{115}$$

and therefore any R^{-3} contributions to the interaction could be derived either semi-classically or relativistically.

The conditions under which these terms might contribute are summarized in Table 6.1.

TABLE 6.1. *Conditions for the Existence of an R^{-3} Interaction Energy due to Magnetic Dipole Interaction*

L_a	L_b	S_a	S_b	$\langle \phi_0 \mid H_{SS,3} \mid \phi_0 \rangle$	$\langle \phi_0 \mid H_{SL,3} \mid \phi_0 \rangle$	$\langle \phi_0 \mid H_{LL,3} \mid \phi_0 \rangle$
0	0	$\neq 0$	$\neq 0$	$\neq 0$	0	non-zero only if L_a and $L_b \neq 0$
$\neq 0$	0	$\neq 0$	$\neq 0$	$\neq 0$	$\neq 0$	
0	$\neq 0$	$\neq 0$	$\neq 0$	$\neq 0$	$\neq 0$	
$\neq 0$	0	0	$\neq 0$	0	$\neq 0$	
0	$\neq 0$	$\neq 0$	0	0	$\neq 0$	

For hydrogen atom interactions Meath (1966) finds

$$\Delta E({}'\Sigma_{m_s=0}) = \frac{0.40\alpha^2}{R^4} - \frac{6.50}{R_6} + \dots$$

$$\Delta E({}^3\Sigma_{m_s=0}) = \frac{\alpha^2}{R^3} + \frac{0.40\alpha^2}{R^4} - \frac{6.50}{R^6} + \dots$$

$$\Delta E({}^3\Sigma_{m_s=\pm 1}) = \frac{-\alpha^2}{2R^3} + \frac{0.40\alpha^2}{R^4} - \frac{6.50}{R^6} + \dots \qquad (116)$$

in atomic units.

The sum of all R^{-3} terms is zero. This follows from a general theorem proved by Meath (1966). For the triplet state with $M_s = 0$ the R^{-3} term is 10% of the dispersion interaction at $22a_0$. The R^{-3} interaction is not retarded at large separations.

Mavroyannis and Stephen (1962) have considered additional effects of the magnetic moments. As perturbation they used the equivalent of two terms of eqn. (C36) of Appendix C, namely

$$H_{\text{int}} = (-\boldsymbol{p}_a \cdot \mathfrak{E}^{\perp}(a) - \boldsymbol{p}_b \cdot \mathfrak{E}^{\perp}(b)) - (\boldsymbol{M}_a \cdot \boldsymbol{B}(a) + \boldsymbol{M}_b \cdot \boldsymbol{B}(b)). \qquad (117)$$

For small separations, one then finds three additional terms in the interaction energy,

$$U_2 = -\frac{4}{3R^6}\sum_{n,m}{}' \frac{(p_a)_{0n}\cdot(M_a)_{n0}(p_b)_{0m}\cdot(M_b)_{m0}}{(\Delta E_a^n+\Delta E_b^m)}$$

$$U_3 = \frac{2}{9R^4}\sum_{n\ m}{}' \frac{\Delta E_a^n \Delta E_b^m}{\hbar^2 c^2}\left[\frac{|(p_a)_{0n}|^2|(M_b)_{0m}|}{\Delta E_a^n+\Delta E_b^m} + \frac{|(p_b)_{0n}|^2|(M_a)_{0m}|^2}{\Delta E_a^n+\Delta E_b^m}\right]$$

and

$$U_4 = -\frac{2}{3R^6}\sum_{n,m}{}' \frac{|(M_a)_{0n}|^2|(M_b)_{0m}|^2}{\Delta E_a^n+\Delta E_b^m}.$$

Notice that U_3 varies as R^{-4}, but at any separation it is smaller than U_2 by a factor of about $(R/\lambda)^2$, where λ is the wavelength of some average transition in the molecule ($R \ll \lambda$). U_2 differs from zero only when the molecule has no center of inversion, i.e. when the molecule is optically active. This interaction is usually rather weak since the $\boldsymbol{M}$ matrix elements are generally smaller than the electric dipole elements by a factor of about l/λ, where l is the size of the molecule. If one could produce a molecule in the form of a right- or left-handed helix with conjugated double bonds, the ratio would become

$$\frac{|M_{0n}|}{|p|_{0n}} \simeq \frac{l\ \text{(radius of the helix)}}{\lambda\ \text{(length of the helix)}}$$

as may be seen by using the particle-in-a-box model for the electrons, as discussed in Chapter 7. Hence, this ratio can be made greater by increasing the radius relative to the length. Such an ideal situation is unlikely, and usually the sum of the U terms is negligible when compared with the dispersion interaction.

At large separations U_3 and U_4 behave as R^{-7}, while U_2 behaves as R^{-9}.

CHAPTER 7

Forces between Molecules

7.1. Introduction

Forces between molecular systems have already been encountered in earlier chapters; but there we were not interested in asymmetries, nor in any features which arise from molecular shapes. When computing, for instance, the dispersion forces between H_2 molecules we found them depending on the average polarizabilities, because we assumed the molecules to rotate rapidly enough to make such an average meaningful. The only departure from this procedure occurred in section 2.3 where attention was given to the effects of anisotropic polarizabilities. The time has now come to examine the shape-dependent forces between molecules more carefully and to study in particular the first-order repulsive interactions between them when they are not rotating, so that an average over all orientation can be more accurately computed.

A full quantum mechanical calculation of this sort has so far been performed only for the simplest system, the H_2 molecule. The last and most mature evaluation of the problem of two interacting H_2 molecules was published by Mason and Hirschfelder (1957b). Their work is based to some extent on concepts and formulas presented earlier by Margenau (1943)† and by Evett and Margenau (1953), who approximated certain three- and four-center exchange integrals (see below) in crude ways, whereas Mason and Hirschfelder used more refined methods

† This paper contains a computing error which led to an unreasonable choice of the screening constant Z for the wave function used. In the following reference the work was carried to a greater number of decimal places and the error became apparent and was corrected.

in the evaluation of these integrals. Earlier, in 1942, de Boer (1942) independently published a simpler calculation in which all these integrals were neglected; nevertheless, his results agreed fairly closely with data derived from experiments and also with those of the second-named calculation. Mason and Hirschfelder probed this matter and concluded that the contribution of multi-center exchange integrals is large, and that their cancellation in de Boer's work was fortuitous. For this reason we shall follow in the first part of our account the analysis of Margenau (1943) and Evett and Margenau (1953) which contains explicitly all the terms that appear in the H_2–H_2 problem.

7.2. Interaction between Hydrogen Molecules

The interaction is composed of three parts: exchange forces, long-range dispersion forces, and the forces due to the permanent quadrupoles carried by H_2 molecules.

7.2.1. Exchange Forces

A simple, manageable state function for H_2 is an antisymmetrized product of two hydrogen type functions but with a screening constant $Z = 1.116$ (Wang, 1928). It will here be employed in the four-center problem. Let the four protons in the two interacting molecules be labeled *a, b, c, d*. The symbols will also be used for hydrogen orbitals of an electron about these protons, so that $b(3)$ means

$$\left(\frac{Z}{\pi a_0^3}\right)^{1/2} \exp\left(\frac{-Zr_{b3}}{a_0}\right) = b(3) \tag{1}$$

with r_{b3} standing for the distance between proton b and electron number 3. This function satisfies the Schroedinger equation

$$\left(\frac{\hbar^2}{2m}\nabla_3^2+\frac{Ze^2}{r_{b3}}\right) b(3) = \frac{1}{2} Z^2 \frac{e^2}{a_0} b(3). \tag{2}$$

Spin functions α, β are introduced in the usual way, but we write $a(i)\alpha(i)$ simply as $a_+(i)$. The basic orbital encountered is

$$\varphi_1 = a_+(1)b_-(2)c_+(3)d_-(4) \tag{3}$$

and from it we construct the normalized Slater determinant

$$\psi_1 = \mathcal{A}\varphi_1 . \tag{4}$$

Numerous different assignments of spin functions to the coordinate functions a to d can be made, and these produce a considerable number of φ_i with corresponding ψ_i, constructed in conformity with (4). But only six of these are needed since we are interested only in singlet states of the four-atom system, i.e. in states for which the total z-component of spin, $\Sigma\sigma_z$, equals zero. The only functions of type (4) which can combine to give zero spin are

$$\begin{aligned} \psi_1 &= \mathcal{A}(a_+b_-c_+d_-) \\ \psi_2 &= \mathcal{A}(a_+b_+c_-d_-) \\ \psi_3 &= \mathcal{A}(a_+b_-c_-d_+) \\ \psi_4 &= \mathcal{A}(a_-b_+c_+d_-) \\ \psi_5 &= \mathcal{A}(a_-b_+c_-d_+) \\ \psi_6 &= \mathcal{A}(a_-b_-c_+d_+). \end{aligned} \tag{5}$$

From these the following two independent singlet functions can be formed:

$$\begin{aligned} \Psi_A &= \tfrac{1}{2}(\psi_1-\psi_3-\psi_4+\psi_5) \\ \Psi_B &= \tfrac{1}{2}(\psi_2-\psi_3-\psi_4+\psi_6). \end{aligned} \tag{6}$$

One sees this when the spin functions of Ψ_A and Ψ_B are factored out, for they combine to the characteristic $(\alpha\beta-\beta\alpha)$-combination of pairs. It will also be observed, through the same procedure, that Ψ_A represents bonds between atoms a and b, as well as c and d, while the bonds of Ψ_B join a to c, and b to d.†

Strictly, then, the interaction state under consideration is a linear combination of Ψ_A and Ψ_B, and the variational method leads to the secular equation

$$\begin{vmatrix} H_{AA}-E\Delta_{AA} & H_{AB}-E\Delta_{AB} \\ H_{AB}-E\Delta_{AB} & H_{BB}-E\Delta_{BB} \end{vmatrix} = 0. \tag{7}$$

But it is clear on physical grounds that, unless the bonds are broken, only one of the functions, Ψ_A and Ψ_B, can play the dominant role in our problem. For weak (thermal) impacts between molecules this can not happen; and our attention will be restricted to that situation. Analytically, neglect of one of the

† The construction of functions having desired symmetries and the concept of bond eigenfunctions are clearly set forth by Eyring, Walter, and Kimball (1944, chap. 13).

functions is justified by virtue of the smallness of H_{AB} and Δ_{AB} in comparison with H_{AA} and Δ_{AA}; the foregoing references [Margenau (1943) and Mason and Hirschfelder (1957b)] indicate in detail why this inequality holds at all distances of approach which are larger than the kinetic theory diameter. Equation (7) therefore reduces to

$$H_{AA}-E\Delta_{AA} = 0. \tag{8}$$

When the matrix elements H_{AA} and Δ_{AA} are expanded via (6) they become linear combinations of elements H_{ij} and Δ_{ij}, respectively; the subscripts i, j are meant to refer to the individual functions listed in (5). The evaluation of H_{ij} and Δ_{ij} proceeds as follows.

Because of the antisymmetry of all ψ and the symmetry of H,

$$H_{ij} = 4! \int \varphi_i H \mathcal{A} \varphi_j d\tau \tag{9}$$

and the integral includes a summation over spins. But an integral like $\int \varphi_i H P_\mu \varphi_j \, d\tau$, where P_μ is any permutation among the four electrons, can always be reduced to the form

$$I_k = \int P_k \varphi_1 H \varphi_1 d\tau,$$

φ_1 being defined in (3). Hence every H_{ij} is a simple sum of exchange integrals of the form I_k. These fall into five classes in accordance with the properties of the symmetric group on four elements. We shall enumerate and label them by reference to the permutation P_k which appears in I_k. But in the calculation all spin function disappear, for they either sum to 1 or cause the integral to vanish. Henceforth, then, we shall mean by I_k the following

$$I_k = \int P_k[a(1)b(2)c(3)d(4)]Ha(1)b(2)c(3)d(4)d\tau_1 d\tau_2 d\tau_3 d\tau_4. \tag{10}$$

The exchange integrals I_k are the essential ingredients of the subsequent analysis. Rather than carry through a cumbersome notation in which the subscript k is replaced by a specific permutation symbol we shall use the abbreviations listed in Table 7.1. Since for every I_k defined by (10) there also occurs an I'_k, a multiple overlap integral which differs from (10) by the absence of H, we include the names of the I'_k along with the I_k.

TABLE 7.1. *List of Symbols used for*
(1) $I_k = \int (P_k\varphi_1)H\varphi_1 d\tau$ and (2) $I'_k = \int (P_k\varphi_1)\varphi_1 d\tau$.
Symbol for integral is given opposite permutation P_k (which is written in the form of cycles).[a] Unprimed symbols refer to (10), primed symbols are defined by (10) with omission of H from the integral.

P_k	I_k	I'_k	P_k	I_k	I'_k	P_k	I_k	I'_k	P_k	I_k	I'_k	P_k	I_k	I'_k
$(a)(b)(c)(d)$	ε	ε'	(ab)	u_1	u_1'	(abc)	v_1	v_1'	$(ab)(cd)$	w_1	w_1'	$(abcd)$	z_1	z_1'
			(ac)	u_2	u_2'	(bac)	v_1	v_1'	$(ac)(bd)$	w_2	w_2'	$(dcba)$	z_1	z_1'
			(ad)	u_3	u_3'	(abd)	v_2	v_2'	$(ad)(bc)$	w_3	w_3'	$(bacd)$	z_2	z_2'
			(bc)	u_4	u_4'	(bad)	v_2	v_2'				$(abdc)$	z_2	z_2'
			(bd)	u_5	u_5'	(acd)	v_3	v_3'				$(acbd)$	z_3	z_3'
			(cd)	u_6	u_6'	(cad)	v_3	v_3'				$(dbca)$	z_3	z_3'
						(bcd)	v_4	v_4'						
						(cbd)	v_4	v_4'						

[a] The symbol $(abcd)$ means that a is to be replaced by b, b by c, c by d and d by a in whatever follows. When a single letter occurs in parentheses, no change in ordering is indicated: e.g. (bac) (d) symbolizes $(c1)$ $a(2)$ $b(3)$ $d(4)$.

It will be noticed that an integral belonging to a given permutation is equal to that of its reciprocal permutation; this follows from the Hermitean character of H. There are no further equalities unless the nuclei are disposed symmetrically in space.

The elements H_{ij}, constructed from the I_k via (9), are presented in Table 7.2.

TABLE 7.2. *Matrix Elements of H_{ij} in Terms of Exchange Integrals*

	1	2	3	4	5	6
1	$\varepsilon-u_2-u_5+w_2$	$-u_4+v_1+v_4-z_2$	$-u_6+v_4+v_3-z_3$	$-u_1+v_1+v_2-z_3$	$w_1-2z_1+w_3$	$-u_3+v_2+v$
2		$\varepsilon-u_1-u_6+w_1$	$-u_5+v_2+v_4-z_1$	$-u_2+v_1+v_3-z_1$	$-u_3+v_3+v_2-z_2$	w_2-2z_3+
3			$\varepsilon-u_3-u_4+w_3$	$w_1+w_2-2z_2$	$-u_1+v_2+v_1-z_3$	$-u_2+v_1+v$
4				$\varepsilon-u_3-u_4+w_3$	$-u_6+v_4+v_3-z_3$	$-u_5+v_2+v$
5					$\varepsilon-u_2-u_5+w_2$	$-u_4+v_1+v$
6						$\varepsilon-u_1-u_6$

When H_{AA} is compounded from these, there results

$$H_{AA} = \varepsilon+u_1+u_6-\tfrac{1}{2}(u_2+u_3+u_4+u_5)-(v_1+v_2+v_3+v_4) + w_1+w_2+w_3-(z_1+z_2-2z_3) \quad (11)$$

and Δ_{AA} is given by an identical combination of primed symbols. At this point it becomes necessary to introduce the explicit four-atom Hamiltonian, which will allow us to decompose the exchange integrals I_k further into *elementary* integrals corresponding to the individual terms of H.

Let us again use the abbreviations

$$\alpha_i = \frac{e^2}{r_{ai}}, \quad \beta_i = \frac{e^2}{r_{bi}}, \quad \text{etc.,} \quad \varrho_{ij} = \frac{e^2}{r_{ij}} \quad (12)$$

so that

$$H = -\frac{\hbar^2}{2m}\sum_1^4 \nabla_i^2-\sum_1^4(\alpha_i+\beta_i+\gamma_i+\delta_i)+\sum_{i>j}\varrho_{ij}+E_N,$$

provided E_N stands for the repulsive Coulomb energy between the four nuclei. In view of (1) and (2) we now obtain

$$Ha(1)b(2)c(3)d(4) = \Big\{-2\frac{Z^2e^2}{a_0} -(1-Z)(\alpha_1+\beta_2+\gamma_3+\delta_4)-(\alpha_2+\alpha_3+\alpha_4+\beta_1+\beta_3+\beta_4+\gamma_1+\gamma_2+\gamma_4 +\delta_1+\delta_2+\delta_3)+\sum_{i>j}\varrho_{ij}+E_N\Big\}a(1)b(2)c(3)d(4). \quad (13)$$

To label the elementary integrals in a self-revealing manner we write, as in Chapter 5,

$$(a\gamma d) \equiv \int a(1)\frac{e^2}{r_1}d(1)d\tau_1,$$

$$(ab\varrho cd) \equiv \int a(1)b(2)\frac{e^2}{r_{12}}c(1)d(2)d\tau_1 d\tau_2,$$

$$\Delta_{ab} \equiv (a|b) = \int a(1)b(1)d\tau_1, \quad \text{etc.}$$

The following symmetries have already been noted:

$$(a\gamma d) = (d\gamma a),$$

$$(a\alpha b) = (a\beta b),$$

$$(ab\varrho cd) = (cb\varrho ad) = (ad\varrho cb) = (cd\varrho ab) = (ba\varrho dc).$$

Thus a Roman letter may be shifted freely from one side of the Greek letter to the other, provided its ordinal place among the Roman letters is unchanged, but the order of the Roman letters on one side of a Greek letter may not be changed if that on the other remains unaltered.

The elementary exchange integrals fall into four categories in accordance with the number of nuclei whose positions affect their value. All *one-center integrals, like* $\{a\alpha a\}$, $\{b\beta b\}$, etc., are of course equal and independent of the position of the nuclei. Among the *two-center integrals* one may distinguish four different types, exemplified by the following specimens:

$$(a\beta a), \quad (a\beta b), \quad (ab\varrho ab), \quad (ab\varrho ba).$$

These all occur in the problem of the hydrogen molecule and are well known.

Three-center integrals fall into three classes, characterized by

$$(a\beta c), \quad (ac\varrho bc), \quad (ac\varrho cb).$$

Since they are of rather complicated structure and, in their exact forms, quite unsuggestive, it is useful to have a quick way of estimating their magnitude. The rough and ready procedure outlined here was employed in Margenau (1943). It gives surprisingly adequate results but is difficult to justify with rigor. We shall comment further on this point in due course. Consider, for example, $(a\beta c)$. The product function $a(1)c(1)$ is largest in the region halfway between the nuclei a and c; therefore, if b is far from a and c,

$$(a\beta c) \approx \Delta_{ac} e^2 / R_{b,ac},$$

where $R_{b,ac}$ is the distance from b to the midpoint between a and c. In a similar way,

$$(ac\varrho bc) \approx (a\gamma b) \approx \Delta_{ab} e^2 / R_{c,ab},$$

$$(ac\varrho cb) \approx \Delta_{ac}\Delta_{bc} e^2 / R_{ac,bc}. \tag{14}$$

These approximations are quite accurate when the distances R involved in the formulas are several times as large as the arguments of the $\triangle$ functions. The integrals do not, however, become infinite as $R \to 0$; they take simple limiting forms which are either two-center integrals or else very manageable expressions easily obtainable from earlier work (Gordadse, 1935; Hirschfelder, Eyring, and Rosen, 1936; Coulson, 1937).

The exact calculation of forms like $(ac\varrho bc)$ and $(ac\varrho cb)$ entails rather formidable labor. Fortunately, however, it is possible to avoid most of it by a simple reduction process now to be described.

If the function c^2 is contracted more and more about its nucleus, until finally it becomes a δ-function located at the nucleus, the integral $(ac\varrho bc)$ turns into $(a\gamma b)$. Alternately, if the function ab is approximated by a δ-function located midway between nuclei a and b, the same integral reduces to the two-center integral† $(c\overline{\alpha\beta}c)\Delta_{ab}$. When one distance is larger than the other—as is generally true in the interaction of two molecules—these independent reductions can always be made and lead to nearly equal numerical results. This we regard as a test of the validity of the reductions, and we take

$$(ac\varrho bc) \simeq \tfrac{1}{2}[(a\gamma b)+\Delta_{ab}(c\overline{\alpha\beta}c)].$$

Similar reasoning shows that

$$(ac\varrho cb) \simeq \tfrac{1}{2}[\Delta_{ac}(c\alpha\gamma b)+\Delta_{cb}(a\gamma\overline{\beta c})].$$

The remaining integrals are known. To give an example of the extent to which this procedure is reliable we consider the case in which the nuclei a, b, c, are collinear, b is between a and c and the distance $a-c$ is three times the distance $c-b$, the latter being the internuclear distance in H_2. The two terms in the above expansion of $(ac\varrho cb)$ then have the values 0.1448 and 0.1417 a.u. respectively. For larger distances $b-c$, the reduction works still better.

All *four-center exchange integrals* are of the form $(ab\varrho cd)$. Their exact calculation is likewise extremely tedious and requires numerical computation. But here again the scheme just described can be used. The reduction can be effected in two ways, and if the results do not differ beyond a tolerated limit of error, their mean is taken as the value of the four-center integral. Thus

$$(ab\varrho cd) \approx \tfrac{1}{2}[\Delta_{ac}(b\overline{\alpha\gamma}d)+\Delta_{bd}(a\overline{\beta\delta}c)].$$

When a and c belong to the same molecule, the two terms on the right become very nearly equal, and each approaches $\Delta_R^2(e^2/R)$ in the limit in which R, the distance between molecular centers, is large.

† $\overline{\alpha\beta}$ here stands for the reciprocal of the electron distance from the mid-point between a and b.

We omit the detailed formulas for the residual integrals, all of which are available in closed form. As far as our present purposes are concerned, the reader will be relieved to know that an even simpler scheme than the alternate reductions here studied, a scheme which will be described a little later, leads to significant results for the interaction energy of molecular hydrogen.

When the exchange integrals, ε, u, ..., z_3 are expressed in terms of their elements, certain combinations often occur. These will first be singled out. We define

$$\begin{aligned} A &= (a\alpha a) = Ze^2/a_0, \\ D &= -\tfrac{1}{2}Z^2e^2/a_0, \\ B_{ab} &= e^2/R_{ab} - (a\beta a), \\ C_{ab} &= e^2/R_{ab} + (ab\varrho ab) - 2(a\beta a), \\ T_{ab} &= (a\beta b)/\Delta_{ab}, \\ X_{ab} &= e^2/R_{ab} + (ab\varrho ba)/\Delta_{ab}^2 - 2T_{ab}. \end{aligned} \tag{15}$$

Here C_{ab} will be recognized as the "Coulomb energy," X_{ab} as the "exchange energy" between two hydrogen-like atoms at a and b. All these expressions are functions of only R_{ab}, the distance between a and b, and most of them vanish exponentially at large distances. In addition we need the following more complicated combinations:

$$X'_{ab} = 2[(ac\varrho bc) + (ad\varrho bd) - (a\gamma b) - (a\delta b)]/\Delta_{ab},$$

$$U_{abc} = \frac{e^2}{R_{ab}} + \frac{e^2}{R_{ac}} + \frac{e^2}{R_{bc}} - (b\alpha c)/\Delta_{bc} - (a\beta c)/\Delta_{ac} - (a\gamma b)/\Delta_{ab}$$
$$+ (ab\varrho bc)/\Delta_{ab}\Delta_{bc} + (ac\varrho ba)/\Delta_{ab}\Delta_{ac} + (bc\varrho ca)/\Delta_{bc}\Delta_{ac},$$

$$U'_{abc} = \frac{(ad\varrho bd) - (a\delta b)}{\Delta_{ab}} + \frac{(bd\varrho cd) - (b\delta c)}{\Delta_{bc}} + \frac{(ad\varrho cd) - (a\delta c)}{\Delta_{ac}},$$

$$V_{ab,cd} = \frac{e^2}{R_{ac}} + \frac{e^2}{R_{ad}} + \frac{e^2}{R_{bc}} + \frac{e^2}{R_{bd}} + 4\frac{(ac\varrho bd)}{\Delta_{ab}\Delta_{cd}}$$
$$-2[(a\gamma b) + (a\delta b)]/\Delta_{ab} - 2[(c\alpha d) + (c\beta d)]/\Delta_{cd},$$

$$W_{abcd} = E_N + \frac{(bc\varrho ab)}{\Delta_{ab}\Delta_{bc}} + \frac{(bd\varrho ac)}{\Delta_{ab}\Delta_{cd}} + \frac{(ba\varrho ad)}{\Delta_{ab}\Delta_{ad}} + \frac{(bc\varrho cd)}{\Delta_{bc}\Delta_{cd}} + \frac{(ca\varrho bd)}{\Delta_{bc}\Delta_{ad}}$$
$$+ \frac{(da\varrho cd)}{\Delta_{cd}\Delta_{ad}} - [(a\gamma b) + (a\delta b)]/\Delta_{ab} - [(b\alpha c) + (b\delta c)]/\Delta_{bc}$$
$$-[(c\alpha d) + (c\beta d)]/\Delta_{cd} - [(a\beta d) + (a\gamma d)]/\Delta_{ad}. \tag{16}$$

By permutation of subscripts, other functions can be constructed from this list. Not all of these are different; for instance it will be seen on inspection that there are four different U functions (U_{abc}, U_{abd}, U_{acd}, U_{bcd}), three different V functions ($V_{ab,cd}$, $V_{ac,bd}$, $V_{ad,bc}$) and three different W functions (W_{abcd}, W_{bacd}, W_{acbd}). Only X' and U' have the property of vanishing when a reduction of the type (11) is performed, and this may be taken as an indication that they are generally small. They disappear when the nuclei in each molecule coalesce.

The function $V_{ab,\,cd}$ has an interesting significance. When it is "reduced", it represents the electrostatic interaction between two linear quadrupoles, one consisting of protons at a and b with two electrons at the midpoint between them, the other of protons at c and d and two electrons at their midpoint. We shall show later that this is indeed its true significance, and that the value of V_{abcd} is well approximated by a quadrupole interaction of this sort.

With the use of these abbreviations, the exchange integrals take on symmetrical—even if somewhat complicated—forms, viz.:

$$\varepsilon = 4D+4(Z-1)A+C_{ab}+C_{ac}+C_{ad}+C_{bc}+C_{bd}+C_{cd},$$

$$u_1 = \Delta^2_{ab}[4D+2(Z-1)(A+T_{ab})+X_{ab}+B_{ac}+B_{ad}+B_{bc}+B_{bd}+C_{cd} +X'_{ab}],$$

$$u_2 = \Delta^2_{ac}[4D+2(Z-1)(A+T_{ac})+X_{ac}+B_{ab}+B_{ad}+B_{bc}+B_{cd}+C_{bd} +X'_{ac}],$$

$$u_3 = \Delta^2_{ad}[4D+2(Z-1)(A+T_{ad})+X_{ad}+B_{ab}+B_{ac}+B_{bd}+B_{cd}+C_{bc} +X'_{ad}],$$

$$u_4 = \Delta^2_{bc}[4D+2(Z-1)(A+T_{bc})+X_{bc}+B_{ab}+B_{ac}+B_{bd}+B_{cd}+C_{ad} +X'_{bc}],$$

$$u_5 = \Delta^2_{bd}[4D+2(Z-1)(A+T_{bd})+X_{bd}+B_{ab}+B_{ad}+B_{bc}+B_{cd}+C_{ac} +X'_{bd}],$$

$$u_6 = \Delta^2_{cd}[4D+2(Z-1)(A+T_{cd})+X_{cd}+B_{ac}+B_{ad}+B_{bc}+B_{bd}+C_{ab} +X'_{cd}],$$

$$v_1 = \Delta_{ab}\Delta_{ac}\Delta_{bc}[4D+B_{ad}+B_{bd}+B_{cd}-(2-Z)(T_{ab}+T_{ac}+T_{bc}) +(Z-1)A+U_{abc}+U'_{abc}],$$

$$v_2 = \Delta_{ab}\Delta_{bd}\Delta_{ad}[4D+B_{ac}+B_{bc}+B_{cd}-(2-Z)(T_{ab}+T_{bd}+T_{ad}) +(Z-1)A+U_{abd}+U'_{abd}],$$

$$v_3 = \Delta_{ac}\Delta_{cd}\Delta_{ad}[4D+B_{ab}+B_{bc}+B_{bd}-(2-Z)(T_{ac}+T_{cd}+T_{ad}) \\ +(Z-1)A+U_{acd}+U'_{acd}],$$

$$v_4 = \Delta_{bc}\Delta_{cd}\Delta_{db}[4D+B_{ab}+B_{ac}+B_{ad}-(2-Z)(T_{bc}+T_{cd}+T_{bd}) \\ +(Z-1)(A+U_{bcd}+U'_{bcd})].$$

$$w_1 = \Delta^2_{ab}\Delta^2_{cd}[4D+X_{ab}+X_{cd}+2(Z-1)(T_{ab}+T_{cd})+V_{ab,\,cd}],$$

$$w_2 = \Delta^2_{ac}\Delta^2_{bd}[4D+X_{ac}+X_{bd}+2(Z-1)(T_{ac}+T_{bd})+V_{ac,\,bd}],$$

$$w_3 = \Delta^2_{ad}\Delta^2_{bc}[4D+X_{ad}+X_{bc}+2(Z-1)(T_{ad}+T_{bc})+V_{ad,\,bc}],$$

$$z_1 = \Delta_{ab}\Delta_{bc}\Delta_{cd}\Delta_{ad}[4D-(2-Z)(T_{ab}+T_{bc}+T_{cd}+T_{ad})+W_{abcd}],$$

$$z_2 = \Delta_{ac}\Delta_{ab}\Delta_{cd}\Delta_{bd}[4D-(2-Z)(T_{ac}+T_{ab}+T_{cd}+T_{bd})+W_{bacd}],$$

$$z_3 = \Delta_{ac}\Delta_{bd}\Delta_{bc}\Delta_{ad}[4D-(2-Z)(T_{ac}+T_{bd}+T_{bc}+T_{ad})+W_{acbd}]. \tag{17}$$

Thus far our development has been quite general. We now make specific assumptions about the position of the atoms, viz. protons a and b belong to one molecule, protons c and d to the other; the distances $a-b$ and $c-d$ are equal ($1.406a_0$). Functions with subscripts ab and cd may therefore be written without subscripts, the understanding being that they are to be evaluated for the internuclear distance of the H_2 molecule. It is also possible to distinguish different orders of magnitude among the various constituents of (17). All quantities without subscripts (or with subscripts ab, cd) are large, all others small. Whether a term can be totally neglected can only be decided by inspection of the Δ-functions which multiply it. This overlap integral has the well-known simple form

$$\Delta_{ab} = \Delta(s) = \left(1+s+\frac{1}{3}s^2\right)e^{-s},$$

s being $(Z/a_0)R_{ab}$. The ratio Δ_{ac}/Δ_{ab} is therefore always fairly small, and it is safe to neglect its fourth power, though not the second. We are thus enabled to simplify the list (17) as follows:

$$\varepsilon = 4D+4(Z-1)A+2C+C_{ac}+C_{ad}+C_{bc}+C_{bd},$$

$$u_1 = \Delta^2[4D+X+C+2(Z-1)(A+T)+B_{ac}+B_{ad}+B_{bc}+B_{bd}],$$

$$u_2 = \Delta^2_{ac}[4D+2B+2(Z-1)A+X_{ac}+X'_{ac}+2(Z-1)T_{ac}],$$

$$u_3 = \Delta^2_{ad}[4D+2B+2(Z-1)A+X_{ad}+X'_{ad}+2(Z-1)T_{ad}],$$

$$u_4 = \Delta_{bc}^2[4D+2B+2(Z-1)A+X_{bc}+X'_{bc}+2(Z-1)T_{bc}],$$
$$u_5 = \Delta_{bd}^2[4D+2B+2(Z-1)A+X_{bd}+X'_{bd}+2(Z-1)T_{bd}],$$
$$u_6 = u_1,$$
$$v_1 = \Delta\Delta_{ac}\Delta_{bc}[4D+B-(2-Z)T+(Z-1)A+U_{abc}+U'_{abc}$$
$$-(2-Z)(T_{ac}+T_{bc})],$$
$$v_2 = \Delta\Delta_{bd}\Delta_{ad}[4D+B-(2-Z)T+(Z-1)A+U_{abd}+U'_{abd} \qquad (17b)$$
$$-(2-Z)(T_{bd}+T_{ad})],$$
$$v_3 = \Delta\Delta_{ac}\Delta_{ad}[4D+B-(2-Z)T+(Z-1)A+U_{acd}+U'_{acd}$$
$$-(2-Z)(T_{ac}+T_{ad})],$$
$$v_4 = \Delta\Delta_{bc}\Delta_{bd}[4D+B-(2-Z)T+(Z-1)A+U_{bcd}+U'_{bcd}$$
$$-(2-Z)(T_{bc}+T_{bd})],$$

$$w_1 = \Delta^4[4D+2X+4(Z-1)T+V_{ab,\,cd}],$$
$$w_2 = w_3 = 0,$$

$$z_1 = \Delta^2\Delta_{bc}\Delta_{ad}[4D-2(2-Z)T-(2-Z)(T_{bc}+T_{ad})+W_{abcd}],$$
$$z_2 = \Delta^2\Delta_{ac}\Delta_{bd}[4D-2(2-Z)T-(2-Z)(T_{ac}+T_{bd})+W_{bacd}],$$
$$z_3 = 0.$$

In u_1 the term X' had been dropped because of its smallness, and in u_2 to u_5, the C functions are neglected. Numerical comparison with terms retained justifies this curtailment.

The primed functions, defined in Table 7.1, are easily obtained from this list; for they are simply the product of Δ functions which appears in the equations for their unprimed mates. Thus, for example

$$\varepsilon' = 1,$$
$$u_1' = \Delta^2,$$
$$u_2' = \Delta_{ac}^2, \quad \text{etc.},$$
$$v_1 = \Delta\Delta_{ac}\Delta_{bc}, \quad \text{etc.} \qquad (18)$$

We are now ready to compute E from (8). Let us first recall the results of the Heitler–London treatment of a single molecule. In our present notation

$$E(H_2) = 2D+(1+\Delta^2)^{-1}\{C+2(Z-1)A+\Delta^2[X+2(Z-1)T]\}$$
$$= 2D+E_m \qquad (19)$$

The term $2D$ represents the energy of two hydrogen atoms of nuclear charge Ze; the remainder is the molecular binding energy. Wang has found the value of Z which minimizes (19) to be 1.166.

The interaction energy ΔE is related to the E which appears in (8) by

$$\Delta E = E - 4D - 2E_m$$

since $4D$ is the atomic energy and E_m is given by (19). When we solve (8), $H_{AA} - E\Delta_{AA} = 0$, in zeroth approximation, that is to say by neglecting all terms in the exchange integrals (17b) which carry subscripts, we obtain exactly $2E_m$ as given in (19). If we solve (8) with retention of all terms we find

$$\begin{aligned}\Delta E = (1+\Delta^2)^{-2}\{&[E_m - B - (Z-1)A](\Delta_{ac}^2 + \Delta_{ad}^2 + \Delta_{bc}^2 + \Delta_{bd}^2)\\ &+ C_{ac} + C_{ad} + C_{bc} + C_{bd} + 2\Delta^2(B_{ac} + B_{ad} + B_{bc} + B_{bd})\\ &- \tfrac{1}{2}\Delta_{ac}^2[X_{ac} + X'_{ac} + 2(Z-1)T_{ac}]\\ &- \tfrac{1}{2}\Delta_{ad}^2[X_{ad} + X'_{ad} + 2(Z-1)T_{ad}] - \tfrac{1}{2}\Delta_{bc}^2[X_{bc} + X'_{bc} + 2(Z-1)T_{bc}]\\ &- \tfrac{1}{2}\Delta_{bd}^2[X_{bd} + X'_{bd} + 2(Z-1)T_{bd}] + \Delta(\Delta_{ac}\Delta_{bc} + \Delta_{ad}\Delta_{bd} + \Delta_{ac}\Delta_{ad}\\ &+ \Delta_{bc}\Delta_{bd})[2E_m - B - (Z-1)A + (2-Z)T]\\ &- \Delta\Delta_{ac}\Delta_{bc}[U_{abc} + U'_{abc} - (2-Z)(T_{ac} + T_{bc})]\\ &- \Delta\Delta_{ad}\Delta_{bd}[U_{abd} + U'_{abd} - (2-Z)(T_{ad} + T_{bd})] - \Delta\Delta_{ac}\Delta_{ad}[U_{acd}\\ &+ U'_{acd} - (2-Z)(T_{ac} + T_{ad})]\\ &- \Delta\Delta_{bc}\Delta_{bd}[U_{bcd} + U'_{bcd} - (2-Z)(T_{bc} + T_{bd})] + \Delta^2(\Delta_{bc}\Delta_{ad}\\ &+ \Delta_{ac}\Delta_{bd})[2E_m + 2(2-Z)T]\\ &- \Delta^2\Delta_{bc}\Delta_{ad}[W_{abcd} - (2-Z)(T_{bc} + T_{ad}) - \Delta^2\Delta_{ac}\Delta_{bd}[W_{bacd}\\ &- (2-Z)(T_{ac} + T_{bd})] + \Delta^4 V_{ab,\,cd}.\}\end{aligned} \qquad (20)$$

In the numerical evaluation of this expression a few further simplifications were found justifiable: terms which varied slowly as R_{ab} and R_{cd} approach zero were replaced by their limiting values, Z was given its "normal" value, 1.166, on the reasonable assumption that the small charge distortion occasioned by intermolecular forces would alter it but slightly.

Calculations were performed for four different orientations of the two molecules (Fig. 13).

In case (d) both molecular axes are perpendicular to R and perpendicular to each other. The exchange energies for these posi-

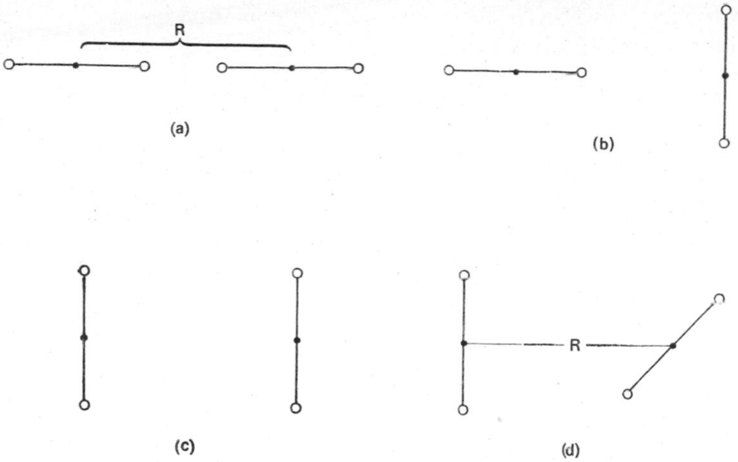

FIG. 13.

TABLE 7.3. *Exchange Energy as a Function of the Intermolecular Separation (R) for Z = 1.166*

R (in Å)	(in millivolts)			
	Position *a*	Position *b*	Position *c*	Position *d*
2.27	154.2	150.7	138.46	139.90
2.72	38.08	34.20	30.45	30.37
3.18	8.364	7.304	6.306	6.262
3.63	1.680	1.402	1.202	1.151

tions are listed as functions of R in Table 7.3: these values exclude, however, the last term of (20).

This term, $[\Delta^4/(1+\Delta^2)^2]V_{ab,\,cd}$ has already been identified as the quadrupole energy on the evidence that $V_{ab,\,cd}$ "reduces" to the electrostatic interaction between two distributions of charges (illustrated in Fig. 14) when the approximation (14) is made. In this approximation the quadrupole energy is given by

$$\Delta E'_Q = \frac{3Q^2}{4R^5} f(\theta_1, \theta_2, \varphi),$$

provided f is the function characteristic of quadrupole interactions introduced in Chapter 2, and

$$Q = 2e^2\left(\frac{d}{2}\right)^2.$$

Our calculation shows that the quadrupole moment for *point* charges must be multiplied by a "diffuseness factor", $\Delta^2/(1+\Delta^2)$, because of the extension of the electron clouds.

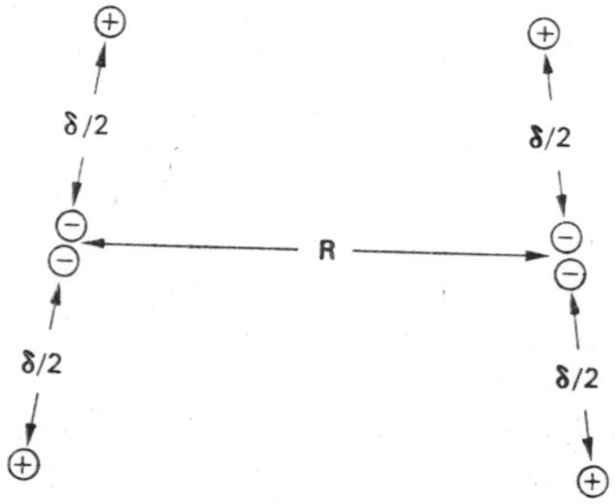

FIG. 14.

Actually, our result, $(\Delta^2/(1+\Delta^2))^2 V_{ab,\,cd}$ is not quite correct in the form $(\Delta^2/(1+\Delta^2))^2 \Delta E'_Q$, which is obtained by the method of reduction. The accurate result, derived in the paper cited, differs from it by a factor 16/25 (4/5 for the quadrupole moment). Thus

$$\Delta E_Q = \frac{3}{25}\left(\frac{\Delta^2}{1+\Delta^2}\right)^2 \frac{e^2 d^4}{R^5}\left[\begin{matrix}1-5\cos^2\theta_1-5\cos^2\theta_2-15\cos^2\theta_1\cos^2\theta_2\\ +2(4\cos\theta_1\cos\theta_2+\sin\theta_1\sin\theta_2\cos\varphi)^2\end{matrix}\right] \tag{21}$$

where the bracket contains the function $f(\theta_1, \theta_2, \varphi)$ and d is the distance between atoms in the H_2 molecule. Numerically, $\Delta = 0.685$, $d = 1.406a_0$, and

$$\Delta E_Q = 0.048\frac{e_2 a_0^4}{R^5} f. \tag{21a}$$

H_2 is a slightly anisotropic molecule, and its dispersion energy is given in terms of its polarizabilities by eqn. (52) of Chapter 2. Following Mason and Hirschfelder (1957) we shall write it in the form

$$\Delta E_{dd} = \overline{\Delta E_{dd}} g(\varkappa, \theta_1, \theta_2, \varphi),$$

where the bar indicates an average over the dipole–dipole energy with respect to orientations. The function g, which depends on angles and on a quantity called the "anisotropy of the polarizability",

$$\varkappa \equiv (\alpha_{\parallel} - \alpha_{\perp})/(\alpha_{\parallel} + \alpha_{\perp})$$

in accordance with eqn. (25) of Chapter 2, has the forms

(a) $1 + 2\varkappa + 3\varkappa^2$

(b) $1 - \frac{\varkappa}{2} - \frac{3}{2}\varkappa^2$

(c) $1 - \varkappa + \frac{3}{2}\varkappa^2$

(d) $1 - \varkappa$

in the four configurations introduced above. The value (Ishiguro *et. al.*, 1952) of $\varkappa$ for H_2 is 0.1173, and

$$\overline{\Delta E_{dd}} = -10.9 \frac{e^2}{a_0} \left(\frac{a_0}{R}\right)^6 \tag{22}$$

as shown in Margenau (1943). A formula for the average dipole–quadrupole interaction is likewise given in Britton and Bean (1955); it is

$$\overline{\Delta E_{dq}} = -116 \frac{e^2}{a_0} \left(\frac{a_0}{R}\right)^8 . \tag{23}$$

Since its effect is small the orientation dependence of ΔE_{dq} will be neglected. Both (22) and (23) are semi-empirical results which are based on known f-values for H_2.

To obtain the total molecular interaction we add ΔE (20), ΔE_Q (21), ΔE_{dd} (22) and $\overline{\Delta E_{dq}}$ (23) for the four different molecular orientations. The results are shown in Figs. 15 and 16.

Figure 15 is taken from the older work of Evett and Margenau (1953), in which all exchange integrals were approximated in the manner described. Also, the orientation dependence of E_{dd} was neglected. Figure 16 presents the results of Mason and Hirschfelder (1957b), who performed a more accurate and elaborate calculation of certain exchange integrals while approximating others.[†] Both are essentially *a priori* quantum mechanical derivations of the molecular interactions from first principles, and it is difficult to say which is the more accurate; in fact, the existing agreement might be regarded as satisfactory.

An empirical curve for the total interaction energy has been derived from transport properties and second virial coefficients by Mason and Rice (1954). It represents an average over molecular orientations and is plotted as curve 1 in Fig. 17. For comparison,

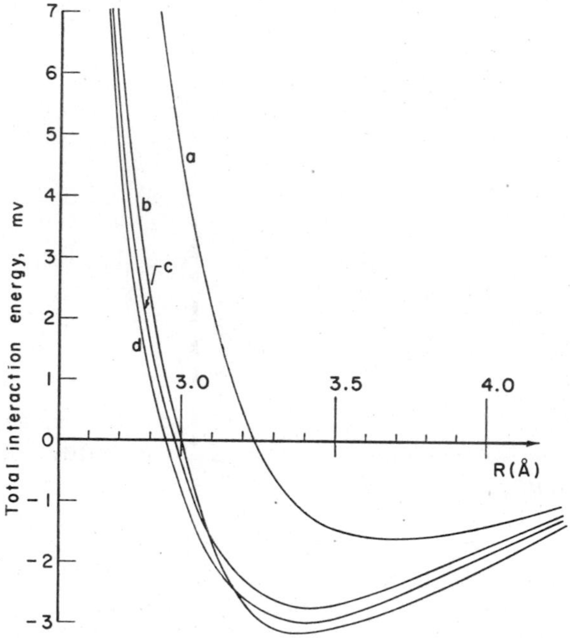

FIG. 15. Interaction energy of two hydrogen molecules, after Evett and Margenau (1953). Orientations *a–d* are explained on p. 235.

† Mason and Hirschfelder (1957b) used the "Mulliken approximation" in evaluating the large exchange integrals of two, three, and four centers. Small integrals were approximated by multipole expansions. The Mulliken approximation assumes that

$$a(1)b(2) = \frac{1}{2}\Delta_{ab}[a(1)a(2)+b(1)b(2)].$$

It has been tested extensively by Barker and Eyring (1954, 1955) and Cizek (1963). The latter paper presents further methods of approximation.

properly weighted averages over the curves of Fig. 15 and those of Fig. 16 are also drawn. They are labeled 2 and 3 respectively.

Mention was made at the beginning of this chapter of the calculations of de Boer (1942). Like those of Evett and Margenau (1953), they are in good agreement with the results of Mason and Rice (1954). This good fortune induced Mason and Hirschfelder (1957b) to examine whether perhaps in calculations of this

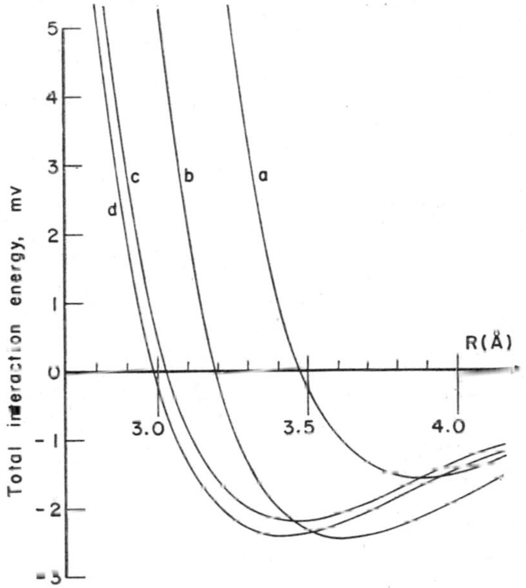

FIG. 16. Interaction energy of two hydrogen molecules, after Mason and Hirschfelder (1957b).

sort it is generally legitimate to follow de Boer's example and ignore three- and four-center exchange integrals. Hence these authors computed the total interaction without them and obtained curve 4 of Fig. 17. Evidently, the effect of these integrals is considerable when all terms are properly included.

The tables and graphs which form the results of the present analysis are not very useful in the calculation of observable effects. It is interesting, therefore, that they can be fairly well simulated by an analytic formula published by Takayanagi (1957), who employs as an approximation to the curves of Fig. 15 (Evett–

Margenau) the expression

$$\Delta E = De^{-\alpha(R-R_0)} - 2De^{-\frac{\alpha}{2}(R-R_0)} + \beta De^{-\alpha(R-R_0)}[P_2(\cos\chi_1)+P_2(\cos\chi_2)].$$

The constants are

$$D = 1.1\times10^{-4} \text{ a.u.}$$
$$R_0 = 6.4a_0$$
$$\alpha = 1.87a_0^{-1}, \quad \beta = 0.075$$

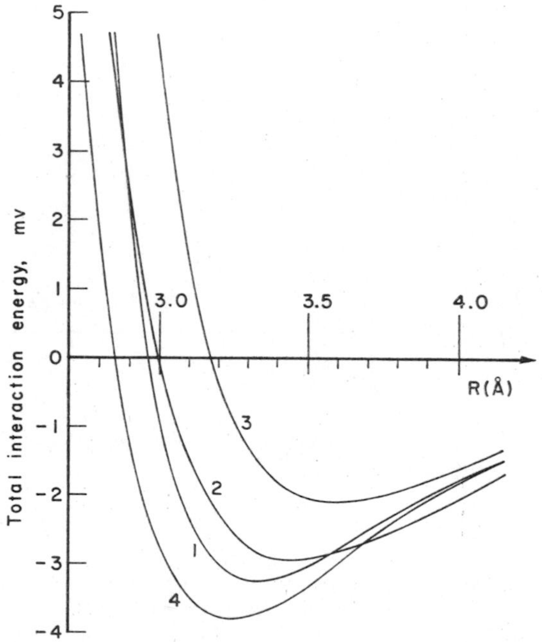

FIG. 17. Interaction energy of two hydrogen molecules, averaged over all orientations. 1, empirical, after Mason and Rice (1954). 2, from Evett and Margenau (1953). 3, from Mason and Hirschfelder (1957b). 4, result obtained with neglect of multiple exchange integrals.

and $\chi_{1,\,2}$ are the angles which the molecular axes of molecules 1, 2 make with the intermolecular axis $\boldsymbol{R}$. Takayanagi's formula is applicable, of course, only in the range where dispersion and quadrupole forces are insignificant; it fits Fig. 15 within a few tenths of a millivolt in the important region between 3 and 4 Å.

Since it does not depend on the azimuthal orientation of one molecule relative to the other it does not distinguish between curves *c* and *d* of Fig. 15.

The dependence of intermolecular forces on orientation is decisive in a study of the collision cross-section of molecules for rotational and vibrational excitation. Calculations of these have been made for H_2 on the basis of Takayanagi's analytic interaction and have yielded reasonable agreement with observations (Davison, 1962; Roberts, 1963).†

We close this section with a brief comment on a problem not covered by its title. When the distance between the two atoms in each molecule is set equal to zero, the interaction energy calculated here should be that between two helium atoms (provided Z is correctly chosen). Our formulas do, in fact, reduce to those for the helium–helium case when this simplification is made: eq. (20) takes on the form

$$\Delta E(\mathrm{He-He}) = (1-\Delta^2)^{-2}\{4C - 2\Delta^2[B+C+X+2(2-Z)A \\ -2(aa\rho aa) - 2(3-Z)T + 4(aa\rho ac)/\Delta]\},$$

where all functions without a subscript now refer to the interatomic distance R. This is nothing other than eqn. (89) of Chapter 3 in our present notation.

7.3. Interaction between a Hydrogen Molecule and a Hydrogen Atom

The formalism of the previous section can be employed directly in the evaluation of the forces between H_2 and H (Margenau, 1944, 1957b).‡

† A good general survey of the entire field will be found in the review articles by Takayanaga (1963, 1965, vol. 1, p. 149).

‡ An earlier, semi-empirical approach to the general interaction between three hydrogen atoms, arbitrarily disposed, was made in a well-known article by Hirschfelder, Eyring, and Topley (1936). It is reviewed in Hirschfelder, Curtiss, and Bird (1954). Our treatment is limited to molecule plus atom. More recent work on the "potential energy surface" of three hydrogen atoms, which should in principle lead to results identical with our first-order $\overline{H}-3E(H)$ because it ignores long-range forces, has been published by Hirschfelder, Eyring, and Rosen, (1936, 1938); Snow and Eyring (1957); Ransil (1957); Kimball and Trulio (1958); Boys and Shavitt (1959); Porter and Karplus (1964). The last reference presents a painstaking and elaborate analysis of the 3H problem judiciously using empirical parameters to approxi-

We assume the three interacting atoms to be located at points a, b, and c, and we use these same letters to denote electronic orbitals centered about these points, as before. The functions needed to construct a doublet state corresponding to a bond between atoms a and b are

$$\psi_1 = \mathcal{A}(a_+ b_- c_+)$$

and

$$\psi_2 = \mathcal{A}(a_- b_+ c_+).$$

The doublet function representing the molecule $a-b$ interacting with atom c is

$$\Psi_A = \sqrt{\left(\frac{1}{2}\right)}\,(\psi_1 - \psi_2)$$

as may be verified by noting that Ψ_A is *symmetric* with respect to an exchange of the bonded nuclei at a and b. Again the total energy is

$$E = H_{AA}/\Delta_{AA} \qquad (24)$$

and

$$H_{AA} = \frac{1}{2}(H_{11} + H_{22} - 2H_{12})$$

$$\Delta_{AA} = \frac{1}{2}(\Delta_{11} + \Delta_{22} - 2\Delta_{12}).$$

The Hamiltonian H acting upon the "normal" product of orbitals is

$$H\{a(1)b(2)c(3)\} = \{3D - (1-Z)(\alpha_1 + \beta_2 + \gamma_3) - (\alpha_2 + \alpha_3 + \beta_1 + \beta_3 + \gamma_1 + \gamma_2) + \varrho_{12} + \varrho_{13} + \varrho_{23} + E_N\}a(1)b(2)c(3)$$

mate some of the integrals that are difficult to calculate. For a comparative comment on the work mentioned above and the results here developed, see the end of this section.

In the most recent calculations on the 3H system all integrals are evaluated exactly and electron correlation is included. The first paper reporting this work, begun by Conroy, has been published (Conroy and Bruner, 1965; linear 3H system). They encounter a small minimum for $R_{ab} = R_{bc}$; this, however, was due to an inadequate numerical integration procedure. In later (1967) work by the same authors this feature disappeared. In general their results are similar but of lower energy than those of Porter and Karplus.

provided again $D = -(1/2)Z^2e^2/a_0$ and E_N is now the electrostatic energy of the *three* nuclei. Table 7.1 reduces to

$$\varepsilon = (abc\,|\,H\,|\,abc), \quad u_1 = (bac\,|\,H\,|\,abc), \quad u_2 = (cba\,|\,H\,|\,abc)$$
$$u_4 = (abc\,|\,H\,|\,abc)$$
$$v_1 = (bca\,|\,H\,|\,abc)$$

with corresponding expressions for the primed elements (of unity instead of H). Then

$$H_{11} = \varepsilon - u_2, \quad H_{12} = -u_1 + v_1, \quad H_{22} = \varepsilon - u_4$$

and

$$H_{AA} = \varepsilon_1 + u_1 - \frac{1}{2}(u_2 + u_4) - v_1$$

$$\Delta_{AA} = \varepsilon_1' + u_1' - \frac{1}{2}(u_2' + u_4') - v_1'. \qquad (25)$$

The exchange integrals now have the following forms:

$$\varepsilon = 3D + 3(Z-1)A + C_{ab} + C_{ac} + C_{bc}$$
$$u_1 = \Delta_{ab}^2\{3D + X_{ab} + B_{ac} + B_{bc} + (Z-1)(A + 2T_{ab} + X_{ab})\}$$
$$u_2 = P_{bc}u_1$$
$$u_4 = P_{ac}u_1$$
$$v_1 = \Delta_{ab}\Delta_{ac}\Delta_{bc}\{3D + (Z-2)(T_{ab} + T_{ac} + T_{bc}) + U_{abc}\}. \qquad (26)$$

An expression like $P_{bc}u_1$ with the symbols b and c interchanged. In this instance

$$X_{ab}' = 2[(ac\varrho bc) - (a\gamma b)]/\Delta_{ab}.$$

Henceforth quantities bearing subscripts a, b, i.e. pertaining to the molecule, will again be written without indices.

It now becomes necessary to find the energy of molecule and atom without interaction and to subtract it from E given by (24). This energy, $E(H_2)+E(H)$, is obtained in this formalism by evaluating all integrals (26) when R_{ac} and R_{bc} approach infinity. In that situation

$$\varepsilon = 3D + C + 3(Z-1)A \qquad \varepsilon' = 1$$
$$u_1 = \Delta^2\{3D + X + (Z-1)(A + 2T)\} \qquad u_1' = \Delta^2$$
$$u_2 = u_4 = v_1 = 0 \qquad u_2' = u_4' = v_1' = 0.$$

If these are inserted in (25) and (24) we find

$$E(\mathrm{H})+E(\mathrm{H_2}) = D+(Z-1)A+(1+\Delta^2)^{-1}\{C+2(Z-1)A \\ +\Delta^2[X+2(Z-1)T]\}+2D. \qquad (27)$$

The last two terms on the right will be recognized as $E(\mathrm{H_2})$ by looking at (19); the first two represent $E(\mathrm{H})$;

$$\int aH_{\mathrm{atom}}a\,d\tau = D+(Z-1)A.$$

The first-order exchange energy we have now calculated is

$$\Delta E = E-E(\mathrm{H})-E(\mathrm{H_2}), \qquad (28)$$

where E is given by (24) and the other terms by (27).

There now remains the evaluation of ΔE_{dd} and ΔE_{dq}. In both of these we neglect the anisotropy of H_2 since its effect is small. The energy ΔE_Q is absent because the atom has no permanent polarity. Thus, with the use of known f-values and formulas, eqn. (27) of Chapter 2, one obtains the results

$$\Delta E_{dd} = -8.40\frac{e^2}{a_0}\left(\frac{a_0}{R}\right)^6 \qquad (29)$$

and

$$\Delta E_{dq} = -148\frac{e^2}{a_0}\left(\frac{a_0}{R}\right)^8. \qquad (30)$$

The total interaction energy is compounded from (28), (29), and (30).

A question arises as to the proper choice of Z. Suppose, following Margenau (1944), we select two values, $Z = 1.071$ and $Z = 1.20$. The first of these lies between the atomic and the molecular Z, the second is clearly unsuitable. It serves merely to indicate the sensitivity of the results to the choice of Z. It turns out that both values produce minima at the same intermolecular distances for different orientations; the depths of the minima differ slightly but the orientation dependence is more marked for the larger Z. We employ again the "method of reduction" of exchange integrals described earlier, thereby introducing unknown errors. Mason and Hirschfelder (1957b) chose $Z = 1.166$, the value for the molecule, and used more reliable methods in the calculation of some of their integrals. Computations were performed for the two orientations of the molecule depicted in Fig. 18.

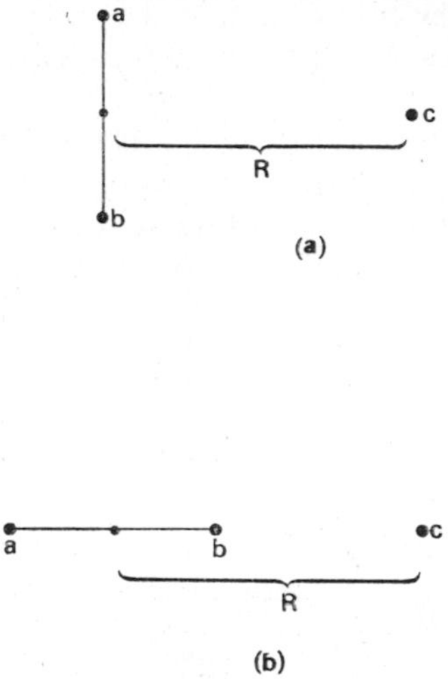

FIG. 18.

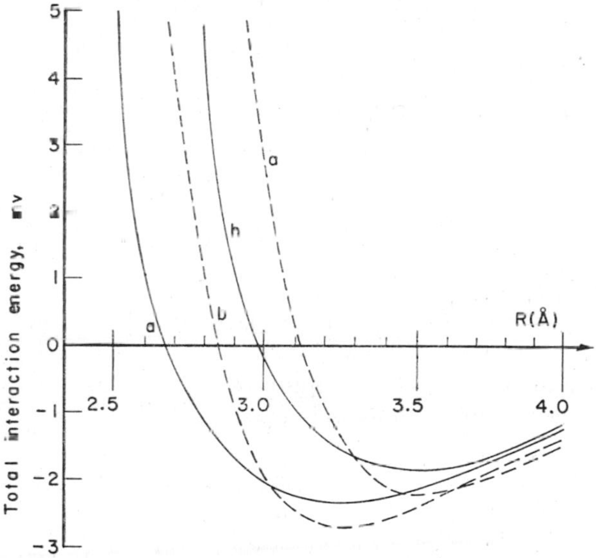

FIG. 19. Total interaction energy between H_2 and H. Curves *a* represent approach of H along bisector of molecular axis, curves *b* along axis. Dashed curves refer to the results of Mason and Hirschfelder (1957b), solid curves to those of Margenau (1944).

Figure 19 shows the results obtained for the total interaction by two sets of authors. Except for the repulsive region, the agreement is gratifying.†

Figure 19 exhibits a feature which is unexpected in view of the work of Hirschfelder, Eyring, and Topley (1936) and all the references listed on p. 241. According to these authors the path of easiest approach of an H atom to an H_2 molecule should be along the molecular axis, whereas Fig. 19 shows it to be perpendicular to the axis. The source of this curious discrepancy does not seem to be known.

7.4. H_2–He

Fundamental consideration has been given to one further example, the interaction between a hydrogen molecule and a helium atom. Roberts (1963) computes the *repulsive* energy; he follows the methods we have outlined and, in the absence of four-center integrals, succeeds in evaluating all occurring integrals exactly, using a digital computer. He finds numerical values which are quite accurately simulated by the function

$$\Delta E = Ce^{-\varkappa R}[1+\delta P_2(\cos\chi)]$$

with

$$C = 17.283 \text{ a.u.},$$

$$\varkappa = 2.027 a_0^{-1},$$

$$\delta = 0.375,$$

$$\chi = \text{angle between } \boldsymbol{R} \text{ and axis of } H_2.$$

P_2 is again the second Legendre polynomial.

Comparison of Roberts's potential with experimental results is possible because the interaction averaged over χ was derived by Amdur and Malmauskas (unpublished but given by Roberts, 1963) from gas diffusion experiments. The agreement obtained is illustrated in Fig. 20.

Occasionally the search for simple models useful in the computation of repulsive interactions has ended in the suggestion

† In Mason and Hirschfelder (1957b, fig. 3) the results of Margenau for $Z = 1.2$ are compared with Mason and Hirschfelder's results for $Z = 1.166$, and there is some disagreement (as one should expect). In Fig. 19 we compare the results for the proper values of Z.

that these be regarded as the sum of all exchange forces between the *atoms* composing the molecules (see, for instance, Brout, 1954). In the case of a diatomic molecule, the name dumbbell model has been applied to this conjecture. While for the H_2–H_2 potential this model is a rough though often tolerable approxi-

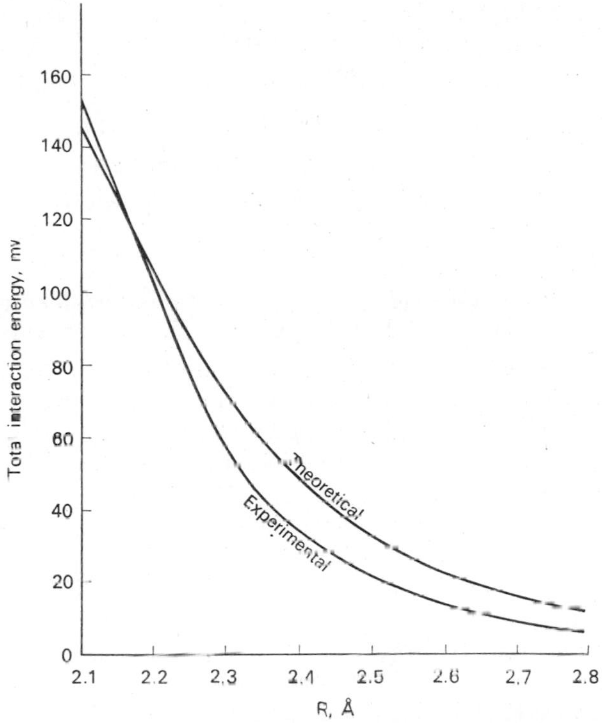

FIG. 20. Repulsive interaction energy, H_2–He, averaged over orientations of the H_2 molecules. "Experimental" curve is taken from Amdur and Malinauskas (see Roberts, 1963); "Theoretical" from Roberts (1963).

mation (Nagamiya and Kishi, 1951), Roberts shows that, in the present instance, the error it introduces at the smallest values of R in Fig. 20 can be as high as 60%; it yields a repulsion which greatly exaggerates the dependence on χ.

Roberts assumed the separation of the H_2-nuclei to be fixed (except for one slightly increased value of the bond length,

$d = 1.486$). In problems involving vibrational excitation it is important to know the dependence of ΔE not only upon angles, but also upon the distance d between the hydrogen atoms. Krauss and Mies (1965) attacked this problem using elaborate *HF* functions which lead to numerical calculations not reproducible here. They arrive at a result which is approximated analytically by the formula

$$\Delta E = Ce^{-\alpha_0 r + \alpha_1 dr}[A(\chi) + B(\chi)r]$$

where

$$C = 198.378 \text{ a.u.},$$

$$\alpha_0 = 1.66176 a_0^{-1}, \quad \alpha_1 = 0.3206 a_0^{-2},$$

$$A(\chi) = 1.10041[1 + 0.18250 P_2(\cos \chi)]$$

$$B(\chi) = -0.52151[1 - 0.27506 P_2(\cos \chi)]$$

$$r = d - d_0 = d - 1.406 a_0.$$

7.5. General Long-range Interaction between Molecules having Permanent Multipoles

Formulas for the interaction of various multipoles have been presented in Chapter 2, where special attention was given to the alignment and induction forces between dipoles and quadrupoles. For two discrete charge distributions, none of whose charges overlap, eqns. (8) and (9) of Chapter 2 express the mutual potential energy in forms convenient for quantum mechanical calculations. A form of V which places in evidence its dependence on the various multipoles has been developed by Buehler and Hirschfelder (1951, 1952). Since it is derived in Hirschfelder, Curtiss, and Bird (1954), we shall merely state it here for reference. It can also be obtained directly from eqn. (9) of Chapter 2.

For a discrete distribution of charges q_i at the points r_i, θ_i, φ_i we define spherical-harmonic, rather than Cartesian, multipole moments Q_n^m by the formula

$$Q_n^m = \sum_i q_i r_i^n P_n^m(\cos \theta_i) e^{im\varphi_i} \tag{31}$$

or, if the distribution is continuous with charge density $\varrho(r, \theta, \varphi)$,

$$Q_n^m = \int \varrho(r, \theta, \varphi) r^n P_n^m e^{im\varphi} r^2 \, dr \sin \theta \, d\theta \, d\varphi. \tag{32}$$

The P_n^m are associated Legendre polynomials, $P_n^{-m} = P_n^m$, and Q_n^m is the moment of a multipole of order 2^n. This quantity, with $2n+1$ components for any given n, carries precisely the information which matters in physical problems; the strict multipole tensor $\vec{\Theta} = \sum_i q_i \boldsymbol{r}_i \boldsymbol{r}_i \ldots \boldsymbol{r}_i$ is redundant for $n > 1$ because of symmetry relations between its components which arise from Poisson's equation. Notice also that the Q_n^m are complex.

If there are two charge distributions with origins at $\boldsymbol{a}$ and $\boldsymbol{b}$, each equipped with its own system of axes (defining r_a, θ_a, φ_a and r_b, θ_b, φ_b) and a distance R_{ab} apart, we define two sets of multipole moments, one for each distribution, in conformity with (31) or (32), thus obtaining $Q_{n_a}^m$ and $Q_{n_b}^m$. In terms of these, the potential energy is given by the formula

$$V = \sum_{n_a=0}^{\infty} \sum_{n_b=0}^{\infty} \sum_{m=-n}^{n} \frac{(-1)^n b^m (n_a+n_b)!}{(n_a+|m|)!(n_b+|m|!)} \frac{Q_{n_a}^{m*} Q_{n_b}^m}{R_{ab}^{n_a+n_b+1}} \tag{33}$$

provided, as stated,

$$R_{ab} > r_i + r_j \quad \text{for every} \quad i, j.$$

The symbol n appearing in (33) is the smaller of n_a and n_b. This result represents the *static* interaction, over and above exchange, dispersion, and polarization effects (these are quantum mechanical in nature), which occurs universally between polar molecules.

Formulas valid when R_{ab} is not greater than the sum of r_1 and r_2, so that the charge distributions overlap, are also given by Hirschfelder, Curtiss, and Bird (1954).

7.6. Forces between Long-chain Molecules

Thus far, the present chapter has dealt with only the simplest molecules. It may seem strange that molecules of intermediate complexity are omitted from our account, for we are now turning to the calculation of intermolecular forces between organic molecules. As so often happens in science, one is able to treat very simple and very complex situations, the latter usually by qualitative methods which rely on the regularities of large numbers and sizes. This is to some extent the case here.† Very little is

† The long-range forces between H_2O molecules, but not their shapes, have been calculated by Margenau and Meyers (1944). The results of this article may serve as a reminder of the relative magnitudes of the different constituents of the long-range force between polar molecules. The coefficients of R^{-6} for the static multipole (orientation) effect, the dispersion and the induction effect for water molecules are in the ratio of 20:3.5:1.

known quantitatively of the forces between ordinary molecules, and we are forced to jump to the study of complex ones. Of these we select examples which introduce ideas not hitherto encountered.

It is clear that the forces between macromolecules can not be computed by starting from the Schrödinger equation, as was done in the previous section. Exchange forces, in particular, can only be estimated from the nature of the atoms which compose the total structure. But a fairly general treatment of the attractive forces is possible, and the approach is different for two types of macromolecules.

The distinction is between saturated, paraffin-like molecules in which the charge distribution is effectively localized in small groups of atoms, on the one hand, and highly conjugated molecules, like hydrocarbon polyenes, on the other. The latter permit an easy transfer of charge along the entire chain and require a treatment somewhat different from that given earlier in the book. We first deal with saturated chains.

7.6.1 *Saturated-chain Molecules*

The two structures which interact are linear arrangements of atomic or submolecular units, each of which gives rise to dispersion forces. The question of additivity of the effects between these local centers has been investigated by Longuet–Higgins and Salem (1961) and answered affirmatively in terms of second order of perturbation theory. In third order there are important nonadditive effects involving two local centers on one structure and one center on the other (1963). Assuming additivity, Muller (1936, 1941) as well as Pitzer and Catalano (1956) have calculated crystal energies and intra-molecular forces in paraffins. Intermolecular forces were computed on this basis by Hamaker (1937) for two small particles (crystal spheres), by de Boer (1936) for two infinite solids interacting across two plane surfaces. Other forces computed in this way are those between circular disks (Dube and Das Gupta, 1939; Bouwkamp, 1947), between thick rectangular rods (Vold, 1954), cylinders (Sparnaay, 1959), and thin rods (de Rocco and Hoover, 1960). In principle, these calculations involve no features except those already discussed in Chapter 2. Hence we present here only a typical calculation of the dispersion energy between two parallel linear chains, a case which illustrates the method, and we follow the clear and circumspect presentation of Salem (1962). Attention is confined to the R^{-6} part of the attractive forces.

Each center of force composing one chain interacts with every center composing the other via the potential $-C/r^6$, r being the distance between two centers. The determination of C is not an easy matter, since the effective centers are not usually known atoms or small molecules for which this constant can be inferred from observational data. But let us defer this problem.

The model for the interaction is illustrated in Fig. 21.

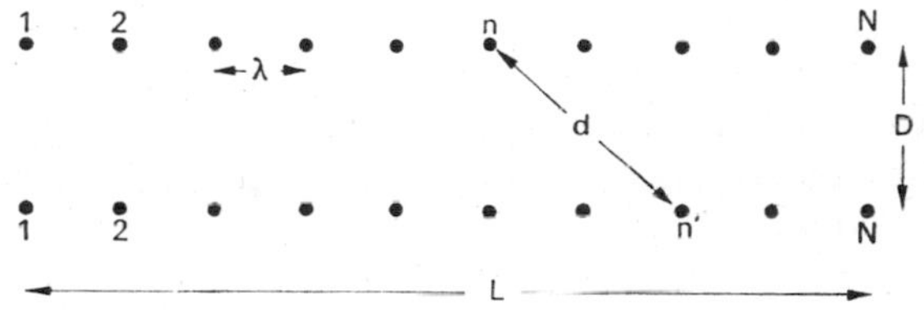

FIG. 21.

There are N "units", or force centers, in each chain, λ cm apart, so that $N\lambda = L$. The distance between the nth unit in one and the n'th in the other chain is

$$d = [D^2+(n-n')^2\lambda^2]^{1/2}$$

provided the chains are separated by D cm. Hence the total interaction energy is

$$\Delta E = -C \sum_{n,\, n'=1}^{N} [D^2+(n-n')^2\lambda^2]^{-3}.$$

Now for every value of $x \equiv |n-n'|$ there are $2(N-x)$ interactions of equal energy, with the exception of $n = n'$, $x = 0$, for which only N equal interactions exist. Hence

$$\Delta E = -C\left[\frac{N}{D^6}+2\sum_{x=1}^{N-1}\frac{N-x}{(D^2+\lambda^2x^2)^3}\right].$$

But N is large, and we shall limit our result to instances where $D \gg \lambda$. One may then replace the summation by an integration over the variable $y = x/N$. Thus, if $\varrho \equiv L/D$, we find

$$\Delta E = -\frac{2C}{\lambda^6 N^4}\int_0^1 \frac{1-y}{(y^2+\varrho^{-2})^3}\,dy$$
$$= -\frac{C}{4\lambda^2 D^4}\,\varrho\left(3\tan^{-1}\varrho+\frac{\varrho}{1+\varrho^2}\right) \tag{34}$$

by an elementary integration.

For large D, small ϱ, the result reduces to

$$\Delta E \rightarrow -\frac{C\varrho^2}{\lambda^2 D^4} = -\frac{CN^2}{D^6}, \quad D \rightarrow \infty. \tag{35}$$

The interaction energy is the sum of those between N units in one chain and N in the other, all of them a distance D apart.

At distances D much smaller than the length of the chain (but large compared with λ) formula (34) reduces to

$$\Delta E \rightarrow -\frac{3\pi}{8}\frac{C}{\lambda^2}\frac{L}{D^5} = -\frac{3\pi}{8}\frac{C}{\lambda}\frac{N}{D^5}, \quad D \ll L. \tag{36}$$

In this interesting limit, then, ΔE is inversely proportional to the 5th power of the intermolecular distance and increases linearly with the length of the chain.

By a slight modification of the foregoing analysis Salem easily calculates the interaction between two identical circular rings which lie in parallel planes perpendicular to the line of centers, a distance D apart. Suppose we regard the second ring as a mirror image of the first. The distance from a given unit to its image is then D, and the distance between this image unit and any arbitrary unit in the image is $d \sin\theta$, provided d is the diameter of the circle and θ the angular opening between the two units of the image. But D and $d \sin\theta$ are mutually perpendicular; hence the separation between a given unit in the first ring and an arbitrary unit in the second is $D^2+d^2\sin^2\theta$ or $D^2[1+(L/\pi D)^2\sin^2\theta]$. From this one obtains at once

$$\Delta E = -\frac{CN^2}{\pi D^6}\int_{-\pi/2}^{\pi/2}\frac{d\theta}{[1+(L/\pi D)^2\sin^2\theta]^3}.$$

A contour integration (see for example, Margenau and Murphy,

1964, vol. 2, p. 93) then leads to

$$\Delta E = -\frac{3\pi^2 C}{8\lambda^2 D^4} \frac{\varrho_a^2}{(1+\varrho_a^2)^{1/2}} \left[1 + \frac{2}{3}(1+\varrho_a^2)^{-1} + (1+\varrho_a^2)^{-2} \right] \tag{37}$$

$$\text{with} \quad \varrho_a \equiv \frac{L}{\pi D}.$$

The dependence on D, which is contained in ϱ, is somewhat complicated; but at small separations, $\pi D \ll L$, one recovers formula (36).

To complete the account we now turn briefly to an identification of the units, the interacting foci to which the formula $\Delta E = -C/r^6$ applies. In this we continue to record the aforementioned work of Salem, who treats the example of paraffin chains. Two such objects are depicted in Fig. 22.

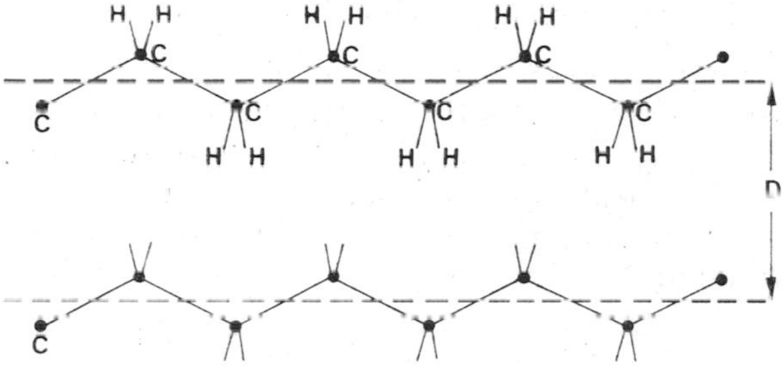

FIG. 22.

The chemical unit of each chain is the CH_2 group. Nevertheless it would be ill advised to treat it as the center of van der Waals forces, the reason being that the polarizable electrons which give rise to dispersion forces are not near the centers of mass but are located between them in the space pervaded by what chemists call the bond. The concept of bond polarizability is now a common and fruitful one (see Vickery and Denbigh, 1949), and it is advantageously applied to the present problem.

The active centers in the situation schematized in Fig. 22 are therefore the regions between two carbon atoms and those between carbon and hydrogen. To take account of this circumstance Salem selects as his unit the arrangement (henceforth called U) depicted in Fig. 23.

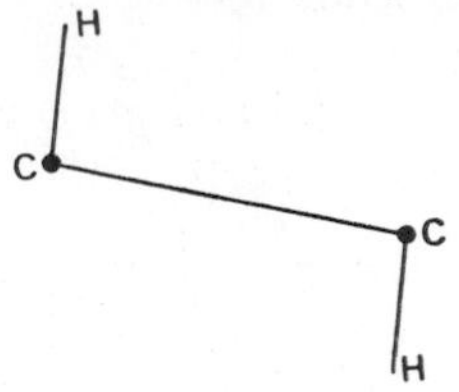

FIG. 23.

It is chemically equivalent to the CH_2 group because each carbon atom counts for only $\frac{1}{2}$ in forming the C—C bond in Fig. 23.

The energy between two such units w_{uu} is then made up of bond–bond attractions in the following way:

$$w_{uu} = w_{\mathrm{CC,\,CC}} + 4w_{\mathrm{CC,\,CH}} + 4w_{\mathrm{CH,\,CH}}$$

To evaluate it one needs to know the dispersion energy between two bonds, e.g. $w_{\mathrm{CC,\,CC}}$ (Salem, 1960). This calculation makes use of known values of bond polarizabilities, such as P_{CC} and P_{CH}, in terms of which each w can be expressed.

The requisite formula follows from the elementary definition of polarizability

$$P_k = \frac{2}{3} e \sum_{k'}^{\infty} \frac{\langle k | \sum_i \boldsymbol{r}_i | k' \rangle \langle k' | \sum_i \boldsymbol{r}_i | k \rangle}{E_{k'} + E_k} \tag{38}$$

and eqn. (26) of Chapter 2:

$$\Delta E = -\frac{2e^4}{3R^6} \frac{\overline{\left(\sum_i \boldsymbol{r}_i^{(1)}\right)^2}\ \overline{\left(\sum_j \boldsymbol{r}_j^{(2)}\right)^2}}{\bar{E}_1 + \bar{E}_2}. \tag{39}$$

The last expression differs from eqn. (26) of Chapter 2 only in being summed over all magnetic quantum numbers, i.e. in being averaged over all directions. $\bar{E}_1$ and $\bar{E}_2$ are certain average energy differences which are not well defined. The summations over i and j extend over all electrons in the two molecules, and the index k in (38) designates the molecular state for which P is being computed.

If we make the same approximation in (38) as was used in (39) to arrive at the average energies in the denominator we obtain, employing matrix multiplication,

$$P_k = \frac{2}{3}\,\frac{e^2}{\bar{E}}\,\langle k|\big(\sum_i r_i\big)^2|k\rangle \equiv \frac{2}{3}\,\frac{e^2}{\bar{E}}\,S \tag{40}$$

for short. Suppose now, contrary to fact but certainly as a reasonable approximation, that the average energies appearing in (39) and (40) are equal. Noting that we need to distinguish between α's, S's and $\bar{E}$'s for the two molecules, we equip them with subscripts, eliminate the $\bar{E}$'s and thus arrive at

$$\Delta E = \frac{e^2}{R^6}\,\frac{P_1P_2}{P_1S_1^{-1}+P_2S_2^{-1}}\,. \tag{41}$$

The subscript k has now been omitted, and the P's refer to the polarizabilities of the interacting structures, in our case the electron bonds. Such bond polarizabilities are known, chiefly from the work of Denbigh (1940) and Vickery and Denbigh (1949). If they are substituted in (41), ΔE becomes the bond dispersion energy and may be equated to $w_{CC,\,CC}$, $w_{CC,\,CH}$ and $w_{CH,\,CH}$ in the formula for $w_{n,\,n'}$.

For applications of these principles we refer to Salem's article which also contains the details encountered in evaluating the various sums S. For the CH_2 units of our example, the constant C has the value

$$w_{CH_2-CH_2} = -\frac{C}{r^6},\; C = 58.1\,(\text{Å}^6)\,\text{ev} = 1340\,(\text{Å}^6)\quad \text{kcal/mole},$$

the average polarizability (average over directions) being $P_{CH_2} = 1.84\,(\text{Å})^3$. With this value formula (3) becomes

$$\Delta E = -53.7\,\frac{N}{D^5}\,\text{ev} = -1240\,\frac{N}{D^5}\quad \text{kcal/mole}$$

provided again that D is expressed in Å. The spacing λ is 1.27 Å. This work has been criticized by Zwanzig (1963) on two grounds: nonadditivity and anisotropic effects are not negligible. Using formulas discussed in Chapter 5, Zwanzig evaluated the nonadditive corrections. However, after also evaluating the anisotropic contributions he finds results not very different from those of Salem.

7.6.2. Large Conjugated Double-bond Molecules

The long-range forces between large conjugated double-bond molecules like ethylene, acetylene, and the higher polyenes are unusually strong, suggesting that high-mobility, easily displaceable electrons are responsible for them. Closer inspection of their charge distribution shows that such electrons are indeed present. In the terminology of molecular physics, the conjugated structures contain two kinds of orbitals. First, there are the σ-molecular orbitals which are localized in individual atoms or in specific bonds. Their charge distributions permit a description along the lines of the preceding section. In addition, conjugated-bond molecules involve π orbitals in which an electron is not bound to a particular site but is free to course over the entire length of the bond system.† Such electrons can hardly be said to form a dipole in connection with any given nucleus; rather they form an extended charge oscillator of length comparable with the entire chain. London (1942), in a first suggestive treatment of the interaction problem, visualized these charges as suitably placed *monopoles* and provided a simple method for calculating the dispersion forces. The need for a new approach is clear from the fact that second-order perturbation theory, which was used for the most part in Chapter 2, breaks down unless the matrix elements of the Coulomb energy V are much smaller than the differences between unperturbed energy levels,

$$|\langle i|V|j\rangle| \ll |E_j - E_i|.$$

This relation is not satisfied when the states $|i\rangle$ and $|j\rangle$ make contributions to the V-elements over the entire length of a molecule. Or, in a similar vein, it is apparent that a multipole expansion makes little sense when the charge distribution reaches a linear size comparable with the distance between two chains.

After London's simple and illuminating contribution, Coulson and Davies (1952) published a more rigorous treatment of the π-bond problem in which realistic forms of the orbitals were used and integrals were accurately evaluated. Because of the more specific nature of their work we forego its full inclusion here. A simple, comprehensive, and systematic treatment of the problem, which opens perspectives on the approach of Coulson and

† A good discussion of these bonds in ethylene is found in Kauzmann (1957).

Davies as well as on that of London, is found in a paper by Haugh and Hirschfelder (1955) and we shall base our exposition upon it.

Let $\Psi_i^{(1)}$ represent the electronic state function of molecule 1 in its ith excited state, with $i = 0$ denoting the state of lowest energy, the state in which the interaction to be calculated occurs; $\Psi_j^{(2)}$ is defined analogously. The classical interaction energy is V. This, we recall, consists of terms depending only on the constant R, others which depend on the distance of one electron from one of the nuclei, and terms like e^2/r_{st}, s and t denoting electrons in the two different molecules. The first group can be ignored for they have vanishing matrix elements between orthogonal electronic states. The second remains finite if only one electron alters its state in the matrix element; such transitions, however, are known to make small contributions to the van der Waals forces; indeed, they vanish if the multipole expansion is used to represent V. Hence we are left with only the electron–electron interactions in V, and we may write as the significant perturbation

$$V = \sum_{s>t}' \frac{e^2}{r_{st}}, \tag{42}$$

where s and t take values from one to the total number of electrons in molecules 1 and 2 respectively. We call these numbers $n^{(1)}$ and $n^{(2)}$ and assume them to be even (for simplicity, since the forces can not depend on the parity of n). The dispersion energy, the second-order perturbation as usual, is

$$\Delta E = -\sum_{l,k}' \frac{|\langle \Psi_0^{(1)}\Psi_0^{(2)} | V | \Psi_l^{(1)}\Psi_k^{(2)}\rangle|^2}{E_l^{(1)}+E_k^{(2)}-E_0^{(1)}-E_0^{(2)}}. \tag{43}$$

For each Ψ we write an antisymmetrized product of real molecular orbitals, with spin functions properly assigned. These orbitals shall be written as $\varphi_1^{(1)}, \varphi_2^{(1)} \ldots$; subscript + and − indicate multiplication by a spin function α or β, and the one-electron energies associated with the various φ_i shall be ε_i. If the ground state of the molecule is a singlet, only transitions to singlet states l and k can appear in (43), and we may restrict our attention to them. This means that every orbital must appear twice, once with a positive and once with a negative subscript before antisymmetrization. For the ground state of one molecule we thus obtain

$$\Psi_0 = \mathcal{A}[\varphi_{1+}\varphi_{1-}\varphi_{2+}\varphi_{2-}\ldots\varphi_{i+}\varphi_{i-}\ldots\varphi_{n/2+}\varphi_{n/2-}]. \tag{44}$$

We need two such functions, one for molecule 1 and one for molecule 2. In order to designate the molecule in reference, it will be necessary to add superscript 1 or 2 to every Ψ, φ, and n. We forego this now and write the superscripts only when they matter.

An excited state, in which one electron has been promoted from φ_i with energy ε_i to a state φ_j with a *higher* energy ε_j, takes the form

$$\Psi_l \sqrt{(\tfrac{1}{2})}\{\mathcal{A}(\varphi_{1+}\varphi_{1-} \ldots \varphi_{i+}\varphi_{j-} \ldots \varphi_{n/2+}\varphi_{n/2-}) + \mathcal{A}(\varphi_{1+}\varphi_{1-} \ldots \varphi_{j+}\varphi_{i-} \ldots \varphi_{n/2+}\varphi_{n/2-})\}, \tag{45}$$

the extra factor $\sqrt{1/2}$ being needed to insure correct normalization. Again, superscripts have been omitted. To every index l of the total molecular state there corresponds in this way the promotion of one electron from its normal state i to a higher state j.

The constituents of V, whose matrix elements appear in (43), are two-particle terms of the form e^2/r_{st}, with s and t labeling electrons in two different molecules. To calculate $\langle 00 | e^2/r_{st} | lk \rangle$ we integrate, of course, over all $n^{(1)}$ electron coordinates of molecule 1 and over all those of molecule 2. But only coordinates s and t are affected by e^2/r_{st}, hence the integrations over all coordinates but s in molecule 1 and over all but t in molecule 2 can be performed directly. We thus find

$$\int \Psi_0^{(1)} \Psi_l^{(1)} d\tau_1 \ldots d\tau_{s-1}\, d\tau_{s+1} \ldots d\tau_{n^{(1)}} = \frac{1}{\sqrt{(2)} n^{(1)}} [\varphi_{i+}^{(1)}(s)\varphi_{j+}^{(1)}(s) + \varphi_{i-}^{(1)}(s)\varphi_{j-}^{(1)}(s)]. \tag{46}$$

The integration on the left is meant to include summations over spins. A similar expression, with s replaced by t and every superscript 1 by 2, results from the integration over all electrons save number t in molecule 2. The integral which remains to be performed to obtain the matrix element of e^2/r_{st} is

$$\iint \varphi_i^{(1)}(s)\varphi_j^{(1)}(s) \frac{e^2}{r_{st}} \varphi_{i'}^{(2)}(t)\varphi_{j'}^{(2)}(t)\, d\mathbf{r}_s\, d\mathbf{r}_t. \tag{47}$$

It extends only over space coordinates and will be abbreviated to $v_{ii',jj'}$. Straightforward substitution of these results, with attention to (42), converts (43) into

$$\Delta E = -4 \sum_{i=1}^{n^{(1)}/2} \sum_{j=(n^{(1)}/2)+1}^{\infty} \sum_{i'=1}^{n^{(2)}/2} \sum_{j'=(n^{(2)}/2)+1}^{\infty} \frac{v_{ii',jj'}^2}{\varepsilon_j^{(1)} - \varepsilon_i^{(1)} + \varepsilon_{j'}^{(2)} - \varepsilon_{i'}^{(2)}}. \tag{48}$$

The strange asymmetry in the summations occurs because we have labeled the orbitals *occupied* in the ground state of one molecule successively $\varphi_1, \varphi_2, \ldots, \varphi_{n/2}$, whereas the *excited* states, j and j', are unlimited.

We now consider the element (47). If r_{st}^{-1} were expanded in a multipole series, (47) and (48) would yield the usual result. This treatment is proper for σ electrons. For π electrons, London's method becomes appropriate. He defined a charge distribution corresponding to a transition of one electron from φ_i to φ_j as follows:

$$\varrho_{ij}^{(1)}(s) \equiv \sqrt{(2)}e\varphi_i^{(1)}(s)\varphi_j^{(1)}(s) \tag{49}$$

and similarly

$$\varrho_{i'j'}^{(2)}(t) \equiv \sqrt{(2)}e\varphi_{i'}^{(2)}(t)\varphi_{j'}^{(2)}(t)$$

and in terms of these (47) becomes

$$v_{ii',jj'} = \frac{1}{2}\iint \varrho_{ij}^{(1)}(s)\frac{1}{r_{st}}\varrho_{i'j'}^{(2)}(t)\,dr_s\,dr_t. \tag{50}$$

The fictitious charge distribution ϱ_{ij} will have different signs in different regions of space, and it is tempting to simplify further calculations by finding those regions in which ϱ_{ij} is positive and those in which it is negative, and then replacing the continuous ϱ-distribution by an equivalent set of point charges, positive in the former and negative in the latter parts of space. When these point charges, called monopoles, are given proper magnitudes and positions, they can be made to simulate the continuous distribution of "transition charge density" and lead to a correct result for the interaction energy.

Following Haugh and Hirschfelder, we divide space into cells which are bounded by the nodal surfaces of ϱ_{ij}, enumerating them by the index k. A similar partitioning is made with respect to $\varrho_{i'j'}$, with index k'. We place in cell k a point charge $g_{ij}^{(1)}(k)$ equal in magnitude and sign to the transition charge contained in it:

$$g_{ij}^{(1)}(k) = \int_k \varrho_{ij}^{(1)}\,dr; \quad g_{i'j'}^{(2)}(k') = \int_{k'} \varrho_{i'j'}^{(2)}\,dr,$$

where the integrations cover the cells k and k'. Furthermore, these monopoles g shall be located at points $\boldsymbol{R}$ where they produce the same dipole moment as the true distribution in each cell:

$$g_{ij}^{(1)}(k)\boldsymbol{R}_{ij}^{(1)}(k) = \int_k \boldsymbol{r}\varrho_{ij}^{(1)}\,dr; \quad g_{i'j'}^{(2)}(k')\boldsymbol{R}_{i'j'}^{(2)}(k') = \int_{k'} \boldsymbol{r}\varrho_{i'j'}^{(2)}\,dr.$$

With these specifications (50) becomes a discrete sum,

$$v_{ii',jj'} = \frac{1}{2}\sum_{kk'}\int_k\int_{k'} \varrho_{ij}^{(1)}(s)\,\frac{1}{r_{st}}\,\varrho_{i'j'}^{(2)}(t)\,d\boldsymbol{r}_s\,d\boldsymbol{r}_t$$

$$= \frac{1}{2}\sum_{kk'}\int_k\int_{k'} g_{ij}^{(1)}(k)\delta[\boldsymbol{r}_s-\boldsymbol{R}_{ij}^{(1)}(k)]\,\frac{1}{r_{st}}\,g_{i'j'}^{(2)}(k')\delta[\boldsymbol{r}_t-\boldsymbol{R}_{i'j'}^{(2)}(k')]\,d\boldsymbol{r}_s\,d\boldsymbol{r}_t$$

$$= \frac{1}{2}\sum_{kk'} g_{ij}^{(1)}(k)g_{i'j'}^{(2)}(k')(|\boldsymbol{R}_{ij}^{(1)}(k)-\boldsymbol{R}_{i'j'}^{(2)}(k')|)^{-1}. \tag{51}$$

This expression, inserted in (48), gives the dispersion energy in the London "monopole" approximation.† Beyond this point the problem reduces to the more specific and cumbersome task of determining the size and disposition of the discrete charges. London contented himself with the computation of an easy but impractical example, the transition of a hydrogen atom from $1s$ to $2p$, where, because of the nature of the p-function, k takes on the values 1 and 2. Here $g(1) = -g(2) = 0.21e$, while the distance between these point charges $|\boldsymbol{R}(1)-\boldsymbol{R}(2)| = 3.55a_0$. Haugh and Hirschfelder, however, actually calculate the cellular structure and the point charges for four linear molecules, from ethylene to hexatriene. We will insert a word here about the details of this procedure.

It is evident that exact molecular orbitals, even if they were available, would occasion difficulties which the approximate character of the method hardly justifies. On the other hand, the very simplest kind of orbital led to reasonable results. It is constructed by assigning to a π electron a one-dimensional motion, completely unhindered, along the axis of the molecule. The state function for this kind of behavior is known from the problem of the "one-dimensional box". For N conjugated double bonds and a molecule of length l,

$$\varphi_n = \left(\frac{2}{l}\right)^{1/2}\sin\frac{n\pi x}{l}, \quad n \text{ (our former } i) = 1, 2, \ldots, 2N.$$

† The term "monopole" is perhaps misleading. If φ_i and φ_j are orthogonal orbitals, (49) shows at once that $\int \varrho_{ij}\,d\boldsymbol{r} = 0$. Hence the sum of the positive discrete charges must equal the sum of the negative charges. It might be better to speak of multipoles of finite extension, or simply of the point-charge approximation.

The associated energies are

$$\varepsilon_n = \frac{n^2h^2}{8ml^2}.$$

The free-particle model has been applied to other physico-chemical problems (see especially Kuhn, 1948a, b, 1949; Scherr, 1952) and meets with fair success when l is taken to be $(2N+1)$ times D, the length of the $C—C$ bond. Thus, for $N = 1$ (ethylene), $l = 4.2$ Å.

Ethylene represents the simplest example, and the only one we choose here to illustrate the method. The important transition is from $n = 1$ to $n = 2$; higher values of n have small transition densities with the ground state ($n = 1$). To this transition correspond point charges of magnitude $\pm 0.60e$, 1.78 Å apart. They represent an extended dipole of 5.14 debye (5.14×10^{-18} e.s.u.). The energy difference, $\varepsilon_2 - \varepsilon_1 = 6.39$ ev. With these data, (48) and (51) provide the energy of interaction between the π electrons, which is part of the total interaction between the ethylene molecules. This energy, denoted by $E_{\pi\pi}$, depends on the relative orientations of the molecules. For large distances of separation R, such that R is much greater than the separation between the point charges, the monopoles behave somewhat like dipoles and give rise to a simple formula:

$$E_{\pi\pi} = -\frac{21.25 \text{ ev}}{R^6}[-2\cos\theta_1\cos\theta_2 + \sin\theta_1\sin\theta_2\cos(\varphi_1-\varphi_2)]^2$$

wherein, as in Chapter 2, the θ's are polar angles relative to the line joining the molecular centers, the φ's are azimuths, and R is expressed in Å. For smaller separations the tables in Haugh and Hirschfelder (1955) must be consulted.

The total interaction consists of three parts:

$$\Delta E = E_{\sigma\sigma} + E_{\sigma\pi} + E_{\pi\pi}.$$

The calculation of $E_{\sigma\sigma}$ proceeds along the lines outlined previously and involves bond polarizabilities. $E_{\sigma\pi}$ is computed by allowing the sigma electrons situated in the bonds to interact with the π-electron concentrations. For ΔE, Haugh and Hirschfelder (1955) present an interesting table which contains numerical values pertaining to a distance of separation R of 10 Å and to many different orientations. When the ethylene molecules are collinear (— —), $\Delta E = -5.21 \times 10^{-4}$ ev. The smallest interaction energy is found for the configuration —|, ΔE being -1.04×10^{-4} ev. The mean over all orientations is 1.81×10^{-4} ev. This value agrees in order of magnitude with inferences from data on viscosity. Exact agreement is not to be expected because a theory of viscosity capable of handling the complexities of the present ΔE with respect to its dependence on distance and on orientation does not exist.

Important extensions of the basic theory here given were made by Sternlicht (1964), who introduced electron correlation into the calculations.

Among recent, more concrete applications is the work of Craig *et al.* (1966) who study the crystal structure of benzene in terms of present knowledge concerning the various components of the intermolecular forces. They are able to explain observed variations of the crystal structure with temperature by showing that the repulsive forces, which they assume to act simply between individual hydrogen-atoms, are highly dependent on the orientation of the molecules.

Considerable work has been done to determine the configuration of various long-chain molecules using our knowledge of the interaction between small parts of these chains.†

This chapter has dealt with van der Waals interactions between large molecules in a narrow sense, excluding forces between charged configurations. There are, however, also important electrostatic interactions between heterocyclic molecules since they do have electrostatic multipole moments. De Voe and Tinoco (1962) have shown that these electrostatic interactions play an important role in stabilizing the double stranded DNA molecule. The multipole moments produced by the charges on the various atoms can be approximately evaluated by using very simple π-electron molecular orbital theory. In addition, two-body (see Sinanoglu,

† Polypeptides have been studied by Brant and Flory (1965), and vinyl polymers by Abe, Jernigan and Flory (1966). Recent work on polyoxymethylene chains, however, seems to suggest that the elementary view set forth in these articles does not always lead to a prediction of the correct configuration. The interactions must evidently be treated often in some more complicated way (P. J. Flory, private communication).

Abdulnur, and Kestner, 1964, p. 301) as well as many-body dispersion interactions are important.

Certain other interesting types of interactions can only be mentioned in this book. For example, when hydroxyl or amino-groups in one molecule are near oxygen or nitrogen atoms in another molecule, the hydrogen atom interacts strongly with both molecules leading to what is commonly called "hydrogen bonding". The latest review of this rather complicated subject has been written by Löwdin (1966). Another example is the charge fluctuation interaction (Kirkwood and Shumaker, 1952; Jehle, Parke, and Solyers, 1964) between protein molecules dissolved in water at their isoelectric point (i.e. at the pH where the molecule is electrically neutral). At any one instant the molecule is charged and thus electrostatic interactions between two molecules exist.

In a recent publication by V. Magnasco and G. F. Musso, *J. Chem. Phys.* **46**, 4015 (1967), the repulsive energy in the H_2-H_2 interaction is computed numerically. Professor Magnasco kindly prepared a comparison between the older results presented in this chapter and his own, permitting us to publish the table below, in which EM refers to Evett and Margenau, *Phys. Rev.* **90**, 1021 (1953); MH to Mason and Hirschfelder, *J. Chem. Phys.* **26**, 756 (1957) and MM to the publication by Magnasco and Musso, cited above. The data refer to case c on p. 235.

D(Å)	EM	MH	MM
2.23	...	...	0.00589
2.27	0.00509	0.00585	...
2.60	...	...	0.00161
2.73	0.00112	0.00118	...
2.98	...	...	0.00038
3.18	0.000232	0.000227	...

The agreement between the old and the new results, Professor Musso says, should be judged excellent in the whole region where results can be compared (2.2–3.0Å). To quote him: "This agreement is even more gratifying in view of the fact that our calculation is based on *exact values* for the multi-center integrals, whereas those of EM and MH are based upon different approximations to the multi-center integrals."

CHAPTER 8

Interactions between Excited Atoms

IN PREVIOUS chapters all of the interacting atoms and molecules were assumed in their ground states. We now consider the interactions between two atoms, at least one of which is in an excited state.

The Hamiltonian of our system is the sum of the exact Hamiltonians of the two atoms and a perturbation V:

$$H = H_0 + V = H_a + H_b + V. \tag{1}$$

We shall suppose the exact eigenfunctions of the atoms (a and b) to be known. Actually, accurate functions are available for only a few of the lower states of helium and for all states of hydrogen. These exact functions will be denoted in $|ai\rangle$ and $|bi\rangle$, where i specifies the excited state.

Perturbation theory will be used in this chapter, although we shall later indicate ways in which the variational method may be modified so as to apply to excited states, even when the ground state is not accurately known.

When overlap is small—and this is the only case to be treated in this chapter—we can neglect the exclusion principle and write the zero-order excited states of our system as

$$|ij\rangle \equiv |ai\rangle\,|bj\rangle. \tag{2}$$

Neglect of overlap can be serious; but we shall postpone discussion of this point until the end of the chapter. We shall further assume that V can be expanded in a multipole series, as in Chapter 2. Later the effect of the finite velocity of light on these interactions at large separations (retardation effects) will also be taken into account.

For the interaction of ground state atoms the results of Chapter 2 are

$$\Delta E(R) = E_1 - \sum_{ij\neq 0} \frac{|V_{00,\,ij}|^2}{E_a^i + E_b^j - E_a^0 - E_b^0}, \tag{3}$$

where

$$V_{ij,\,kl} \equiv \langle ij | V | kl \rangle \quad \text{and} \quad V_{00,\,00} = 0. \tag{4}$$

When one of the atoms is excited, both the first- and second-order perturbations must be studied carefully.

8.1. First-order Perturbation Energy

The first-order perturbation energy E_1 is given by the matrix element of the perturbation over the zero-order eigenfunctions of the state considered if that state is nondegenerate, i.e. if

$$H_0 | ij \rangle \neq H_0 | kl \rangle. \tag{5}$$

In that case, if atom a is in state i and atom b in state j,

$$E_1 = V_{ij,ij}. \tag{6}$$

As one example we might consider a hydrogen atom a in its first excited state and a helium atom b in its ground state. Then

$$|a1\rangle |b0\rangle = |10\rangle \tag{7}$$

and there exists no other zero-order eigenfunction of the same energy. Hence, from our discussion in Chapter 2, we see that

$$V_{10,\,10} = 0. \tag{8}$$

This leads to a general rule: *In the interaction of two unlike atoms there are no first order energy corrections* when overlap is neglected.

As a counter example, consider the interaction of two hydrogen atoms, one in its first excited state and the other in its ground state. Now states $|10\rangle$ and $|01\rangle$ have the same energy. Therefore, (6) no longer applies. In general, there *are* first-order contributions to $\Delta E(R)$ when two like atoms in different states of excitation interact.

There is an especially important class of degeneracies which will concern us most. In the last example the state of excitation of the two atoms is such that radiation from one can be absorbed by the other. We shall refer to this condition as resonance.

Mathematically this means that the transition dipole matrix element between the ground and excited state of the *atom* is finite, i.e. the oscillator strength (see Chapter 2) is not zero for this transition. Such resonant states will lead to the most important contributions of longest range in the first-order energy.

To solve the first-order problem we use degenerate perturbation theory.† Consider the case where m zero-order eigenfunctions are degenerate. They will be designated as $\psi_1^0 \ldots \psi_m^0$ and assumed to be orthonormal, i.e.

$$\langle \psi_i^0 | \psi_j^0 \rangle = \delta_{ij} \quad (i, j = 1 \ldots m). \tag{9}$$

There will then be m new linear combinations which will not be degenerate through terms of first order in the perturbation

$$\varphi_k = \sum_{i=1}^{m} C_{ki} \psi_i^0, \quad (k = 1, 2 \ldots m). \tag{10}$$

To carry out the perturbation theory we substitute

$$\begin{aligned} E_k &= E^0 + E_1^{(k)} + E_2^{(k)}, \\ \psi_k &= \varphi_k + \psi_1^{(k)} + \psi_2^{(k)}. \end{aligned} \tag{11}$$

$E_1^{(k)}$ is the first order energy, $E_2^{(k)}$ the second order energy, E^0 being the energy of the degenerate states. When the Schrödinger equation is written and only first-order terms are retained, we have

$$(H_0 - E^0)\psi_1^{(k)} = \sum_{j=1}^{m} C_{kj}(E_1 - V)\psi^0. \tag{12}$$

Upon expanding $\psi_1^{(k)}$ in the complete set of functions ψ_j^0, i.e.

$$\psi_1^{(k)} = \sum_j A_{kj} \psi_j^0, \tag{13}$$

we obtain

$$\sum_j (E_j^0 - E^0) A_{kj} \psi_j^0 = \sum_{j=1}^{m} E_1 C_{kj} \psi_j^0 - \sum_j \Big(\sum_{l=1}^{m} C_{kl} V_{jl}\Big) \psi_j^0. \tag{14}$$

The degeneracy causes difficulty only for $j \leqslant m$, that is to say when $E_j^0 = E^0$. In that case it is required that, for every j,

$$\sum_{l=1}^{n} V_{jl} C_{kl} - E_1 C_{kj} = 0. \tag{15}$$

† For further discussion of degenerate perturbation theory, see Bates (1961, vol. 1, p. 194) and Eyring, Walter, and Kimball (1944, chap. 7).

This equation has a solution only if

$$|\vec{V}-E_1\vec{I}| = 0, \tag{16}$$

where we have employed matrix notation. If the original states were not orthonormal the unit matrix would be replaced by the overlap matrix, $\langle \psi_i^0 | \psi_j^0 \rangle$. Equation (16) will form the basis of all first-order calculations on excited-state interactions. The eigenfunctions corresponding to the new first-order eigenvalues are found by requiring that

$$\sum_{i=1}^{m} C_{ki}^* C_{ki} = 1. \tag{17}$$

As an example, consider again the case of two like atoms, one in its first excited state, $|a1\rangle$, and the other in the ground state $|b0|\rangle$

$$\psi_1^0 = |10\rangle, \quad \psi_2^0 = |01\rangle. \tag{18}$$

both of the same energy.

According to (16) we must solve

$$\begin{vmatrix} V_{10,\,10}-E_1 & V_{10,\,01} \\ V_{01,\,10} & V_{01,\,01}-E_1 \end{vmatrix} = 0 \tag{19}$$

and obtain

$$E_1 = \pm|V_{01,\,10}| \tag{20}$$

since $V_{10,10} = 0$ for like (and unlike) atoms. The eigenfunctions are easily found by calculation or inspection† to be

$$\psi_1 - \frac{1}{\sqrt{2}}(\psi_1^0 | \psi_2^0)$$

$$E_1^{(1)} - |V_{01,\,10}|, \tag{21}$$

† Note that

$$-V_{10,\,01}C_{11}+V_{10,\,01}C_{12} = 0$$

and

$$V_{10,\,01}C_{21}+V_{10,01}C_{22} = 0$$

so that upon normalization

$$C_{11} = +C_{12} = \frac{1}{\sqrt{2}},$$

$$C_{21} = -C_{22} = \frac{1}{\sqrt{2}}.$$

and
$$\varphi_2 = \frac{1}{\sqrt{2}}(\psi_1^0 - \psi_2^0)$$
$$E_1^{(2)} = -|V_{01,\,10}|. \tag{22}$$

Specifically, for two hydrogen atoms, one in a 1s state, the other in a 2p resonant state ($m = 0, \pm 1$) and

$$V = \frac{1}{R^3}[x_a x_b + y_a y_b - 2z_a z_b]. \tag{23}$$

where x, y, and z are components of the dipole moment in atomic units. From (21) and (22) we have

$$E_1 = \pm |\langle \varphi_1 | V | \varphi_2 \rangle|, \tag{24}$$

where $\psi_1^0 = |10\rangle \equiv |2p_{za} 1s_b\rangle$, for example if $m = 0$. Performing the integration over angles we easily see that

$$E_1(m = \pm 1) = \pm \frac{|r_{01}|^2}{3R^3}, \tag{25}$$

$$E_1(m = 0) = \pm \frac{2|r_{01}|^2}{3R^3}, \tag{26}$$

provided $|r_{01}|$ is the radial matrix element

$$r_{01} = r_{10} = \langle 2p_a | r | 1s_a \rangle \tag{27}$$

and $2p_a$ is the radial factor of the 2p wave function.

A novel feature emerges at this point. While asymptotic van der Waals' forces between atoms in their ground states are always attractive, "resonance forces" of long range (R^{-3} dependence) are equally likely tobe repulsive or attractive even for large R.

In the treatment of the interaction of two hydrogen atoms we considered the various excitations, $m = 0, \pm 1$, separately. The reason we were permitted to do so is this: the symmetry of the atomic orbitals (represented by the magnetic quantum number m) did not allow the various m values to combine with one another. In other words, the i of (10) must belong to one definite m value. Because of this even if we had used the six possible 1s and 2p orbital combinations in place of (18) the resulting matrix equation would have had the following block form:

$$\begin{bmatrix} m=0 & 0 & 0 \\ 0 & m=1 & 0 \\ 0 & 0 & m=-1 \end{bmatrix} = 0 \qquad (28)$$

and the solutions would have been exactly those listed in (25) and (26) for each 2×2 matrix.

This suggests that we use further elements of symmetry to reduce the general matrix equation, (16). For when the functions $\psi_{j_i}^0$ are enumerated according to molecular symmetries, the matrix in (16) again takes on a blocked form, only states with the same molecular symmetry appearing in each block. Symmetry for diatomic molecules is denoted by the symbol

$$^{2S+1}\Lambda^{\pm}_{g,u} \qquad (29)$$

The left superscript $2S+1$ refers to the spin multiplicity. In the interaction of two "spin one-half" atoms, hydrogen or alkali atoms, for example, we can have $S = 0$ or 1, singlets or triplets. Λ is the symbol representing the electronic orbital angular momentum along the internuclear axis:

$$\Lambda = |M_L|. \qquad (30)$$

The symbols Σ, Π, Δ, Φ, ..., are used to designate states with $\Lambda = 0, 1, 2, 3, \ldots$. For $\Lambda > 0$ all states are doubly degenerate. The subscripts g and u refer to symmetry under reflection in a plane of symmetry midway between the nuclei and at right angles to the internuclear line. If the state function is invariant under this symmetry operation it is of g or gerade (even) symmetry; if it changes sign it is u or ungerade. The superscript + or − is determined by the behavior of the state function under reflection in a plane of symmetry which passes through both nuclei. If it does not change sign, it is plus; if it does, we use the negative sign. Thus all functions involving only atomic s orbitals are plus.

To return to the hydrogen interactions, the element of symmetry used in (28) to reduce the matrix equation from one 6×6 equation to three 2×2 equations was M_L. The 2×2 equations were solved algebraically to find the eigenfunctions. However, we notice that the resulting eigenfunctions (21) and (22) are simply eigenfunctions of g and u symmetry operators. Thus if we had used the complete molecular symmetry classification at the start, our matrix would have been immediately diagonal with the eigenfunctions as listed. For example, the eigenfunctions of the $2p_a(m = 1)$ orbital denoted as 1 and the $1s_b$ orbital denoted as 0, would be

$$|10\rangle_- = \frac{1}{\sqrt{2}}(|10\rangle - |01\rangle) \quad \text{or} \quad {}^3\Pi_u \tag{31}$$

if it were a triplet state. Thus far we have not specified the spin symmetry.

Usually one is not greatly interested in the eigenfunctions but only in the energy values. In that case Fontana (1961b) has shown how the interactions can be computed with ease. Further interest in the eigenfunctions will be satisfied by the papers of Mulliken (1960) and Linder and Hirschfelder (1958), for they present in detail the eigenfunctions of proper symmetry for states involving $2s$ and $2p$ hydrogen orbitals.

To evaluate (16) easily we need a general expression for the matrix elements of V. Using eqn. (9) of Chapter 2, one finds the matrix elements of V between two states specified by p and p' (they do not include the g and u symmetry elements) to be

$$\begin{aligned}(p'|V|p) &= \sum_{a,b,\alpha} \frac{(-1)^b 4\pi(a+b)!}{R^{a+b+1}[(2a+1)(2b+1)(a-\alpha)!(a+\alpha)!(b+\alpha)!(b-\alpha)!]^{1/2}} \\ &\quad \times (q_1'|Y_a^{\alpha *}(\theta_1,\varphi_1)|q_1)(q_2'|Y_\alpha^{-\alpha *}(\theta_2,\varphi_2)|q_2) \\ &\quad \times (n_1'l_1'|r_1^a|n_1l_1)(n_2'l_2'|r_2^b|n_2l_2)\,, \end{aligned} \tag{32}$$

where q and q' are the quantum numbers l and m while (nl) and $(n'l')$ specify the radial quantum states of the two atoms; α occurs because of the tensor contraction (see Chapter 2).

When spin-orbit coupling effects are small ($V \gg H_{LS}$) and L and S are good quantum numbers individually, the matrix elements over the spherical harmonics can be expressed in terms of the Clebsch–Gordan C coefficients. This was done in Chapter 2 following eqn. (32) of that chapter, and the result was

$$(l'm' \mid Y_a^{\alpha *}(\theta\varphi) \mid lm) = (l'm' \mid Y_a^{-\alpha}(\theta, \varphi) \mid lm)(-1)^{\alpha}$$

$$= (-1)^{\alpha} \left[\frac{(2l+1)(2a+1)}{4\pi(2l'+1)} \right]^{1/2} C(lal'; m, -\alpha, m)\, C(lal'; 000). \quad (33)$$

Thus the matrix elements wich occur in (32) exist only if $C(l_2'al_1;000)\,C(l_1'bl_2';000)$ is finite, i.e. only if $a = l_2 + l_1, l_2' + l_1 - 2, \ldots$ and $b = l_1' + l_2, l_1' + l_2 - 2, \ldots$. When $a = b = 1$ we have the resonant interactions studied before. For example, in the interaction of two atoms, each in a state with $l = 1$ and resulting in a state of total molecular symmetry $^1\Delta_g$, the interaction is proportional to $C(1a1;000)\,C(1a1;000)$. By arguments employed earlier the interaction varies as R^{-2a-1} or R^{-s}.

Any terms with $a = 0$ or $b = 0$ do not contribute to the interaction since they are canceled by effects arising from the electron–nuclear attractions. For the $^0\Pi_u$ state the results vary as R^{-3} ($a = 0+1$, $b = 0+1$) as found before.

In using (32) we have not specified symmetry with respect to reflection in the mirror plane between the two nuclei, i.e. the g and u classification. To do so will further simplify (16). This is easily accomplished for the molecular sigma states. However, in general, unless the matrix is originally constructed with states of proper g and u symmetry, it is simpler to discover the effects of this symmetry by solving the matrix equation.

Fontana (1961b) has evaluated the interaction of variously excited hydrogen and hydrogen-like atoms in this way. His results supplement and verify or correct those listed by Mulliken (1960) and by Linder and Hirschfelder (1958) in references mentioned earlier. The values in Mulliken's article are listed in terms of dipole matrix elements and therefore are very general.

We present in Table 8.1 the results of Fontana and Linder and Hirschfelder for the interaction of two hydrogen atoms, both in various p states.

When the $2s$ orbitals are also included, then, since they are degenerate with $2p$ (in the non-relativistic hydrogen atom) the matrix in (16) has 22×22 elements. It reduces to six 1×1 diagonal blocks when the plus and minus symmetry is used, plus four 2×2 and two 4×4 matrices. Fontana has also offered procedures

TABLE 8.1. *Interaction of Two Hydrogen Atoms*

State	Orbitals included[a]	First-order perturbation theory $\Delta E(R)$
${}^1\Delta_g$, ${}^3\Delta_u$	$2p_1$, $2p_1$	$216/R^5$
${}^1\Pi_g$, ${}^3\Pi_u$	$2p_0$, $2p_1$	$432/R^5 - 62.208/R^7$
${}^1\Pi_u$, ${}^3\Pi_g$	$2p_0$, $2p_1$	0
${}^1\Sigma_u^-$, ${}^3\Sigma_g^-$	$2p_1$, $2p_{-1}$	0
${}^1\Sigma_g^+$, ${}^3\Sigma_u^+$	$2p_0$, $2p_0$	$9\sqrt{(6)}/R^3 + [2592 + 972\sqrt{(6)}]/R^5$
${}^1\Sigma_g^+$, ${}^3\Sigma_u^+$	$2p_1$, $2p_{-1}$	$-9\sqrt{(6)}/R^3 + [2592 - 972\sqrt{(6)}]/R^5$

[a] The subscripts designate the *m* quantum number of the orbital

by which the number and size of these small matrices can be determined.

When both atoms are highly excited somewhat strange new phenomena occur since many multipole contributions can enter with different signs. Fontana (1961b) finds that the $\Phi_u(2P_1, 3D_2)$ state exhibits a large maximum at about $10a_0$. Other states can become repulsive at short separations while remaining attractive at larger distances. This could lead to bound states of a purely *electrostatic* origin. Linder and Hirschfelder (1958) have exhibited a number of examples of intermolecular potential curves which contain "humps" or secondary maxima. Several ${}^3\Sigma_u^+$ and ${}^1\Sigma_g^+$ states do show small maxima in the region of 10–$12a_0$. Most interesting is the ${}^3\Sigma_u^+$ state formed from a sum of triplet $2s$ and $2p_0$ (or $2p_z$) functions of ${}^3\Sigma_u^+$ symmetry. This combination produces a negative potential for $R < 10a_0$, but reaches a maximum of about 0.5 kcal/mole (0.8×10^{-3} a.u., 0.022 ev, or 175 cm^{-1}) at $11.5a_0$.

When the classical potential V is very small, another perturbation must be considered, namely the spin–orbit coupling, H_{LS}. If $H_{LS} > V$, L (the angular momentum) and S (the spin) are no longer good quantum numbers. One must then designate states by the quantum numbers J and M_J ($\boldsymbol{J} = \boldsymbol{L} + \boldsymbol{S}$, $M_J = J_z$). This removes some of the degeneracy previously encountered, especially that between triplet and singlet states. Since spin was not present in the simple Hamiltonian, eqn. (1), such states previously had the same energy. The matrix elements of (32) are now changed since the state function in this coupled form must be written (Fontana, 1962)

$$\Psi_{jm} = \sum_{m_l} C(lsj;\, m_l,\, m-m_l,\, m)\Psi_{l,\, m_l}\Psi_{s,m-m_l},$$

where l and s are the angular momentum and spin quantum numbers respectively. If the state function Ψ_{jm} is orthonormal to every other state $\Psi_{jm'}$, we can write the essential matrix elements as follows:

$$\begin{aligned}(j'm' \mid Y_a^{\alpha *} \mid jm) &= \sum_{m'_l}\sum_{m_l} C(l's'j';\, m'_l,\, m'-m'_l,\, m').\\ &\quad C(lsj;\, m_l,\, m-m_l,\, m)\,(l'm'_l \mid Y_a^{\alpha *} \mid lm_l).\\ &\quad \delta(m'-m'_l,\, m-m_l)\; \delta(s's).\end{aligned} \tag{34}$$

Using (33) and $m_l - \alpha = m'_l$, we eliminate one sum above to obtain

$$\begin{aligned}(j'm' \mid Y_a^{\alpha *} \mid jm) &= \sum_{m_l} C(l's'j';\, m_l-\alpha,\, m'-m_l+\alpha,\, m')\\ &\quad \times C(lsj;\, m_l,\, m-m_l,\, m)C(lal;\, m_l,\, -\alpha,\, m_l-\alpha)\\ &\quad \times C(lal';\, 000)(-1)^{\alpha}\left[\frac{(2l+1)(2a+1)}{(2l'+1)}\right]^{1/2}.\end{aligned} \tag{35}$$

This sum can also be evaluated by introducing Racah W coefficients (Racah, 1942, 1943; Rose, 1957) or their close relatives, the Wigner "6j" coefficients to obtain

$$(j'm' \mid Y_a^{\alpha *} \mid jm) = (-1)^{s-l-j'+a+\alpha}\left[\frac{2j'+1}{2j+1}\right]^{1/2} C(jaj';\, m,\, -\alpha,\, m').$$

$$W = (ljl'j';\, sa)\left[\frac{(2l+1)(2a+1)}{4\pi(2l'+1)}\right]^{1/2} C(lal';\, 000). \tag{36}$$

Tables of W are available (Biedenharn, 1952).

The relation of (36) to (33) for the uncoupled result (where L and S are good quantum numbers independently) can be obtained by setting $S = 0$, since then $j' = l'$, $j = l$, and

$$W(lll'l';\, 0a) = (-1)^{a-l-l'}[(2l+1)(2l'+1)]^{-1/2}. \tag{37}$$

Equation (36) vanishes unless $m_1+m_2 = m'_1+m'_2$. This is the condition that the total angular momentum J be quantized along the internuclear axis. States can also be classified by their parity under inversion (denoted by g or u) in the center of symmetry.

Below we list the results of Fontana for the interaction of two hydrogen atoms, one in the ground state and the other in the $2p$ state. We show both the uncoupled results where the terms are classified by L and M (Σ, Π, Δ, Φ, etc.) and the coupled results discussed above, where states are labeled by J and M (1, 2, 3, . . .). Where possible the j and j_z values for the individual atoms are listed as a double subscript.

Uncoupled

$$
\begin{aligned}
{}^1\Pi_g(1s, 2p_1) &= -\frac{C}{3R^3} \\
&= {}^3\Pi_u \\
{}^1\Pi_u(1s, 2p_1) &= \frac{C}{3R^3} \\
&= {}^3\Pi_g \\
{}^1\Sigma_g^+(1s, 2p_0) &= \frac{2C}{3R^3} \\
&= {}^3\Sigma_u^+ \\
{}^1\Sigma_u^+(1s, 2p_0) &= \frac{-2C}{3R^3} \\
&= {}^3\Sigma_g^+
\end{aligned}
\tag{38}
$$

The constant $C = 1.664787$, a value obtained by computing the matrix elements for the hydrogen atom.

Coupled

(a) $(1s_{1/2},\ 2p_{3/2})$

$$
\begin{aligned}
&2_g(1s_{1/2\,1/2},\ 2p_{3/2\,3/2}) = -2_u(1s_{1/2\,1/2},\ 2p_{3/2\,3/2}) = -C/3R^3, \\
&1_g(1s_{1/2\,1/2},\ 2p_{3/2\,3/2}) = -1_u(1s_{1/2\,1/2},\ 2p_{3/2\,3/2}) = (2-\sqrt{7})C/9R^3, \\
&1_g(1s_{1/2\,1/2},\ 2p_{3/2\,3/2}) = -1u(1s_{1/2\,1/2},\ 2p_{3/2\,1/2}) = (2+\sqrt{7})C/9R^3, \\
&0_g(1s_{1/2},\ 2p_{3/2})_A = -0_u(1s_{1/2},\ 2p_{3/2})_A = -5C/9R^3, \\
&0_g = (1s_{1/2},\ 2p_{3/2})_B = -0_u(1s_{1/2},\ 2p_{3/2})_B = 3C/9R^3.
\end{aligned}
\tag{39}
$$

A and B denote the two solutions of a 2×2 secular equation.

(b) $(1s_{1/2},\ 2p_{1/2})$,

$$
\begin{aligned}
&1_g(1s_{1/2\,1/2},\ 2p_{1/2\,1/2}) = -1_u(1s_{1/2\,1/2},\ 2p_{1/2\,1/2}) = 2C/9R^3, \\
&0_g(1s_{1/2},\ 2p_{1/2})_A = 0_u(1s_{1/2},\ 2p_{1/2})_A = 0, \\
&0_g(1s_{1/2},\ 2p_{1/2})_B = -0_u(1s_{1/2},\ 2p_{1/2})_B = -4C/9R^3.
\end{aligned}
\tag{40}
$$

Further results (Fontana, 1962) are available for states which arise from the interaction of two hydrogen atoms or two alkali atoms in which the effective nuclear charge for different atomic levels may be different.

As mentioned, the uncoupled approximation is correct when $V \gg H_{LS}$, the coupled form when $V \ll H_{LS}$. The intermediate case is very complicated. Fontana (1962) has considered the general case for the interaction of two hydrogen atoms, one in the $1s$ state (or $1s_{1/2,\,-1/2}$), the other in the $2p_1$ state (or $2p_{3/2,\,3/2}$). The uncoupled results $[\pi_u(1s, 2p)]$ are accurate to within 10% or better for $R < 15a_0$ (provided the multiple results alone yield the correct interaction); the coupled results $[1_u(1s_{1/2-1/2}\, 2p_{3/2\, 3/2})]$ are accurate to 10% for $R > 40a_0$. Over the range of greatest interest the uncoupled result is the more accurate.

8.1.1. *Exchange Contributions in First-order Interactions*

The discussion in this chapter has so far neglected all exchange effects arising from the Pauli principle. Estimates of the effect of electron exchange or antisymmetry have been made by Linder and Hirschfelder (1958) and by Mulliken (1960). For $2s$ and $2p$ excited hydrogen atoms one can neglect exchange at distance over $8a_0$ in most cases with an error less than 20%. In some cases such as the ${}^1\Pi_u$, ${}^3\Pi_g$, ${}^3\Sigma_u^-$, and ${}^3\Sigma_g^-$ states one can almost never neglect it. Exchange contributions are greater for the triplet than for the singlet states, as might be expected. Specific examples of the separations at which the error in ignoring exchange is less than 20% are $15a_0$ for the ${}^1\Delta_g$ state, $13.5a_0$ for the ${}^3\Delta_u$ state, $10a_0$ for the ${}^1\Pi_u$, $12a_0$ for the ${}^1\Pi_g$, $7.5a_0$ for the ${}^3\Pi_u$, and $7a_0$ for the ${}^3\Sigma_g^+$ state. In Chapter 4 we found only very small corrections due to exchange in the interaction of ground state hydrogen atoms even at $8a_0$. However, when excited states are involved, charge distributions are much more diffuse, i.e. their overlap integral for a given R is much larger than it is for two $1s$ orbitals. For instance at $8a_0\, \langle 1s_a | 1s_b \rangle = 0.0102$, while $\langle 2pz_a | 2pz_b \rangle = 0.319$ (for hydrogen).

8.1.2. *Radiation Corrections to the First-order Resonant Interaction of Like Atoms*

We recall from Chapter 6 that, to include radiation effects, we must replace the matrix elements $\langle q' | V | q \rangle$ by $\langle q' | p_a \cdot \overleftrightarrow{T}_{ab} \cdot p_b | q \rangle$, where T_{ab} is the general frequency-dependent tensor coupling the dipole moments of the two atoms. Since, as discussed in Chapter 6 [eqn. (76) and the following pages—see also Chapters 2 and 5], T_{ab} is complex, we must actually use $1/2\,(T^*_{ab}+T_{ab})$. In this case we find in terms of the wavelength $\lambdabar = \hbar c/E$, E being the energy of the excited state relative to the ground state:

$$\Delta E = -\langle \boldsymbol{p}_a\rangle \cdot \left[(\overleftrightarrow{\boldsymbol{I}} - \hat{\boldsymbol{R}}_a\hat{\boldsymbol{R}}_b)\frac{\cos R\lambdabar}{\lambdabar^2 R} - (\overleftrightarrow{\boldsymbol{I}} - 3\hat{\boldsymbol{R}}_a\hat{\boldsymbol{R}}_b)\left\{\frac{\sin R/\lambdabar}{\lambdabar R^2} + \frac{\cos R/\lambdabar}{R^3}\right\}\right] \cdot \langle \boldsymbol{p}_b\rangle \tag{41}$$

Here $\overleftrightarrow{\boldsymbol{I}}$ is the unit tensor and $\hat{\boldsymbol{R}}$ is a unit vector along the internuclear axis. $\langle \boldsymbol{p}_a\rangle$ and $\langle \boldsymbol{p}_b\rangle$ are matrix elements of the transition dipole moment between the atomic states involved. For small separations, this reduces to the standard result:

$$\Delta E \sim +\langle \boldsymbol{p}_a\rangle \cdot \frac{(\overleftrightarrow{\boldsymbol{I}} - 3\hat{\boldsymbol{R}}_a\hat{\boldsymbol{R}}_b)}{R^3} \cdot \langle \boldsymbol{p}_b\rangle \quad \text{(small } R) \tag{42}$$

which is equivalent to (25) and (26). At large separations we obtain

$$\Delta E \sim -\langle \boldsymbol{p}_a\rangle \cdot (\overleftrightarrow{\boldsymbol{I}} - \hat{\boldsymbol{R}}_a\hat{\boldsymbol{R}}_b) \cdot \langle \boldsymbol{p}_b\rangle \frac{\cos R/\lambdabar}{\lambdabar^2 R} \quad \text{(large } R). \tag{43}$$

The result is simply the interaction of a dipole $\langle \boldsymbol{p}_a\rangle$ with the wave zone (large R) electric field of the oscillating dipole $\langle \boldsymbol{p}_b\rangle$ (Panofsky and Phillips, 1955).

It can be derived directly by second-order perturbation theory if the radiation field is included. The technique is that used in Chapter 6 where we obtained the interacton between two ground state atoms by fourth-order perturbation theory:

$$\Delta E = \sum_I \frac{\langle i\,|\,\boldsymbol{p}_a\cdot\boldsymbol{E}^{\perp}(a)\,|\,I\rangle\langle I\,|\,\boldsymbol{p}_b\cdot\boldsymbol{E}^{\perp}(b)\,|\,i\rangle}{E_i - E_I} + \sum_{II} \frac{\langle i\,|\,\boldsymbol{p}_b\cdot\boldsymbol{E}^{\perp}(b)\,|\,\mathrm{II}\rangle\langle \mathrm{II}\,|\,\boldsymbol{p}_a\cdot\boldsymbol{E}^{\perp}(a)\,|\,i\rangle}{E_i - E_{II}}. \tag{44}$$

State function $|\,i\rangle$ represents the system with no photons present. In state $|\,I\rangle$ both atoms are in their ground states but a photon of wave vector k and polarization λ is present, leading to $E_i - E_I = E - \hbar ck$, where E is the excitation energy of one atom. State $|II\rangle$ is another intermediate state in which both atoms are excited *in addition* to having a photon in the radiation field.

Using the quantized field of eqn. (10) in Chapter 6, we then obtain

$$\Delta E = \frac{2\pi}{V}\sum_{k,\lambda=1,2}[\langle \boldsymbol{p}_a\rangle\cdot\hat{e}^{(\lambda)}\hat{e}^{(\lambda)}\cdot\langle \boldsymbol{p}_b\rangle]\left[\frac{e^{i\boldsymbol{k}\cdot\boldsymbol{R}}}{E-(\hbar ck)}-\frac{e^{-i\boldsymbol{k}\cdot\boldsymbol{R}}}{E+(\hbar ck)}\right], \quad (45)$$

which, by virtue of eqn. (16) of Chapter 6, becomes

$$\Delta E = 4\pi\langle \boldsymbol{p}_a\rangle\langle \boldsymbol{p}_b\rangle : \int \frac{(\overleftrightarrow{I}-\hat{k}\hat{k})e^{i\boldsymbol{k}\cdot\boldsymbol{R}}}{(E^2-(\hbar ck)^2)}(\lambda ck)^2\frac{d^3k}{(2\pi)^3}. \quad (46)$$

The integration over angles yields

$$\Delta E = \frac{2}{\pi}\langle \boldsymbol{p}_a\rangle\langle \boldsymbol{p}_b\rangle : \int\left[(\overleftrightarrow{I}-\hat{R}\hat{R})\frac{\sin kR}{kR}\right.$$

$$\left.+(\overleftrightarrow{I}-3\hat{R}\hat{R})\left(\frac{\cos kR}{k^2R^2}-\frac{\sin kR}{k^3R^3}\right)\right]\frac{(\hbar c)^2k^4dk}{[E^2-(\hbar ck)^2]}. \quad (47)$$

The k integrations are easily carried out by extending the limits of integration from $-\infty$ to $+\infty$ and calculating the residues. The result is (41).†

If, however, instead of considering a steady-state situation we treat one in which an atom is adiabatically excited at some time, t_0, we must use time-dependent theory. In that case the interaction energy emerges multiplied by a factor $e^{-\gamma R/C}$, where γ^{-1} is the lifetime of the states (Stephen, 1964).

Indeed, the entire concept of an interaction energy is not well defined at large separations since distant atoms exchange photons very infrequently so that the state of excitation of an atom changes slowly with time. These points have been studied in detail by Stephen (1964), Hamilton (1949), and McLachlan (1964b). The latter author also discusses the general problem of resonant energy transfer in which one excited molecule imparts its excitation to another separated from the first by a distance less than the wavelength of the excitation photon divided by 2π. In this analysis an interesting sidelight appears: the second molecule can become excited before a photon reaches it, contrary to what

† This perturbation treatment is taken from a paper by McLone and Power (1964).

one would believe from the so-called principle of causality. From the point of view of virtual excitation within small time intervals, this is understandable. McLachlan shows, however, that all observables such as the dipole moment of one of the molecules exhibit causal behavior.

In this context we shall not pursue the closely related problem of spectral line shapes, lifetimes, and how they are modified by the presence of another atom or atoms, even though some of this information is contained in the imaginary part of $\overleftrightarrow{T}_{ab}$. For reference the reader should consult the papers of Stephen (1964), Hamilton (1949), and McLachlan (1964b), the recent work of Hutchinson and Hameka (1964), Philpott (1966), and various reviews (Breene, Jr., 1957; Ch'en and Takeo, 1957; Baranger, 1962; Margenau and Lewis, 1959). The lifetime of a state with an attractive interaction is long at short distances and equal to the natural lifetime at large separations.

8.2. Second-order Perturbation Energy

As already shown in Chapter 2, contributions to the second-order interaction energy exist for like as well as unlike atoms. To treat the interaction of excited unlike atoms the same procedures as in Chapter 2 are employed except that the zero-order functions are no longer the ground state solutions of the unperturbed Hamiltonian. One can even use the method of relating matrix elements to the oscillator strengths of each transition to arrive at semi-empirical results [see eqn. (27) of Chapter 2], as explained.

To obtain the second-order interaction of two like atoms, the only complication encountered is that one must use zero-order state functions which are linearly combined to remove degeneracy.

As an example consider again the interaction of one atom in the ground state with another in the first excited state. The required zero-order function were

$$|01\rangle_+ \equiv \frac{1}{\sqrt{2}}[|01\rangle + |10\rangle], \tag{48}$$

$$|01\rangle_- \equiv \frac{1}{\sqrt{2}}[|01\rangle - |10\rangle]. \tag{49}$$

These are (21) and (22) written in a more general notation. Similarly we represent all excited states in the form

$$|ij\rangle_+ \equiv \sqrt{\frac{1}{2}}\{|ai\rangle|bj\rangle+|aj\rangle|bi\rangle\}, \tag{50}$$

$$|ij\rangle_- \equiv \sqrt{\frac{1}{2}}\{|ai\rangle|bj\rangle-|aj\rangle|bi\rangle\}. \tag{51}$$

Because of their opposite symmetries with respect to an exchange of a and b, our symmetric V has no matrix elements between sets (50) and (51). The second-order energy therefore splits at once into two:

$$(E_2)_+ = \sum_{ij}{}' \frac{|\langle 01|V|ij\rangle_+|^2}{E_a^0+E_b^1-E_a^i-E_b^j}, \tag{52}$$

$$(E_2)_- = \sum_{ij}{}' \frac{|\langle 01|V|ij\rangle_-|^2}{E_a^0+E_b^1-E_a^i-E_b^j}. \tag{53}$$

The prime on the summations is to indicate that the pair of terms, 01, are to be excluded. Since the state functions (50) and (51) remain unaltered upon an interchange of i and j [except for a trivial sign change of (51)] we must count each pair of indices (ij) only once in the summations. Diagonal terms (ii) cause no trouble because both $\langle 01|V|ii\rangle_+$ and $\langle 01|V|ii\rangle_-$ vanish.†

Let us now suppose, subject to further consideration, that the denominators in the summations can be replaced by a suitable average $\bar{E}$. We may then write, for example,

$$\sum_{ij}{}' \frac{|\langle 01|V|ij\rangle_+|^2}{E_a^0+E_b^1-E_a^i-E_b^j} = \frac{1}{\bar{E}}\{\langle 01|V^2|01\rangle_+ - \langle 01|V|01\rangle_+^2\} \tag{54}$$

employing the law of matrix multiplication. The second term in brackets can be identified from

$$E_1 = \langle 01|V|10\rangle \tag{55}$$

and we obtain the interaction energy *through* second order as

$$(E_1)_+ + (E_2)_+ = E_1\left(1-\frac{E_1}{\bar{E}}\right)+\langle 01|V^2|01\rangle_+/\bar{E}, \tag{56}$$

$$(E_1)_- + (E_2)_- = -E_1\left(1+\frac{E_1}{\bar{E}}\right)+\langle 01|V^2|01\rangle_-/\bar{E}. \tag{57}$$

† This is evident for $\langle 01|V|ii\rangle_-$ since $|ii\rangle = 0$. For $\langle 0|V|ii\rangle_+$ it is true for reasons of parity; if, for instance, X_{0i} is finite, $X_{1i} = 0$, and vice versa.

There remains, then, only the calculation of the diagonal elements of V^2. If it is remembered that terms like $(X^2)_{01}$, $(XY)_{01}$ vanish because of parity (provided that $r_{01} \neq 0$, as we are supposing), we find

$$\langle 01 \mid V^2 \mid 01 \rangle_+ = \langle 01 \mid V^2 \mid 01 \rangle_- = \langle 01 \mid V^2 \mid 01 \rangle = \frac{1}{R^6}\{(X^2)_{00}(X^2)_{11} + (Y^2)_{00}(Y^2)_{11} + 4(Z^2)_{00}(Z^2)_{11} + 2(XY)_{00}(XY)_{11} - 4(XZ)_{00}(XZ)_{11} - 4(YZ)_{00}(YZ)_{11}\}. \tag{58}$$

Were we to compute the dispersion energy for the state $|01\rangle$, we should obtain

$$E_2 = \sum_{i,j}{}' \frac{\langle 01 \mid V \mid ij \rangle}{E_a^0 + E_b^1 - E_a - E_b^j} = \frac{1}{\bar{E}}\{\langle 01 \mid V^2 \mid 01 \rangle - \langle 01 \mid V \mid 01 \rangle^2\} = \frac{\langle 01 \mid V^2 \mid 01 \rangle}{\bar{E}}. \tag{59}$$

The last terms of (56) and (57) are therefore nothing but the ordinary van der Waals energy between the two resonating molecules. Therefore, except for the term $-E_1^2/\bar{E}$, resonance and dispersion forces are additive. The additional term is rather unphysical; it probably suggests that expression (59) is physically meaningless because it distinguishes between resonating and non-resonating members of the interacting pair.

Like E_1, $\langle 01 \mid V^2 \mid 01 \rangle$ depends on the orientation of the two systems, i.e. on M_0 and M_1. A complete analysis of this dependence would start with the irreducible tensor expansion of V used earlier in this Chapter and introduced in Chapter 2. For the example discussed earlier in which $M_0 = 0$ and $M_1 = 1,0$ or -1 the results can be found by elementary means. We choose for the three degenerate substates of the excited atom:

$$\psi_{-1} = \sqrt{\left(\frac{3}{8\pi}\right)} \frac{X - iY}{r} Q(r),$$

$$\psi_0 = \sqrt{\left(\frac{3}{4\pi}\right)} \frac{Z}{r} Q(r),$$

$$\psi_{+1} = \sqrt{\left(\frac{3}{8\pi}\right)} \frac{X + iY}{r} Q(r),$$

where X, Y, Z have the same meaning as in (23) and Q is a function of r. One then readily finds the following results.

If $M_1 = -1$

$$(X^2)_{11} = \int \psi_{-1}^* X^2 \psi_{-1}\, d\tau = \tfrac{2}{5}(r^2)_{11},$$
$$(Y^2)_{11} = \tfrac{2}{5}(r^2)_{11},$$
$$(Z^2)_{11} = \tfrac{1}{5}(r^2)_{11}.$$

The same values are obtained for $M_1 = 1$. But if $M_1 = 0$,

$$(X^2)_{11} = (Y^2)_{11} = \tfrac{1}{5}(r^2)_{11}, \quad (Z^2)_{11} = \tfrac{3}{5}(r^2)_{11}.$$

For the S-state,

$$(X^2)_{00} = (Y^2)_{00} = (Z^2)_{00} = \tfrac{1}{3}(r^2)_{00}, \quad \text{and} \quad (XY)_{00} = (XZ)_{00} = (YZ)_{00} = 0.$$

Hence we obtain from (58)

$$\langle 01 \,|\, V^2 \,|\, 01 \rangle = \frac{e^2}{3R^6}(r^2)_{00}(r^2)_{11} \times \begin{cases} \frac{8}{5} & \textit{if } M = \pm 1. \\ \frac{14}{5} & \textit{if } M = 0. \end{cases} \tag{60}$$

This shows a strong directionality of the van der Waals forces.

Finally, a word must be said about the mean value $\langle \bar{E} \rangle$ which appears in (56) and (57). It has the same value in both formulas. It is *negative*, as usual in the case of dispersion forces, if resonance takes place between the two lowest states of the molecules. However, if energy levels exist between the two involved in resonance, $\bar{E}$ may be *positive*. In that case a useful value of $\bar{E}$ will be very difficult to find, and the theory becomes highly complex unless the approximation which equates it with an average energy is adopted. Yet it is important to realize that the van der Waals forces which accompany resonance, aside from being noncentral, can be repulsive as well as attractive.

In more complex situations where many states are degenerate we can extend the perturbation procedure introduced in the earlier parts of this chapter. From (14), when $j > m$, the coefficients in the first-order state function are

$$(E_j^0 - E^0) A_{kj} = -\sum_{l=1}^{m} C_{kl} V_{jl}. \tag{61}$$

The second-order energy is given by (see, for example, Hirschfelder, Brown, and Epstein, 1964):

$$E_2^{(k)} = \langle \varphi_k \,|\, V \,|\, \psi_1^{(k)} \rangle, \tag{62}$$

$$= \left\langle \sum_{i=1}^{m} C_{ki}\psi_i^0 \;\middle|\, V \,\middle|\, \sum_{j>m}^{\infty} A_{kj}\psi_j^0 \right\rangle, \tag{63}$$

$$E_2^{(k)} = \sum_{j>m} \frac{\sum_{i=1}^{m} C_{ki}^* V_{ij} \sum_{l=1}^{m} C_{kl} V_{jl}}{E_1^0 - E_j^0}. \tag{64}$$

To use this result we must replace all quantum numbers by pairs which define the states of both atoms. For example, let $1 \equiv (01)$, $2 \equiv (10)$. Introducing this notation in (64) and the values of the coefficients found previously, it is straightforward to prove for this particular case, at least, that (64) is equivalent to (52) and (53):

$$E_2^{(1)} = \sum_{j>m} \frac{\langle 01 \,|\, V \,|\, \psi_j^{(0)} \rangle \langle \psi_j^{(0)} \,|\, V \,|\, 01 \rangle}{E^0 - E_j^0}$$

$$= \sum_{(ij)}{}' \frac{\langle 01 \,|\, V \,|\, ij \rangle_+ \; {}_+\langle ij \,|\, V \,|\, 01 \rangle}{E^0 - E_a^i - E_b^j}.$$

If there were no degeneracy (or resonance, in particular), we would have a delta function of i and l in (64) and this would lead to (59) because of (17). This is true for $n = 1$.

Very little is known about the numerical magnitudes of these second-order effects, but they are significantly larger than those between ground state atoms. Consider our former example of one atom in the ground state and another in the first optically excited state with $M = 0$. In this instance, because of (58), the interaction is proportional to a number, about one, times R^{-6} times $|r_{00}|^2 \, |r_{11}|^2$ [see (27) for notation]. For hydrogen the values are $|r_{00}| = {}^3/_2$, $|r_{01}| \cong 0.29$, and $|r_{11}| \cong 6$ a.u. For ground state atoms the interaction is proportional to $\frac{2}{3}\,|r_{00}|^4$. Hence, when one atom is excited, the second-order attraction is about twenty times as large as when both atoms are in the ground state. The diffuse nature of the excited orbitals again leads to significant changes in the interaction.

Overlap and exchange effects are important even at fairly large separations when one or both of the atoms are excited. However, very few specific calculations have been made of these in second-order interactions.

A brief comment may here be added on the possibilities of obtaining $E_2^{(m)}$ and $\psi_1^{(m)}$, both relating to the mth excited state, by variational methods. This can be done simply by minimizing the functional

$$\tilde{E}_2^{(m)} = \{2\langle \psi_m^0 \mid V \mid \psi_1^{(m)} \rangle + \langle \psi_1^{(m)} \mid H_0 - E_m^{(a)} \mid \psi_1^{(m)} \rangle\} \qquad (65)$$

subject to the requirement that $\psi_1^{(m)}$ be orthogonal to the exact eigenfunctions of all $m-1$ lower excited states. When all states of lower energy have different symmetries this requirement is automatically satisfied. If, however, some of the lower excited states are of the same symmetry, this condition is difficult to satisfy, for we usually do not know the exact functions but only approximations to them, such as ψ_m^0. Sinanoglu (1961b) has shown that one can use (65) if a trial function for $\psi_1^{(m)}$, say $\tilde{\psi}_1^{(m)}$, is written in the form

$$\tilde{\psi}_1^{(m)} = \sum_{k=1}^{m-1} \frac{V_{mk}}{E_m - E_k^0} \psi_k^0 + X, \qquad (66)$$

where X is restricted so that

$$\langle X \mid \psi_k^0 \rangle = 0, \qquad k = 1, \ldots, m-1.$$

The first term removes all of the complicating effects. Miller (1966), however, indicates that even if such constraints as (66) are not imposed, a reasonable result is obtained in many cases.

8.2.1. *Retardation Effects in the Second Order*

Retardation has subtle effects also on second-order forces. These have been investigated by McLone and Power (1965), who treat the interaction between two nonidentical systems using procedures identical with those discussed in Chapter 6. Among the new features of their analysis is the occurrence of twice as many diagrams, since the excitation of both systems must be considered in all possible ways. Furthermore, some additional poles appear in the integrals since downward transitions are also possible ($\Delta E < 0$ in the perturbation denominators).

These authors list results for one atom (b) in the ground state and another (a) in the first excited state. The short- and long-range behavior, special cases of McLone and Power's general results, are listed below for various values of the magnetic quantum number, M:

$$\Delta E(R) \underset{R\to 0}{\sim} \frac{4}{R^6(E_a'-E_b')} |\langle \boldsymbol{p}_a \rangle|^2 |\langle \boldsymbol{p}_b \rangle|^2 \quad \text{for} \quad M = 0,$$

$$\Delta E(R) \underset{R\to 0}{\sim} \frac{1}{R^6(E_a'-E_b')} |\langle \boldsymbol{p}_a \rangle|^2 |\langle \boldsymbol{p}_b \rangle|^2 \quad \text{for} \quad M \neq 1,$$

where E_b' is the energy of the first excited state of atom b and E_a' that of the excited state of atom a, each divided by $\hbar c$;

$$\Delta E(R) \underset{R\to \infty}{\sim} \frac{8(E_a')^2 E_b'}{(E_a')^2-(E_b')^2} |\langle \boldsymbol{p}_a \rangle|^2 |\langle \boldsymbol{p}_b \rangle|^2 \frac{\sin {}^2 E_a' R}{R^4} \quad \text{for} \quad M = 0,$$

$$\Delta E(R) \underset{R\to \infty}{\sim} \frac{2(E_a')^2 E_b'}{(E_a')^2-(E_b')^2} |\langle \boldsymbol{p}_a \rangle|^2 |\langle \boldsymbol{p}_b \rangle|^2 \frac{\cos^2 E_a' R}{R^2} \quad \text{for} \quad M \neq 1.$$

The behavior at $R \to 0$ is as found previously, but the long-range formulas exhibit slower changes as functions of R; their range is much longer than suggested by R^{-6}. In fact for $M = \pm 1$ it falls off as R^{-2}, more slowly even than the first-order resonant interaction.

Philpott (1966) has studied the case in which both atoms are excited and has given explicit formulas. At large separations the interactions again behave as R^{-2}.

These long-range potentials can lead to many interesting physical phenomena. In view of them one would expect, for example, that the spectrum of an impurity in a crystal should depend on the size and shape of the sample.

Plasmas contain many excited species of molecules and ions. The existence of very long-range forces, taken together with the possibility of interaction curves which exhibit multiple maxima and minima, vaguely leads one to expect novel phenomena, the precise nature of which can not be foreseen. The field of intermolecular forces between excited molecules is an inviting one which beckons for further investigation.

8.3. Double Minima. Crossing of Potential Curves

Much of the recent work on the interaction of excited atoms has not been included in our discussion here because in many cases the results are more typical of bound states of molecules than of ordinary van der Waals effects, the potentials often having wells several electron volts deep relative to infinite separation.

Furthermore, recent calculations involve very elaborate computations using a large basis of orbital sets. The availability of faster computers with larger memories has caused the number of such calculations to increase rapidly in the last few years. We have included many of these in the bibliography.

One interesting feature which has been found in several calculations is the appearance of *double minima* in potential energy curves of excited states of atoms and molecules. Earlier we mentioned that there was a possibility of a potential well at large separations in some states due to purely electrostatic causes. But as will now be shown double minima also arise because the various potential curves calculated without configuration interaction cross one another at small separations.

In calculating intermolecular potential curves with elementary state functions, one often discovers that function ψ_1 has the same energy as function ψ_2 at some distance R_c. These functions are not eigenfunctions of the exact Hamiltonian but of some approximate one, say, H_0. Consider the effect of the perturbation

$$V = H - H_0$$

on this potential crossing. Using degenerate perturbation theory at the point of crossing, we obtain the energy by solving

$$\begin{vmatrix} E_1+V_{11}-E & V_{12} \\ V_{21} & E_2+V_{22}-E \end{vmatrix} = 0.$$

The solution is

$$E=(1/2)(E_1+E_2+V_{11}+V_{22})+\sqrt{[(1/4)(E_1+E_2+V_{11}+V_{22})^2+V_{12}^2]}.$$

At R_c, $E_1 = E_2$, i.e. the two functions ψ_1 and ψ_2 give the same energy. The V_{ij} are matrix elements of V with respect to ψ_i and ψ_j. If ψ_1 and ψ_2 have different molecular symmetries, $V_{12} = 0$ and thus $E = 0$; the curves cross one another. If, on the other hand, V_{12} is not zero, as is the case when ψ_1 and ψ_2 are of the same symmetry (V has the symmetry of the molecule), then E is not zero and the curves can not cross. These general features are demonstrated in Fig. 24. We can state a theorem due to von Neumann and Wigner (1929) as a result of these considerations:

For an infinitely slow change of internuclear distance two electronic states of the same species can not cross each other.

If the nuclei are moving, V has a time dependence and a more general theory is needed. This has been provided by Landau

(1932, 1965), Zener (1932), and Bates and Lewis (1955). Which particular curve the system follows near the "crossing point" also depends on the velocity. For small velocities the system follows the curves of Fig. 24(b). This illustrates how, in case the

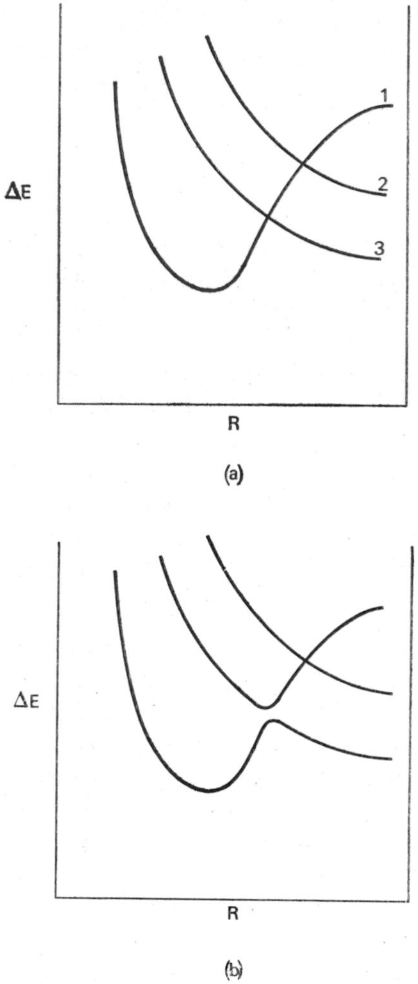

FIG. 24. Intermolecular potential curve crossings; curves 1 and 3 have the same molecular symmetry. (a) Approximate solutions (each based on some zero-order Hamiltonian). (b) Correct solutions using degenerate perturbation theory (or configuration-interaction).

approximate potential curves intersect properly, intermolecular potentials with two or more minima can appear. The same problem can be investigated variationally.

In calculations on excited states of the hydrogen molecule Cade (1961) used a limited basis set ($1s$, $2s$, $2p\sigma$) and found several examples of potentials with double minima. One curve had symmetry $^1\Sigma_g$ and another $^3\Sigma_u$: both of these seemed to involve $2s$ and $2p$ orbital mixing. The second minimum occurred at very large separations. Because Cade used very few orbitals and carried out only a small amount of orbital-exponent variation, he was not certain, especially for the triplet case, that more elaborate calculations would confirm his results.

Davison (1961) performed an extensive machine computation on the lowest two $^1\Sigma_g^+$ states of hydrogen. Using about twenty terms and expressing his orbitals in elliptical coordinates, (see Chapter 3), he found a potential curve with a definite double minimum. Even with this large basis set the outer depression is difficult to calculate. The best results lead to minima at $R = 1.9a_0$ and $4.3a_0$. The ground state minimum is at $1.449a_0$. A single maximum occurs at $3.3a_0$. Relative to this energy maximum of -0.6844 ± 0.004 a.u., the inner well has a depth of 0.0318 a.u. (0.87 ev) and the outer well a depth of 0.0163 a.u. (0.44 ev). This situation arises from a mixing of ($1s$ $2s$) and $(2p\sigma_u)\,^1\Sigma_g^+$ levels. At the inner minimum one molecular orbital behaves as a normal H_2^+ orbital while the other is very diffuse. At the outer minimum the electron density corresponds more closely to H^+H^-. Davidson also predicted higher states with double minima.

Matsumoto, Bender, and Davidson (1967), employing a large basis set, have considered the interaction of helium atoms at small separations. They used up to fifty configurations consisting of $10\sigma_g$, $4\sigma_u$, $8\pi_u$, $4\pi_g$, and $2\delta_g$ functions of the usual two center ellipsoidal type. While they do not report or expect any double minima in the potential, they do find several SCF level crossings. However, since these occur at very small angles it is unlikely that double minima would ever be found.

At $R \leqslant 0.25a_0$ the $1\sigma_g^2\, 2\sigma_g^2$ configuration has the lowest energy. This corresponds at large separations to the interaction of excited helium atoms ($1s$ $2s$). There is crossing of the $1\sigma_g^2\, 1\sigma_u^2$ and $1\sigma_g^2\, 1\pi_u^2$ potentials near $0.6a_0$, and of the $1\sigma_g^2\, 1\sigma_u^2$ and $1\sigma_g^2\, 2\sigma_g^2$

near $0.7a_0$. It is only beyond $0.7a_0$ that the normal $1\sigma_g^2\,1\sigma_u^2$ configuration is dominant. In Table 8.2 we list the results of Matsumoto, Bender, and Davidson for three of the configurations as well as for the complete calculation involving a fifty-term configuration interaction (CI).

TABLE 8.2. *Total Electronic Energy of two Helium Atoms* ($^1\Sigma_g$ *states*) *(in atomic units)*

R (a_0)	$E_{SCF}(1\sigma_g^2 2\sigma_g^2)$	$E_{SCF}(1\sigma_g^2 1\sigma_u^2)$	$E_{SCF}(1\sigma_g^2 1\pi_u^2)$	Fifty-term CI[a]
0.25	−12.7256	−12.206	−12.488	−12.79939
0.50	−10.7393	−10.591	−10.273	−10.82226
0.75	−9.309	−9.5026	−9.212	−9.57500
1.00	−8.272	−8.7928	−8.210	−8.85760

[a] The fifty-term CI result can be approximated by the following potential: $\Delta E(R) = ae^{-br}$, where $a = 247$ ev, $b = 4.23$ Å. These formulas apply to the *lowest*-energy solution of the total Hamiltonian of the system.

CHAPTER 9

Physical Adsorption

PHYSICAL adsorption[†] of atoms on a surface involves relatively weak interactions between the atom and the solid material. It is to be distinguished from chemisorption, where the interaction energies are usually of the order of bond energies, that is, about 50–100 kcal/mole. This distinction is, of course, not always rigorous. We shall consider here only physical adsorption, meaning by this interactions with energies less than about 5 kcal per mole.

The effective interaction between the adsorbed molecules and the surface depends in general on the number of molecules adsorbed, since the single-atom surface interaction is modified when another atom is near the first. This more general problem will be treated later. We turn first to the interaction of a single atom with a bare surface or a surface carrying an adsorbed layer whose atoms are well separated from each other.

9.1. Small Density of Adsorbed Atoms—Neglect of Many-atom Effects

9.1.1. Atoms Adsorbed on Insulators

The case of atoms adsorbed on dielectric materials, i.e. insulators, is easily understood. The electrons in the solid are bound to atoms which in turn are stationed at lattice sites. If the molecule in the crystal is similar to the molecule in free space, a condition usually prevailing in nonmetals, the total interaction is

† For a comprehensive review of all phases of this subject the reader is referred to Brunauer (1943) and Young and Crowell (1962).

obtained by summing the two-body interactions between the adsorbed species and one solid molecule over all molecules in the solid.

In symbols

$$\Delta E = \sum_c \Delta E_{ac}(R). \tag{1}$$

Here the adsorbed molecules are labeled a, the solid molecules c. If ΔE_{ac} is the dispersion energy plus the additional terms (repulsion, R^{-8} attraction, etc.), then, in view of eqn. 29, Chapter 2,

$$\Delta E = \sum_c -\frac{3}{2}\frac{P_a P_c}{R^6}\frac{\bar{E}_a \bar{E}_c}{\bar{E}_a+\bar{E}_c} + \sum_c (\Delta E_{ac})_{\text{remainder}} \tag{2}$$

$\bar{E}_a$ and $\bar{E}_c$ are mean excitation energies of the atoms. Moreover, if the solid is sufficiently far from the adsorbed molecule relative to the lattice spacings in the solid, we can approximate the first term by a continuum model with polarizability $(P_c)_\omega$ per unit volume. For this special case

$$(\Delta E)_{\text{cont model}}$$
$$\cong -\frac{3}{2}(P_c)_\omega P_a \frac{\bar{E}_a \cdot \bar{E}_c}{\bar{E}_a+\bar{E}_c}\int_{R_0}^{\infty} dz \int_{-\infty}^{\infty} dy \int_{-\infty}^{\infty} \frac{dx}{(x^2+y^2+z^2)^3}$$
$$+\sum_c (\Delta E_{ac})_{\text{remainder}} \tag{3}$$

$$\cong -\frac{3}{2}(P_c)_\omega P_a \frac{\bar{E}_a \bar{E}_c}{\bar{E}_a+\bar{E}_c}\left(\frac{\pi}{6R_0^3}\right) + \sum_c (\Delta E_{ac})_{\text{remainder}}. \tag{4}$$

The interaction is therefore effectively proportional to R_0^{-3}, where R_0 is the perpendicular distance from the atom to the surface. The remainder term varies more rapidly with R and only a few solid molecules near the surface contribute to its sum. Steele and Kebbekus (1965) have studied this problem for some simple lattices. They used the Lennard-Jones potential for $\Delta E_{ac}(R)$.

In general the transition from (2) to (3) is not permissible and one must use the discrete summation for all terms.

9.1.2. Atoms Adsorbed on Metals

More interesting but also more complex is the interaction of atoms with a metallic surface.

It was first studied by Lennard-Jones (1932), who assumed that the metal was a perfect conductor with an instantaneous response to any electric field. If this is granted, then a charge q placed at a distance R_0 from the surface of the metal will be attracted to the metal by an image force as if there were a charge $-q$ at a distance R_0 inside the metal. Thus a charge q is attracted to a metallic surface with a force of $-q^2/(2R_0)^2$ or a potential of $-q^2/4R_0$. If the atom contains an instantaneous dipole moment, represented here by charges at x, y, z, the energy of interaction with the surface is

$$+\frac{e^2}{4R_0}+\frac{e^2}{[x^2+y^2+(z^2+2R_0)^2]^{1/2}}-\frac{e^2}{4(z+R_0)}$$
$$=-\frac{e^2(x^2+y^2+2z^2)}{16R_0^3}+O(R_0{}^{-4}). \qquad (5)$$

If we then average over all positions of the electrons of a spherical atom, noting that $\langle x^2\rangle = \langle y^2\rangle = \langle z^2\rangle$, we obtain

$$\Delta E = -\frac{e^2\langle z^2\rangle}{4R_0^3} \qquad (6)$$

or

$$\Delta E = -\frac{e^2\langle r^2\rangle}{12R_0^3},$$

$\langle r^2\rangle$ representing the average value of the square of the radius vector of the adsorbed atom. By means of relations given in Chapter 2 this can be related to the polarizability of the adsorbed species.

In deriving these results Lennard-Jones made several assumptions: (a) he neglected all repulsive and other short-range effects which stabilize the adsorbed layer; and (b), as mentioned, he assumed that the metal could respond infinitely rapidly to any electrostatic perturbation. The adsorbed atoms, e.g. inert gas atoms, are characterized by resonance frequencies in the far ultraviolet. It is well known that electrons in metals can not be regarded as free at such frequencies. In fact, they act as if they were bound systems with frequencies characterized by energies in the 3–25 ev range.

We shall first remedy the second of these shortcomings, later returning to discuss the additional repulsive and many-body effects.

Bardeen (1940) pointed out that by the nature of its derivation, (6), should be an upper limit to this form of interaction. He considered in addition to the image potential the effects due to the relative kinetic energies of the electrons in the metal and in the adsorbed atom. These were appreciable and led to a reduction in the total interaction energy. Only for very high electron densities are these corrections small. Using an approximate form of the London formula, Bardeen estimated the interaction using the free electron model and taking account of the Pauli exclusion principle. He found

$$\Delta E = -\frac{e^2\langle r^2\rangle}{12R_0^3}\left(\frac{Ce^2/2r_s}{\bar{E}_a+Ce^2/2r_s}\right), \tag{7}$$

where the constant C has a value of about 2.6. The image result of Lennard–Jones is seen to result if the mean excitation energy $\bar{E}_a$ is very small (corresponding to low frequencies) or the electron density in the metal is very high, i.e. r_s, the effective radius of an electron inside the metal, must be small. For most metals $\bar{E}_a$ should be taken to be about one-half atomic unit; the image force is then too strong by about a factor of 2. In many cases the factor is even larger.

Another approach (Margenau and Pollard, 1941) proceeds with an even simpler model. It, too, arose from the concern that the response of electrons in a metal at high frequency is not instantaneous. Metals at very high frequencies in fact are more like insulators than metals, as was first pointed out by London (1930). Following Margenau and Pollard (1941) we divide the metal up into infinitesimal volume elements as was done for insulators. The first term in (4) is

$$(\Delta E)_{\substack{\text{cont}\\ \text{model}}} = \frac{(P_c)_\omega e^2\pi\langle r^2\rangle\bar{E}_c}{6R_0^3(\bar{E}_a+\bar{E}_c)}. \tag{8}$$

The polarizability per unit volume of a metal is easily derived. A slab of metal in an electric field $\mathcal{E}$ acquires a surface charge of density $\sigma = \mathcal{E}/2\pi$. If its thickness is d and its area A, its dipole moment is $\sigma Ad = \mathcal{E}\tau/2\pi$, where τ denotes its volume. Thus $(P_c)_\omega = 1/2\pi$ and (8) becomes

$$(\Delta E)_{\substack{\text{cont}\\ \text{model}}} = -\frac{\langle r^2\rangle}{12R_0^3}\left(\frac{\bar{E}_c}{\bar{E}_a+\bar{E}_c}\right). \tag{9}$$

The similarity with the Bardeen result is evident. Equation (9) can also serve to define the meaning of Bardeen's terms more clearly. Later, after a more careful discussion of the electron gas, we return to and refine both results, (7) and (9). At this point we note again the conditions under which the image result is correct, namely, that $\bar{E}_a \ll \bar{E}_c$.

Margenau and Pollard (1941) also write the interaction energy in a more transparent form,

$$\Delta E_{ac} = \frac{-6}{2R^6} e^2 \{\sum_k P_c(\nu_k) |\langle 0 | z_a | k \rangle|^2 + \sum_l P_a(\nu_l) |\langle 0 | z_c | l \rangle|^2\} = 6(S_1 + S_2), \tag{10}$$

which is obtained from the normal second-order energy, e.g. eqn (26) of Chapter 2, by multiplying numerator and denominator by $\Delta E_c^l - \Delta E_a^k$ and identifying the frequency dependent polarizabilities. The states of the atom are labelled by k and those of the solid by l; $\Delta E_c^l = E_c^l - E_c^0$. In (10) $P_c(\nu_k)$ is, for example, the polarizability of an element of the metal at frequency ν_k corresponding to a transition in atom a of energy ΔE_a^k. If we knew the response of a metal at all frequencies we could evaluate the first sum, which we shall call $6S_1$, since the matrix elements are simply related to f values of molecular transitions (see Chapter 2). If only one energy level in the atom is important we can replace S_1 by

$$S_1 = -\frac{e^2\hbar^2}{4mR^6} P_c(\nu) \frac{f_1}{h\nu}. \tag{11}$$

S_2 is more difficult to evaluate. Since ν_l is generally very small relative to ν_k we might remove P_a from the summation and replace it by its static value $P_a(0)$. Then upon relating the matrix elements to the polarizability we find

$$S_2 = -\frac{P_a(0)P_c(0)\bar{E}_c}{4R^6}. \tag{12}$$

Hence the interaction with a small element of the metal is

$$\Delta E(R)_\omega = 6(S_1+S_2) = 6\left\{-\frac{e^2\hbar^2}{4m}\frac{f_1}{h\nu_1}[P_c(\nu_1)]_\omega - \frac{1}{8\pi}P_a(0)\bar{E}_c\right\}\frac{1}{R^6}. \tag{13}$$

After integrating over the entire metal, as in (3), this becomes

$$\Delta E = -\left\{\frac{e^2\hbar}{m}[P_c(\nu_1)]_\omega \frac{f_1}{\nu_1} + \bar{E}_c P_a(0)\right\} / 8R_0^3. \tag{14}$$

Comparison with (7) shows that $\bar{E}_c$ is to be identified with $Ce^2/2r_s$. The second term in (14) is identical to one part of Bardeen's result but it enters here in a more obvious way. Equation (14) differs from previous results in that the response of the metal is taken at frequency ν_1, not at zero frequency. This difference is fundamental. Also the atom spectrum is not approximated by a mean excitation energy and a zero frequency polarizability in the first term. As pointed out earlier, if $\nu_1 = 0$, the second term in (14) is neglected and $[P_c(0)]_\omega$ is replaced by $1/2\pi$, we obtain the image force formula.

As a better approximation to $[P_c(0)]_\omega$ one may use the free electron value

$$[P_c(\nu_1)]_\omega = -\frac{n_0 e^2}{4\pi^2 m \nu_1{}^2}, \tag{15}$$

where n_0 is the actual number of free electrons per unit volume of the metal and m is their true mass. Note that this is much different from $[P_c(0)]_\omega$. With all of these approximations (14) can be rewritten in the form

$$\Delta E = \left[\frac{e^2\hbar}{m}\left(\frac{n_0 e^2}{4\pi^2 m \nu_1^2}\right)\frac{f_1}{\nu_1} - \frac{Ce^2}{2r_s} P_a(0)\right]\frac{1}{8R_0^3}$$

$$= -\left[\frac{Ce^2}{2r_s} P_a(0) - \frac{n_0 e^4 \hbar f_1}{4\pi^2 m^2 \nu_1^3}\right]\frac{1}{8R_0^3}. \tag{16}$$

In this final form the relation of ΔE with the image force formula is of course difficult to discern.

Prosen and Sachs (1942) made the point that the concept of a polarizability, employed by Margenau and Pollard in arriving at (16), is of questionable applicability. To define a polarizability one needs both electrons and positive ions. Hence in order to obtain a meaningful representative volume within the metal one needs to choose elements of several lattice spacings on each side. But the formulas above require the polarizability to be defined at each point. Prosen and Sachs therefore decided to employ the state functions of the metal directly. They considered only the second term in (10), which they wrote in the London

(1937) form

$$\Delta E_{PS} = -\frac{1}{2} P_a(0)\langle \boldsymbol{E}^2 \rangle, \tag{17}$$

where $\boldsymbol{E}$ is the sum of the field strength exerted by the atomic dipole and the ions on the electrons in the metal. Margenau and Pollard (1941) found the first term in (10) to be a half to one-eight of the second term, hence its neglect by Prosen and Sachs may not be serious.

These authors evaluated the interaction for both a non-degenerate and a degenerate electron gas, i.e. a gas in which one does or does not include effects of the Exclusion Principle. Their results differ greatly from those of earlier investigators, since their calculated formulas including degeneracy behave not as R_0^{-3} but as the logarithm of a constant times R_0 divided by R_0^2, a very strange result:

$$(\Delta E)_{PS} = -\frac{e^2 P_a k_m^2}{8\pi^2} \frac{\ln(2k_m R_0)}{R_0^2}, \tag{18}$$

where $k_m = (3\pi^2\varrho)^{1/2}$, ϱ being the electron density in the metal.

Criticism which might be leveled against these results are: they neglect electron interactions and the plasma modes and they involve the use of very approximate state functions for the electrons in the metal. It is important to realize that the effect to be calculated is the difference in the interaction of an adsorbed instantaneous dipole with the ion cores and with the electrons. This difference may be very sensitive to state functions and to the frequency of the dipole oscillations. Prosen and Sachs point out that the inclusion of electron–electron correlations will affect the long-range behavior of the forces.

The objection to the concept of a polarizability at a point in a metal is a valid one and should be considered with greater care. Since experimental results are not precise enough to estimate the exact R_0 dependence, our interest is necessarily confined to the numerical values obtained from the different methods. Most of the recent calculations do assume that a polarizability per unit volume can be defined at each point. They include retardation and will now be sketched, however without including many details of their derivation.

Mavroyannis (1963) applied the S-matrix method of quantum electrodynamics to evaluate the interaction of a neutral molecule with a surface. For a perfect metal (instantaneous response at all frequencies) he finds, employing the method explained in Section 6.1,

$$\Delta E = -\frac{1}{3\pi}\sum_n{}' \int_0^\infty du u^2 \frac{\Delta E_a^n |\langle 0|r|n\rangle|^2}{(\Delta E_a^n)^2+u^2}$$

$$\left[\frac{2}{(2R_0)}+\frac{4}{u(2R_0)^2}+\frac{4}{u^2(2R^2{}_0)^3}\right] \exp(-2R_0 u). \qquad (19)$$

The bracket above multiplied by u^2 is simply the interaction with the image field. In the notation of eqn. (19) of Chapter 6, it is simply $(T_{ab})_{xx}+(T_{ab})_{yy}-(T_{ab})_{zz}$ with the image at $R = 2R_0$.

For extremely large separations, only the first term contributes and we obtain

$$\Delta E \sim -\frac{3\hbar c P_a}{8\pi R^4} \quad (R \gg \lambda). \qquad (20)$$

This is exactly the result of Casimir and Polder (1948). We shall see later that it is correct for most metals. At short distances we may neglect the exponential and retain only the last term in the bracket, the static ($u = 0$) component, and thus obtain the Lennard-Jones formula.

If the adsorbing material is a dielectric the image charge produced by a charge q outside the body is not $-q$ but $-[(\varkappa-1)/(\varkappa+1)]q$ where $\varkappa$ is the dielectric constant (see, for example, Jackson, 1963). When this is introduced into (19) with the use of the dielectric constant at imaginary frequency iu, as explained in Section 6.3,

$$\Delta E = -\frac{1}{3\pi}\sum_n{}' \int_0^\infty du\, u^2 \frac{\Delta E_a^n |\langle 0|r|n\rangle|^2}{(\Delta E_a^n)^2+u^2}$$

$$\left(\frac{\varkappa(iu)-1}{\varkappa(iu)+1}\right)\left[\frac{2}{2R_0}+\frac{4}{u(2R_0)^2}+\frac{4}{u^2(2R_0)^3}\right] \exp(-2uR_0).$$

This substitution has been justified rigorously by Mavroyannis (1963) and McLachlan (1964) proceeding from different points of view.

At relatively short separations the leading term of this expression is

$$\Delta E \sim -\frac{1}{6\pi R_0^3}\sum_n{}' \int_0^\infty du \frac{\Delta E_a^n |\langle 0|r|n\rangle|^2}{(\Delta E_a^n)^2+u^2}\left[\frac{\varkappa(iu)-1}{\varkappa(iu)+1}\right].$$

When the surface is a metal the limiting form of this result for large separations is identical to the image result, eqn. (20), since then the zero frequency dielectric constant enters and $\varkappa(0)$ is much greater than one. Mavroyannis and McLachlan also show that their results are equivalent to the more complicated formulas of Lifshitz and his coworkers (Lifshitz, 1956; Dzyaloshinskii 1959; and Pitaevskii, Dzyaloshkinskii, Lifshitz, and Pitaevskii, 1960, 1961), when the latter are applied to adsorption problems.

We can go a few steps further in applying the dielectric constant method since the dielectric constant of a free electron gas is known, again assuming that the polarizability is defined at all points in the medium. For long wavelengths the dielectric constant is (see, for example, Pines, 1963, p. 148)

$$\varkappa(iu) = 1 + \frac{\omega_p^2}{u^2}, \tag{21}$$

where ω_p is the plasma frequency, $\hbar\omega_p$ the dielectric energy loss of the substance. Specifically, $\omega_p = (4\pi n_0 e^2/m)^{1/2}$, where n_0 is the number of electrons in a unit volume of the metal. Using (21) in (19), we obtain for $R \ll \lambda$,

$$\Delta E = -\frac{1}{12R_0^3}\langle r^2\rangle \frac{\hbar\omega_p/\sqrt{2}}{E_u + \hbar\omega_p/\sqrt{2}}. \tag{22}$$

The relation of this result to Bardeen's (1940, eqn. (7)), is immediate. The ratio c/r_s in that equation is related to ω_p.

Using the mean excitation energy approximation in various forms, (22) may be rewritten in two ways:

$$\Delta E = -\frac{P_a^{1/2}N^{1/2}}{8R_0^3} \frac{\hbar\omega_p/\sqrt{2}}{(N/P_a)^{1/2} + \hbar\omega_p/\sqrt{2}}, \tag{23}$$

$$\Delta E = -\frac{\langle r^2\rangle}{12R_0^3} \frac{\hbar\omega_p/\sqrt{2}}{\dfrac{3}{2}\dfrac{N}{\langle r^2\rangle} + \hbar\omega_p/\sqrt{2}}. \tag{24}$$

N is the number of optically active electrons in the atom.

In Table 9.1 we quote the results of various formulas presented thus far. This table is taken from the paper of Mavroyannis (1963).

Obviously by treating only the leading attractive portions of the interaction we have not evaluated all of its essential features.

TABLE 9.1. *Interaction Energy between an Atom and A Metal*

System	Experimental (cal per mole)	Calculated (cal/mole)					
		Eqn. (23)	Eqn. (24)	Image force [eqn. (6)]	Bardeen (1940) [eqn.(7)]	Margenau and Pollard (1941) [eqn. (16)]	Prosen and Sachs (1942) [eqn. (18)]
Pt–A ($R_0 = 3.30$ Å)	1320	1435	1815	5170	940	1260	2080
Zn–A ($R_0 = 3.24$ Å)	1570	1520	1660	5465	1100	1400	–
Cu–A ($R_0 = 3.18$ Å)	2090	1475	1870	5800	1090	1450	2700

There must be a repulsive contribution for stability of the adsorbed layer. Pollard (1941) evaluated the exchange effects which yield a repulsion. He assumed plane waves for the electrons in the metal, modified so that the functions vanished at the surface. Adsorption of helium and H_2 were the object of his study and he found very large reductions in the attraction near the surface. The potential minimum was also closely related to r_s and appeared at approximately 1.51 r_s.

Recently Wojciechowski (1966) published a comprehensive critical study of physical adsorption. He raised the following points against the theories discussed thus far.

(1) Formulas based on image forces are incorrect because of surface effects (Cutler and Gibbons, 1958) and polarization of the electron gas. The last point is partly accounted for by our earlier use of a dielectric constant for the metal.
(2) The use of a point polarizability is not rigorous, a comment made earlier.
(3) Attractive contributions due to charge transfer have not been included. These are quantum mechanical effects involving state functions resulting from a superposition of configurations (see Chapter 4) in which ionic terms are added.

Such ionic terms reflect conditions in which one electron is transferred from the adsorbed atom to the metal. Their effects can be large (Gundry and Tompkins, 1960a, b; Toya, 1958).

(4) Most calculations ignore the actual structure of a metal and hence the dependence of adsorption on crystallographic planes.

(5) All calculations assume the attractive and repulsive parts to be additive and calculable separately. From our study of the atom–atom interaction, this is certainly not the case.

Wojciechowski has dealt with all of these effects with varying degrees of approximation, applying himself to alkali and alkaline earth atoms interacting with tungsten and copper. His numerical results are to be published shortly.

The above discussion should be considered as representative of some of the work in this field. However, because of the theoretical and practical importance this is a very active research area and thus our treatment is by no means exhaustive. Before closing this section we discuss briefly one current paper which suggests a new approach to certain areas of physical adsorption on metals. Bennett and Falicov (1966), being most concerned with charge transfer and the Pauli Principle, developed a self-consistent field method for evaluating the energy of adsorption and the percentage of charge transfer using a one-electron model of the metal. They neglected all dispersion interactions and assumed the action of an image force for the charge transfer problem. Good agreement with the experimental results for potassium, rubidium, and cesium adsorbed on tungsten was recorded despite these omissions.

Another aspect not treated in this section is ionic adsorption, such interactions being more typically in the energy range characteric for chemisorption.

9.2. Many-atom Effects

It remains for us now to consider how atoms interact with one another while adsorbed. This can lead to increased or decreased effective attractions between the surface and the adsorbed atoms.

The problem was first discussed by Sinanoglu and Pitzer (1960) who used a model for the metal which was similar to that of

Margenau and Pollard (1941). We merely state their results. The effective interaction between atoms a and b is the normal dispersion energy which would obtain in the gas phase plus an additional term

$$[\Delta E(R_{ab})]_{\text{adsorbed}} = [\Delta E(R_{ab})]_{\text{gas}} + \frac{S(R_0)}{R^3_{\text{ab}}}(1-3\cos^2\theta), \quad (25)$$

where R_{ab} is the distance between atoms a and b near the surface and θ is the angle their line of centers makes with the normal. For monolayers, $\theta = 90°$, and the undefined quantity

$$S(R_0) \cong \frac{3}{4} P_a U^T \left(\frac{1.5+6.5\gamma}{2+\gamma}\right),$$

where U^T is the attraction of a single atom with the surface. The number γ is related to the strength of any net electrostatic surface field that may be present. If there is none as we have been assuming in this chapter, $\gamma = 0$. For further details the reader is referred to Sinanoglu and Pitzer (1960), and to the end of this section.

McLachlan (1964), using his field susceptibility methods, arrives at similar expressions for the additional terms, finding

$$\Delta E = \frac{\hbar(2-3\cos 2\theta - 3\cos 2\phi)}{2\pi R^3_{ab}(R')^3} \int_0^\infty P_a(iu)P_b(iu)\left[\frac{\varkappa(iu)-1}{\varkappa(iu)+1}\right] du.$$
$$-\frac{3\hbar}{\pi(R')^6}\int_0^\infty P_a(iu)P_b(iu)\left[\frac{\varkappa(iu)-1}{\varkappa(iu)+1}\right]^2 du. \quad (26)$$

R' is the distance from the image of atom a to b and ϕ is the angle this line makes with the normal to the surface. Of the two terms, the first is generally the larger.

When the atoms are very far apart ($R_{ab} = R'$, $\theta = \phi = 90°$) the total effective interaction is the sum of the two body interaction plus (26):

$$\Delta E = -\frac{3\hbar}{\pi R^6}\int_0^\infty P_a(iu)P_b(iu)\frac{2(\varkappa^2+5)}{3(\varkappa+1)^2} du.$$

If the surface were that of a perfect metal, this could be as small as two-thirds of the two-body dispersion energy. Generally, however, there is not quite so large a reduction because of the weakened response of the metal at higher frequencies.

McLachlan (1964) has extended these calculations to include

the integration of two surfaces. In that case he obtains the general results of Lifshitz.

Krizan and Crowell (1964), Wolfe and Sams (1966), and more recently Everett (1966) have applied (25) and (26) to the interpretation of experimental results. While agreement with experiment is clearly improved by inclusion of many-atom effects experimental problems at present prevent very accurate tests. Several features are already apparent. Electrostatic effects are minor. The suggestion of Sinanoglu and Pitzer to use $\gamma \sim 0.3$ does not agree with the experimental results. If such effects are present it is unlikely that they will exceed 10% of the second-order energy ($\gamma \sim 0.1$) (Sams, 1964). Better agreement is obtained by neglecting electrostatic contributions, setting γ equal to zero. Even then most of the formulas presented tend to overestimate the three-body effects (Wolfe and Sams, 1966; Everett, 1966). Perhaps most disturbing is the fact that these theories often do not predict even the relative order of three-body effects with various adsorbates (Wolfe and Sams, 1966; Everett, 1966). Clearly more experimental and theoretical work is required.

This problem, like many others which remain in the study of intermolecular forces, invites careful investigation.

APPENDIX A

The Sign of $\Delta E(R)$ for Molecular Orbital Calculations Involving Closed Shell Atoms

THE possibility that a molecular orbital theory could yield an attractive interaction between two closed shell atoms has often been discussed. In all cases in which very accurate SCF–MO energies have been obtained ΔE is positive. Donath and Pitzer (1956) in their calculations on the triplet state of H_2, calculations described in Chapter 4, observed that the best limited molecular orbital calculation yielded a repulsive interaction at all separations. Furthermore the coefficient of the term in their function which might seem to lead to something like the attractive energy [the coefficient C_3 of eqn. (13) of Chapter 4] did not behave as R^{-3} at large distances, thus failing to exhibit a feature characteristic of the manner in which the R^{-6} interaction arises [see, for instance, eqn. (11) of Chapter 4]; instead it decreased exponentially. Still there is no general proof that ΔE resulting from this approach can never be negative.

In many cases, however, we can show analytically that the molecular orbital energy does not contain any part of the R^{-6} interaction.† The simplest instance is that of two atoms having like spins.

In the limit of zero overlap, the induced dipole interactions vanish in the MO approximation for the triplet state of H_2. Let us calculate the total energy using Donath and Pitzer's approximate molecular orbital function, eqn. (13) of Chapter 4, which represents all of the essential features, and then pass to

† For discussions of this aspect of the problem, the authors wish to thank J. A. Pople and H. C. Longuet-Higgins.

the limit of very large separations where all overlap integrals can be neglected. We then find

$$\langle \Psi | H | \Psi \rangle = (1-\delta_+^2)(1-\delta_-^2)\langle 1s_a 1s_b | H_a + H_b | 1s_a 1s_b \rangle$$
$$-2\delta_+\delta_-(1-\delta_+^2)^{1/2}(1-\delta_-^2)^{1/2}\left\langle 1s_a 1s_b \left| -\frac{2z_1z_2}{R^3} \right| 2p_a 2p_b \right\rangle$$
$$+\tfrac{1}{4}[\delta_+\sqrt{(1-\delta_-^2)}+\delta_-\sqrt{(1-\delta_+^2)}][\langle 1s_a^2 p_a | H_a+H_b | 1s_a^2 2p_a \rangle$$
$$+\langle 1s_b^2 2p_b | H_a+H_b | 1s_b 2p_b \rangle]$$
$$+\tfrac{1}{4}[\delta_+\sqrt{(1-\delta_-^2)}-\delta_-\sqrt{(1-\delta_+^2)}][\langle 1s_a 2p_b | H_a+H_b | 1s_a 2p_b \rangle$$
$$+\langle 1s_b 2p_a | H_a+H_b | 1s_b 2p_a \rangle$$
$$+2\left\langle 1s_a 2p_b \left| -\frac{2z_1z_2}{R^3} \right| 2p_a 1s_b \right\rangle\Bigg]$$
$$+\delta_+^2\delta_-^2\langle 2p_a 2p_b | H_a+H_b | 2p_a 2p_b \rangle.$$

Provided we take the long range limit of H,

$$H = H_a + H_b + \frac{x_1x_2+y_1y_2-2z_1z_2}{R^3}.$$

Designating the energy of the $2p$ orbitals as E^p, the ground state energy of a hydrogen atom as E^0, and the matrix element

$$\langle 1s_a 1s_b | -2z_1z_2 | 2p_a 2p_b \rangle \quad \text{as } A,$$

we can write (keeping only terms in the coefficients squared)

$$\langle \Psi | H | \Psi \rangle = (1-\delta_+^2-\delta_-^2)(E_a^0+E_b^0)-2\delta_+\delta_-\left(\frac{A}{R^3}\right)$$
$$+\tfrac{1}{2}(\delta_+^2+\delta_-^2)(E_a^p+E_b^p)$$
$$+\tfrac{1}{2}(\delta_-+\delta_+^2)(E_a^0+E_b^0)$$
$$+\tfrac{1}{2}(\delta_+^2-2\delta_+\delta_-+\delta_-^2)(A/R^3).$$

where δ_+ and δ_- are parameters defined after eqn. (6) of Chapter 4. Now

$$\langle \Psi | \Psi \rangle = 1.$$

Thus we have

$$E \leqslant E_a^0+E_b^0+\left[-2\delta_+\delta_-\left(\frac{A}{R^3}\right)+\frac{1}{2}(\delta_+^2+\delta_-^2)(E_a^p+E_b^p-E_a^0-E_b^0)\right.$$
$$\left.-\frac{1}{2}(\delta_+-\delta_-)^2(A/R^3)\right].$$

Since $\delta_+ = \delta_- \equiv \delta$ at large R we obtain

$$E \leqslant E_a^0 + E_b^0 + \delta^2 \left[E_a^p + E_p^b - E_a^0 - E_b^0 - \frac{2A}{R^3} \right].$$

The bracket is positive; hence the optimum value of δ is zero, i.e. no dispersion effects are included in the molecular orbital calculation.

Another noteworthy feature of the calculation by Donath and Pitzer is this. ΔE calculated by the MO method is never negative but is, in fact, equal to the energy resulting from a trial function of the following form:

$$\Psi = c_0 \mathcal{A}(1s_a' 1s_b') + c_1[\mathcal{A}(1s_a 2p_a) + \mathcal{A}(2p_b 1s_b)].$$

The proof that the dispersion interaction is not attainable by molecular orbital calculations for the interaction of two like-spin hydrogen atoms can be extended to the helium interaction in a fairly simple way. This indicates that R^{-6} interactions are not part of molecular orbital calculations involving closed shell atoms.

To prove that ΔE can never be negative in molecular orbital calculations of the interaction of two closed shell atoms is not as simple, since one must show that all covalent and all ionic one- and two-electron substitutions contribute less than the Heitler–London or LCAO–MO term. A general proof of this may be impossible, but we have shown at least that the largest negative contribution plays no role in the MO approximation at large separations.

APPENDIX B

Current Methods for Obtaining $\Delta E(R)$ from Experimental Data

Intermolecular potentials can be obtained directly or indirectly from many types of experimental data. We shall review some of the more successful methods and assess their accuracies, restricting our comments to the interaction of closed shell atoms.

B.1. Long-range Interaction

B.1.1. Thermal Molecular Beam Scattering

As shown by Massey and Mohr (1934) and Landau and Lifshitz, (1958, p. 416), for low energy scattering of atoms the total elastic scattering cross-section from a potential of the form K/r^s is proportional to $(K/\hbar v)^{2/(s-1)}$, where v is the initial relative velocity. From measurements of the velocity dependence of the total cross-section, the value of s is known to be 6 to within $1-2\%$.† The values of the coefficients K determined before 1965 are incorrect because of experimental pumping errors. The ratios of coefficients are not subject to this error. The latest measurements, and the calculated values of K (or total cross-section) are in excellent accord (Rothe and Neynaber, 1965b). This is in agreement with the theoretical results of Fontana and Bernstein (1964), who find that retardation effects and R^{-8} interactions do not contribute substantially to the total cross-section at thermal energies. An extensive review of work in this area was published by Pauly and Toennies (1965, vol. 1, p. 195) and Bernstein and Muckerman (*Advanc. Chem. Phys* (1967) **12**, 389).

† For K–Ar, K–Kr, K–Xe, see Florin (1964); Rothe, Rol, Trujillo, and Neynaber (1962). For atom–molecule collisions see Pauly (1960); Schoomaker (1961).

B.1.2. Low-temperature Transport Data

From the Massey–Mohr result quoted above one can also calculate the viscosity cross-section at low temperatures as a function of K. Munn (1965) has analysed recent low-temperature viscosity data on argon and neon and found good agreement with the latest theoretical sum rule results for K (Bell, 1965a).

B.1.3. Virial Coefficient Data

Attempts to fit a two parameter potential to virial coefficients and then relate the attractive part found in this way to the theoretical R^{-6} coefficient are of limited validity. These procedures oversimplify the interaction and neglect the other contributions to the attraction, e.g. R^{-8} terms. K values so obtained are invariably too large since they include other effects (see Chapter 4).

B.1.4. Limiting Curve of Dissociation (LCD)

Bernstein (1966) has proposed a method by which spectroscopic results for diatomic molecules can be analysed to yield the long-range interaction between the products of dissociation. From predissociation data (Herzberg, 1950, pp. 405–37) one can obtain, for each vibrational level of a molecular state, the maximum rotational quantum number that is occupied, $j_{\max}$, and the energy at which the last observed level occurs. This energy is the dissociation energy plus the height of the centrifugal barrier.

Let $V(r)$ be the potential between the components of the diatomic molecule; then the effective potential is

$$V_{\text{eff}} = V(r) + j(j+1)/2r^2,$$

where j is the rotational quantum number. The top of the barrier occurs at the value $r = r_{\max}$, where

$$r_{\max}^3 \left(\frac{\partial V}{\partial r}\right)_{r=r_{\max}} = j_{\max}(j_{\max}+1).$$

If $V = -C_n r^{-n}$ then

$$V_{\text{eff}}(r_{\max}) = S_n[j_{\max}(j_{\max}+1)]^{n/n-2}$$

with

$$S_n = \frac{n-2}{[n^n C_n^2]^{1/n-2}}.$$

The limiting curve of dissociation satisfies

$$E_{LCD} = E_0 + S_n[j_{max}(j_{max}+1)]^{n/n-2}$$

if r_{max} is large. Thus by plotting the values of log $[j_{max}(j_{max}+1)]$ versus log $(E_{LCD}-E_0)$ one can determine the value of n and the coefficient C_n. This has been done for HgH, HgD, $N_2(C^3\pi_u)$ and $CO(b^3\Sigma^+)$. It is furthermore possible to relate these results to a certain resonance behavior in elastic atomic scattering. In this way Bernstein has shown that resonances in the $X^1\Sigma_g^+$ state (ground state) of H_2 (Buckingham, Fox, and Gal, 1965) directly reveal substantial contributions of the dipole–quadrupole (R^{-8}) interaction to the potential.

This method should prove useful for heteronuclear and excited-atom interactions and as checks on other experimental data.

B.2. Total Interaction

B.2.1. Transport and Virial Data

Rapidly increasing knowledge of intermolecular potentials, many-body effects, and other features connected with condensed media, indicates that simple two- or even three-parameter potentials are not adequate representations of the true interactions. Added to such uncertainties is the fact that the correct two-body potential can only be obtained from low density gas data. The inadequacy of most of the present potentials is made evident by Rowlinson (1966) in a recent article. There exist, however, potential forms which, with some care, can be made to fit the experimental data. Munn (1964) and Kingston (1965), for instance, assumed a long-range behavior of the form $-KR^{-6}$ (taking theoretical K values) and used a multiparameter representation of the short-range forces. Such a potential is capable of yielding adequate transport *and* virial data. Its behavior, even its depth differ radically from the Lennard-Jones 6–12 formula. Neon, argon, krypton, and xenon have been studied in this way. For recent work on helium see Chapter 4 (Bruch and McGee, 1967).

B.2.2. Solid State Data

From the thermal properties of solids Guggenheim and McGlashan (1966), among others, have derived an effective two-body interaction. The accuracy of such a potential is unknown since three- and four-body effects can not be neglected, as was shown in Chapter 5.

B.2.3. X-ray Diffraction of Gases

A unique method to determine intermolecular potentials has been proposed by Mikolaj and Pings (1966), who point out that at low densities one can write an expression for the radial distribution function g of the molecules in a gas in terms of the potential, provided one uses the density expansion given by the Mayer cluster development (see, for example, Hill, 1956):

$$g(R_{12}; T, \varrho) = \exp\{-\Delta E(R_{12})/kT\} \times \{1+\varrho g_1(R_{12}; T)+O(\varrho^2)\},$$

where

$$g_1(R_{12};T) = \int [\exp(-\Delta E(R_{13})/kt)-1] \times [\exp(-\Delta E(R_{23})/kT)-1] dR_3,$$

T being the temperature and ϱ the gas density. Since g_1 is a measure of the three-body effects, the density must be low enough so that the nonadditive three-body potentials do not contribute. From precise X-ray diffraction measurements $\Delta E(R)$ can be found by an iteration procedure. Mikolaj and Pings, working with argon gas near the critical density, found that inclusion of g_1 in the potential had only slight effects on their results. According to estimates of Kestner and Sinanoglu (1963) three-body contributions constitute only a few percent of ΔE. The resulting $\Delta E(R)$ agrees with the best "experimental" potentials previously obtained. If the accuracy can be improved this method promises to be one of the important ways of determining reliable potential curves.

APPENDIX C

The Effective Hamiltonian for the Interaction of Bound Charges with an Electromagnetic Field

---

IN CALCULATING the interaction of atoms or other systems of bound charges with an electromagnetic field in the conventional Coulomb gauge (div $A = 0$, A being the vector potential), certain difficulties arise in the form of singularities which must be properly treated to obtain the correct results. The interaction of two neutral atoms contains terms which cancel the Coulomb interactions at large separations (see, for example, Casimir and Polder, 1948; Power and Zienau, 1959). Such singularities can be removed if the atoms are consistently treated as the sources of the quantized radiation field in the Hamiltonian.

In this Appendix we outline the method used by Power and Zienau (1957a, 1959) and Power (1964) to obtain a useful effective Hamiltonian for the interaction of bound charges with an electromagnetic field.

Let two atoms, A and B, be at rest in positions x_A and x_B. The kinetic energy of each electron will be of the form $(1/2)m\dot{q}^2$ where $\dot{q}$ is the time derivative of the electron coordinate. V_A and V_B include all Coulomb interactions within the atom, those between electrons and nucleus and between the electrons.

The Lagrangian of the two atoms is (see, for example, Goldstein, 1959)

$$L_A = \sum_{\alpha} \tfrac{1}{2} m \dot{q}_{\alpha}^2 - V_A - V_B, \tag{C1}$$

that for the radiation field

$$L_R = \frac{1}{8\pi} \int (E^2 - B^2)\, d\tau, \tag{C2}$$

where $\boldsymbol{E}$ is the total electric and $\boldsymbol{B}$ the magnetic field. The summation in (1) extends over all charges.

All interactions (between atoms and with the radiation fields) are contained in the interaction Lagrangian

$$L_I = -\int \left(\varrho\phi - \frac{1}{c}\, \boldsymbol{J}\cdot\boldsymbol{A}\right) d\tau \tag{C.3}$$

provided ϱ is the charge density, ϕ the scalar potential, $\boldsymbol{A}$ the vector potential, and $\boldsymbol{J}$ the electric current vector.

$$\varrho(\boldsymbol{x}) = \sum_\alpha e_\alpha\, \delta(\boldsymbol{x}-\boldsymbol{x}_\alpha), \tag{C4}$$

$$\boldsymbol{J}(\boldsymbol{x}) = \sum_\alpha e_\alpha \dot{\boldsymbol{q}}_\alpha\, \delta(\boldsymbol{x}-\boldsymbol{x}_\alpha). \tag{C5}$$

The delta functions indicate that the sources of charge and current are points within the atoms.

The electric and magnetic fields are related to the scalar and vector potentials by the equations,

$$\boldsymbol{B} = \nabla \times \boldsymbol{A}, \tag{C6}$$

$$\boldsymbol{E} = -\frac{1}{c}\dot{\boldsymbol{A}} - \nabla\phi. \tag{C7}$$

Thus the total Lagrangian obtained from (Cl), (C2) and (C3) is

$$\begin{aligned} L = {} & \sum_\alpha \left(\frac{1}{2} m\dot{q}_\alpha^2 - V_\alpha\right) + \frac{1}{8\pi}\int \left\{\frac{1}{c^2}\dot{A}^2 - (\nabla\times A)^2\right\} d\tau \\ & + \frac{1}{8\pi}\int |\nabla\phi|^2\, d\tau - \int \varrho(x)\phi(x)\, d\tau \\ & + \frac{1}{c}\int \boldsymbol{J}(x)\cdot\boldsymbol{A}(x)\, d\tau. \end{aligned} \tag{C8}$$

The term $\int \boldsymbol{A}\cdot\nabla\phi\, d\tau$ does not appear because $\boldsymbol{A}$ and $\nabla\phi$ are at right angles. Similarly, only the transverse part $\boldsymbol{J}^\perp$ of $\boldsymbol{J}$ enters.

To derive the standard Hamiltonian we use the coordinates, q_α and $\boldsymbol{A}$, and their conjugate momenta

$$\mathfrak{p}_\alpha = \frac{\partial L}{\partial \dot{\boldsymbol{q}}_\alpha} = m_\alpha \dot{\boldsymbol{q}}_\alpha + \frac{e_\alpha}{c} A(q_\alpha) \tag{C9}$$

and

$$\pi = \frac{\partial L}{\partial \dot{A}} = \frac{1}{4\pi c^2}\dot{A}, \tag{C10}$$

a quantity which is related to the total polarization, as we shall see. Thus

$$H = \sum_\alpha \mathfrak{p}_\alpha \cdot \dot{q}_\alpha + \int \pi \cdot \dot{A}\, d\tau - L \tag{C11}$$

$$= \sum_\alpha \frac{1}{2m_\alpha} \left(\mathfrak{p}_\alpha - \frac{1}{c} e_\alpha A\right)^2 + V_A + V_B + \frac{1}{8\pi} \int [(4\pi c)^2 \pi^2 + (\nabla \times A)^2]\, d\tau + \frac{1}{8\pi} \int |\nabla \phi|^2\, d\tau. \tag{C12}$$

Or, using (C7) we find

$$H = \sum_\alpha \frac{1}{2m_\alpha} \left(\mathfrak{p}_\alpha - \frac{1}{c} e_\alpha A\right)^2 + V_A + V_B + \frac{1}{8\pi} \int (\boldsymbol{E}^2 + \boldsymbol{B}^2) d\tau \tag{C13}$$

$$= \mathrm{H}_{\text{atom}} + \mathrm{H}_{\text{field}} + \mathrm{H}_{\text{int}} \tag{C14}$$

provided

$$H_{\text{int}} = \sum_\alpha \left(\frac{1}{m_\alpha c} \mathfrak{p}_\alpha \cdot A + \frac{1}{2m_\alpha c^2} A \cdot A \right) \tag{C15}$$

and

$$H_{\text{field}} = H_{\text{rad}} + H_{\text{coul}} \tag{C16}$$

$$= \frac{1}{8\pi} \int \{(\boldsymbol{E}^\perp)^2 + \boldsymbol{B}^2\}\, d\tau + \frac{1}{8\pi} \int |\nabla \phi|^2\, d\tau, \tag{C17}$$

where $\boldsymbol{E}^\perp$ is written for the transverse component of the electric field and H_{coul} for the Coulomb interaction between the two atoms. If we use the multipole expansion of Chapter 2, the first term of H_{coul} is the dipole–dipole interaction. The $\mathfrak{p}_\alpha$ can also be expressed as time derivatives of a dipole moment.

(C13) is the form usually adopted to treat the interaction of atoms with radiation fields. However, as mentioned earlier, it has several faults which can be eliminated if the atom is consistently included as the source of radiation in H_{field} as well as in H_{int}.

This can be accomplished in several ways. We can modify our Lagrangian (C8) by subtracting the following quantity, which is zero:

$$\frac{d}{dt} \int A \cdot \boldsymbol{P}\, d\tau \tag{C18}$$

where $P(x) = \sum_{\alpha} e_{\alpha} q_{\alpha} \delta(x - x_{\alpha})$, is the polarization. This changes terms in the Lagrangian from $P \cdot A$ to $\dot{A} \cdot P$ and the resulting Hamiltonian is greatly modified. Since a constant has been added to the Lagrangian, the state functions of the Hamiltonian are only changed by a phase factor; the energy difference remaining unchanged (Göppert–Mayer, 1931; McLachlan, 1963a). Alternatively, it is possible to perform a contact transformation of the Hamiltonian equation (C13). We shall follow this latter procedure since it is more general. The former procedure is explained elsewhere (Power and Zienau, 1957a).

To simplify our analysis we shall define a polarization vector field P which is the source of the electromagnetic field:

$$J = \dot{P}, \tag{C19}$$

$$\varrho = -\nabla \cdot P. \tag{C20}$$

These relations are assumed to hold only within the volume of our atom. We assume that the source is electrically neutral, for otherwise another term must be added.

A multipole expansion of the polarization field operator for the atom yields

$$P(x) = \sum_{\alpha} e q_{\alpha} \, \delta(x - x_{\alpha}) - \frac{1}{2} \sum_{\alpha} e q_{\alpha} (q_{\alpha} \cdot \nabla) \delta(x - x_{\alpha}) + \ldots . \tag{C21}$$

The summations extend over *all* atoms with the delta functions again defining the origin of the multipoles. The longitudinal electric field is given by the longitudinal component of the polarization.

$$E^{\parallel} = -\nabla \phi = -4\pi P^{\parallel}(x) \tag{C22}$$

or by components

$$E_j^{\parallel} = -4\pi e q_j \, \delta_{ij}^{\parallel}(x) + 4\pi Q'_{jk} \partial_j \delta_{ik}^{\parallel}(x) + \ldots . \tag{C23}$$

Here ∂_j denotes a partial derivative with respect to coordinate j. The longitudinal delta functions have their usual properties: If $f_j(r)$ is the jth component of any vector function of r, then

$$f_i^{\parallel}(r) = \int \delta_{ij}^{\parallel}(r - r') f_j(r) \, dr'. \tag{C24}$$

For more details we refer to Power (1964, pp. 73–75) or Belinfante (1946).

The delta functions in (C23) have as their arguments the position of the atoms, since the multipole expansion is about the center of the atom.

Using (C22) we rewrite (C13), obtaining

$$H = \sum_{\alpha} \frac{1}{2m_{\alpha}} [\mathfrak{p}_{\alpha} - eA(q_{\alpha})]^2 + V_A + V_B$$

$$+ \frac{1}{8\pi} \int [|\boldsymbol{E}^{\perp}|^2 + \boldsymbol{B}^2] \, d\tau + 2\pi \int |\boldsymbol{P}^{\parallel}|^2 \, d\tau \tag{C25}$$

where $\boldsymbol{P}^{\parallel}$ is the longitudinal component of $\boldsymbol{P}$.

We now modify H so that the new transverse electric field operator $\mathfrak{E}^{\perp}$ does not contain any static components which lead to the singularities mentioned earlier.† For this purpose, we first modify the electric field and then the rest of the Hamiltonian.

The polarization $\boldsymbol{P}$ as defined thus far is completely longitudinal and vanishes outside the atom. We now add to the $\boldsymbol{P}$ a transverse component which is subject to the restrictions that $\boldsymbol{P}^{\perp}$ plus $\boldsymbol{P}^{\parallel}$ must vanish outside the atom. With this addition the electric field strength becomes

$$\mathfrak{E}^{\perp} = \boldsymbol{E}^{\perp} + 4\pi \boldsymbol{P}^{\perp}(\boldsymbol{x}). \tag{C26}$$

Since this substitution removes the troublesome components in the expectation values of the electric field strength, it is a physically measurable dynamical variable and our Hamiltonian should be written in terms of it rather than $\boldsymbol{E}^{\perp}$. By this transformation the atom *and* its Coulomb field are incorporated into the radiation field. The modified portions of the Hamiltonian become, on substitution of (C26) into (C17),

$$H_{\text{rad}} + 2\pi \int |\boldsymbol{P}^{\parallel}(x)|^2 \, d\tau = \frac{1}{8\pi} \int [|\mathfrak{E}^{\perp}|^2 + B^2] \, d\tau$$

$$- \int \mathfrak{E}^{\perp} \cdot \boldsymbol{P}^{\perp} \, d\tau \tag{C27}$$

$$+ 2\pi \int |\boldsymbol{P}^{\perp}|^2 \, d\tau + 2\pi \int |\boldsymbol{P}^{\parallel}|^2 \, d\tau$$

$$= \frac{1}{8\pi} \int [|\mathfrak{E}^{\perp}|^2 + B^2] \, d\tau - \int \mathfrak{E}^{\perp} \cdot \boldsymbol{P} \, d\tau$$

$$+ 2\pi \int |\boldsymbol{P}|^2 \, d\tau. \tag{C28}$$

† This point, together with other relevant details, is discussed by Power (1964, appendix 4).

Because of our definition of $\boldsymbol{P}$ as the total polarization the last term is very small in the region outside the atoms and of no interest to us. It contains self-energies and contact interactions only and does not contribute to any coupling between the electromagnetic fields and the atoms. It is responsible for Lamb shifts, etc. Since $\boldsymbol{P}$ commutes with all field variables there is actually no need to distinguish between $\mathfrak{E}^{\perp}$ and $\boldsymbol{E}^{\perp}$. To prevent confusion, however, we shall continue to use both symbols in this Appendix.

To transform the remaining parts of the Hamiltonian it becomes necessary to find the canonical transformation S which converts $\boldsymbol{E}^{\perp}$ to $\mathfrak{E}^{\perp}$, so that

$$\mathfrak{E}^{\perp} = S\boldsymbol{E}^{\perp}S^{-1}. \tag{C29}$$

The required choice is

$$S = \exp\left(+\frac{i}{\hbar}\int \boldsymbol{P}^{\perp}\cdot\boldsymbol{A}\, d\tau\right). \tag{C30}$$

A proof is straightforward if the following vector identity

$$e^{A}Be^{-A} = B+[A, B]+\frac{1}{2!}[A,[A,B]]+\dots \tag{C31}$$

and a relation known from quantum electrodynamics (Power, 1964), viz.:

$$\left[-\frac{1}{4\pi}E_i^{\perp}(\boldsymbol{x}),\ A_j(\boldsymbol{y})\right] = i\hbar\ \delta_{ij}^{\perp}(\boldsymbol{x}-\boldsymbol{y}) \tag{C32}$$

are used. Here $[A, B]$ means $AB-BA$.

Transformation (C29) does not change the energy of a state but simply rearranges the variables (Fontana, 1961b).

Upon making a multipole expansion of $\boldsymbol{A}$ and S and using (C21), we find

$$\boldsymbol{A}(q_\alpha) = \boldsymbol{A}(0)+(q_\alpha\cdot\nabla)\boldsymbol{A}(0)+\dots, \tag{C33}$$

$$S = \sum_\alpha \exp\left(-\frac{ie}{\hbar}q_\alpha\cdot\boldsymbol{A}(0)-\frac{1}{2}\frac{ie}{\hbar}(q_\alpha\cdot\nabla)q_\alpha\cdot\boldsymbol{A}(0)+\dots\right), \tag{C34}$$

and thence

$$\begin{aligned} &S^{-1}\sum_\alpha\frac{1}{2m}[\mathfrak{p}_\alpha-e\boldsymbol{A}(q_\alpha)]^2S \\ &=\sum_\alpha\frac{1}{2m}(\mathfrak{p}_\alpha)^2-\sum_\alpha\frac{e}{2m}[q_\alpha\times\mathfrak{p}_\alpha]\cdot[\nabla\times\boldsymbol{A}(0)] \\ &+\sum_\alpha\frac{e^2}{8m}[q_\alpha\times[\nabla\times\boldsymbol{A}(0)]]^2+\dots. \end{aligned} \tag{C35}$$

The second term is the magnetic dipole contribution used in the treatment of the Faraday effect (Power and Shail, 1959).

We now combine all terms and write the total Hamiltonian in the form

$$H = \sum_{\alpha} \frac{1}{2m} (\mathfrak{p}_{\alpha})^2 + \sum_{l} V_l + \frac{1}{8\pi} \int [|\mathfrak{E}^{\perp}|^2 + |\boldsymbol{B}|^2] \, d\tau$$

$$- \sum_{l} \boldsymbol{p}_l \cdot \mathfrak{E}^{\perp} - \sum_{l} \sum_{ij} Q_{ij}^{(l)} \nabla \mathfrak{E}_i^{\perp}$$

$$- \sum_{l} \boldsymbol{M}_l \cdot \boldsymbol{B} + \sum_{\alpha} \frac{e^2}{8m} [\boldsymbol{q}_{\alpha} \times \boldsymbol{B}]^2 + \dots$$

$$+ 2\pi \int |\boldsymbol{P}|^2 \, d\tau. \qquad \text{(C36)}$$

Again, α refers to a sum over individual electrons and l sums over atoms a and b. The following are properties of one atom:

electric dipole moment: $\boldsymbol{p}_l = \sum_{\alpha} e\boldsymbol{q}_{\alpha}$

magnetic dipole moment: $\boldsymbol{M}_l = \sum_{\alpha} \frac{e}{2m} [\boldsymbol{q}_{\alpha} \times \mathfrak{p}_{\alpha}]$

traceless quadrupole tensor (ij component):

$$Q_{ij}^{(l)} = \frac{e}{2} \sum_{\alpha} \left((q_{\alpha})_i (q_{\alpha})_j - \frac{1}{3} q_{\alpha}^2 \, \delta_{ij} \right).$$

Only the traceless quadrupole moment appears since $\nabla \cdot \mathfrak{E}^{\perp} = 0$. No longitudinal components enter and the entire interaction is retarded.

In Chapter 6 we use (C36) in the dipole approximation, neglecting self-energy terms, viz.:

$$H = H_{\text{atom}} + H_{\text{rad}} + H'_{\text{int}}, \qquad \text{(C37)}$$

where H_{atom} and H_{rad} have been defined in (C14) and (C17) and

$$H'_{\text{int}} = -\sum_{l} \boldsymbol{p}_l \cdot \mathfrak{E}^{\perp}. \qquad \text{(C38)}$$

The dipole approximation, we observe, amounts to neglecting all variations of the fields, i.e. all gradients, across the atom. The gradients lead to higher multipole effects.

Equation (C38) is the form of the interaction term used in Chapters 5 and 8 to evaluate the contribution of radiation effects to intermolecular forces between neutral atoms.

Post Scriptum

Current Research, Problems, and Trends in the Theory of Intermolecular Forces

IN VIEW of the intense activity in the fields covered by this book and of the delay between writing and publication, it may be appropriate to add a post scriptum in which current work is briefly sketched and some new references are added.

In the theory dealing with forces at large separations two major developments have occurred. First, there have been applications of time dependent perturbation theory to the dispersion interaction of many electron atoms. The theory is applied in either the complicated coupled form or in the approximate uncoupled form in which perturbation of one orbital does not affect the perturbation of the others. These are discusssed in a recent review by Dalgarno (1967). A modification intermediate between these two versions uses double perturbation techniques to solve the coupling problem (Deal and Kestner, 1966). These same approaches can be used with the many-electron theory (pp. 130 and 140) when the atoms do not overlap. At present such methods are not useful when overlap is important because there is then no simple relationship between the Hamiltonian describing the interaction and the polarizability or some other *one-electron* operator. The major advantage of all of these techniques is that the frequency dependent polarizability, and thus the dispersion interaction based on eq. (84) of Chapter 2, can be obtained correctly to first order in intra-atomic electron correlation by solving only a one-electron equation, provided the zero order solution for the atom is the Hartree–Fock result (see Chapter 4). Furthermore, the intra-atomic electron correlation functions (see Chapter 4) or the exact atomic state functions are not needed. It appears that these methods, which correct those of Karplus and Kolker quoted earlier, can easily be extended to larger systems and all of the attractive interactions. The general connection between time dependent Hartree theory and intermolecular forces is given by McLachlan, Gregory, and Ball (1963). Related methods of response functions and reaction field techniques are discussed in papers by Linder (1967).

Secondly, there is the rather startling observation by Herring (1962, 1966) that the Heitler–London theory predicts the wrong singlet (${}^1\Sigma_g{}^+$)–triplet (${}^3\Sigma_u{}^+$) spacing at large separations. In fact, this difference has the wrong sign for hydrogen atom interactions beyond $50a_0$. The exchange energy, one half of the singlet–triplet difference, should behave asymptotically (Herring and Flicker, 1964) as

$$-.818R^{5/2} \exp(-2R)$$

whereas the Heitler–London result has the asymptotic form

$$\{-(28/35)+(2/15)[0.5772+\ln R]\} R^3 \exp(-2R)$$
$$+\{-2.5+0.8[0.5772+\ln R]\} R^2 \exp(-2R).$$

At distances over $50a_0$ the triplet is the lowest energy state. The cause of this error is a correlation effect, primarily an ionic or charge transfer term (Alexander and Salem, 1967). Herring (1962, 1966) has proposed methods to avoid this difficulty which could have applications to other calculations of interaction energies. This long-range singlet–triplet effect may also be observed in alkali-atom scattering (Davison and Dalgarno, 1968).

The interaction of many electron atoms at large separations involves additional problems if overlap and/or problems such as singlet triplet separations are to be studied. We do not know the correct asymptotic behavior of the electronic wave function of the atoms and, in addition, we do not know what relativistic effects contribute to the problems. Once we have a many-electron wave function it is no longer obvious what the asymptotic behavior should be. And while relativistic effects on the behavior of the outer electrons are probably minor, this is not true of the inner electrons, whose orbitals are greatly contracted in atoms with a large nuclear charge. This leads to changes in electron correlation, which will shift nodes in the orbital functions and alter the screening of the nuclear charge compared with that in a non-relativistic treatment. Whether this could seriously effect the overlap of atoms at large separations remains to be studied.

At intermediate separations many researchers have been searching for methods of calculating the intermolecular potential in powers of the overlap integral. Murrell and Shaw (1967a) have further developed the procedure (Murrell, *et al.* (1965)) described on p. 149 and have calculated the He–He interaction, to be sure, neglecting second order exchange effects, (1967b). Using the latest values for dispersion interactions they find that agreement with the latest analysis of experimental data (Bruch and McGee, 1967) is good. Alternative attempts to derive such theories have been made by Musher and Salem (1966), Salem (1966), Van der Avoird (1967), Jansen (1967), Hirschfelder (1967), and others. Most of these are based on perturbation rather than variational theory. In view of our earlier remarks (p. 150) a variational theory may be thought to be more successful. In much of this work little attention has been given to the practical problems of calculating an interaction between many-electron atoms. For example, no one has adequately calculated the overlap except in a limited Hartree – Fock approximation; the error in this is unknown.

Several calculations have been made on three-body interactions. Williams, Schaad, and Murrell (1967) conclude that three-body overlap effects should be small for most rare-gas solids. Ross and Alder (1966) have found that the effective repulsive potential in a liquid is weaker than in the gaseous phase, at least for typical two-parameter empirical potentials. This can be justified by Thomas–Fermi–Dirac calculations. Lucas (1967) has shown that the n-body induced dipole–dipole interaction converges rapidly in rare gas crystals (see also Doniach, 1963). He finds, for example, that the effective pairwise nonadditive interaction is only eight per cent of the additive terms in solid xenon. The effects of three-body forces on radial distribution functions of liquids have been analysed by Rushbrooke and Silbert (1967), Rowlinson (1967), and Pings (1967). However, little is known experimentally of many body forces. We do not even know in general the analytic form of the angular dependence of short-range three- and four-body interactions, aside from the special cases dealt with in Chapter 5.

The success of the simple Woodward–Hoffmann rules in predicting the stereochemistry of concerted organic reactions (Woodward and Hoffmann, 1965, 1968) has led to the development of semi-empirical perturbation theories for the interaction of large organic systems in their ground and excited states. This work (Fukui, 1966; Fukui and Fujimoto, 1966; Salem, 1968) is capable of predicting organic reactions and holds much promise for biological problems.

As to the future, we can expect even more applications of theory to large organic systems, especially those of biological interest. In this context the study of charge transfer interactions (Mulliken, 1950, 1952; Aono, 1959), a topic not treated in this book, should become increasingly important. The same is likely to be true for hydrogen bonding. A recent paper on this subject is written by Van Duijneveldt and Murrell (1966). The effects of solvent on interaction and ways to modify interactions by chemical substitution are subjects of future interest. There will doubtless continue to be many developments in new experimental techniques, for example, collision-induced adsorption (Futrelle, 1967), new methods of analysis of differential scattering data (Bernstein and Muckerman, 1967), and interpretation of low density X-ray data (Mikolaj and Pings, 1966).

While the earlier theories of intermolecular forces could be content with the use of electrostatics, it now appears that quantum electrodynamics, field theory, many-body techniques, and a host of modern physical tools will also be the equipment of future theoreticians. Of late, computers have certainly been of great importance. While their use is sure to increase, we expect that *a priori* calculation of forces between large molecules is very far off and that judicious use of computers coupled with physical insight and powerful mathematical tools will be more important tasks in the immediate future. Leaving aside such fallible prognoses of coming events, we state in conclusion a simple necessity. Success in this field will require more cooperation between theorists and experimenters. Experimentalists must use theoretical potentials (or parts of potentials) whenever possible and study those systems for which theory can provide some guidance.

Bibliography of Papers dealing with Theoretical Aspects of Intermolecular Forces

ABDULNUR, S.: *see* SINANOGLU, O.

ABE, A., JERNIGAN, R. L. and FLORY, P. J. (1966) *J. Am. Chem. Soc.* **88,** 631.

ABRAHAMSON, R. A. (1963) Repulsive interaction potentials between rare-gas atoms: homonuclear two-center systems, *Phys. Rev.* **130**, 693.

ABRAHAMSON, R. A. (1964) Repulsive interaction potentials between rare-gas atoms: heteronuclear two-center systems, *Phys. Rev.* **133,** A990.

ABRIKOSOV, A. A., GORKOV, L. D. and DZIALOSHINSKII, I. E. (1965) *Quantum Field Theoretical Methods in Statistical Physics*, Pergamon, Oxford.

ABRIKOSOVA, I. I.: *see* DERJAGUIN, B. J.

ALDER, B. J., FERNBACH, S. and ROTENBERG, A. (Eds.) (1962) *Methods in Computational Physics*, Academic Press, New York.

ALDER, B. J. and PAULSON, R. H. (1965) Pair potential that stabilizes cubic argon, *J. Chem. Phys.* **43,** 4172.

ALDER, B. J. and POPLE, J. R. (1957) Third virial coefficients for intermolecular potentials with hard sphere cores, *J. Chem. Phys.* **26,** 325.

ALEXANDER, M. H. and SALEM, L. (1967) *J. Chem. Phys.* **46,** 430.

AMDUR, I. (1949) *J. Chem. Phys.* **17,** 844.

AMDUR, I. (1958) Comments on 'Note on the He–He interaction potential and its determination from Amdur's Scattering Measurements', *J. Chem. Phys.* **28,** 987.

AMDUR, I. and BERTRAND, R. R. (1962) Redetermination of the He–He interaction between 0.55 and 1.0 A, *J. Chem. Phys.* **36,** 1078.

AMDUR, I. and HARKNESS, A. L. (1954) *J. Chem. Phys.* **22,** 664.

AMDUR, I. and JORDAN, J. E. (1966) Elastic scattering of high energy beams: repulsive forces, *Advanc. Chem. Phys.* **10,** 29.

AMDUR, I., JORDAN, J. E. and COLGATE, S. O. (1961) Scattering of high velocity neutral particles, XI: a further study of the He–He potential, *J. Chem. Phys.* **34,** 1525.

AMDUR, I., LONGMIRE, M. S. and MASON, E. A. (1961) Scattering of high velocity neutral particles, XII: He–CH_4, He–CF_4, the CH_4–CH_4 and CF_4–CF_4 interactions, *J. Chem. Phys.* **35,** 895.

AMDUR, I. and MASON, E. A. (1956a) Scattering of high velocity neutral particles, VII: xenon–xenon, *J. Chem. Phys.* **25,** 624.

AMDUR, I. and MASON, E. A. (1956b) Scattering of high velocity neutral particles, VIII: H–He, *J. Chem. Phys.* **25,** 630.

AMDUR, I. and MASON, E. A. (1956c) Scattering of high velocity neutral particles, IX: Ne–A, A–Ne, *J. Chem. Phys.* **25**, 632.

AMDUR, I., MASON, E. A. and JORDAN, J. E. (1957) Scattering of high velocity neutral particles, X: He–N_2, A–N_2: the N_2–N_2 interaction, *J. Chem. Phys.* **27**, 527.

AMDUR, I.: *see* MASON, E. A. and JORDAN, J. E.

AMME, R. C. and LEGVOLD, S. (1960) Vibrational transitions and the intermolecular potential, *J. Chem. Phys.* **33**, 91.

ANDERSON, P. W. (1950) *Phys. Rev.* **80**, 511.

AONO, S. (1959) *Prog. Theoret. Phys.* (Kyoto) **22**, 313.

ARMSTRONG, R. L. (1966) Repulsive interactions and molecular rotation in inert-gas lattices, *J. Chem. Phys.* **44**, 530.

AROESTE, H. and JAMESON, W. J. (1959) Short-range interaction between a hydrogen molecule and a hydrogen atom, *J. Chem. Phys.* **30**, 372.

AROESTE, H.: *see* JAMESON, W. J.

ASHKIN, M.: *see* CARR, W. J.

ASTON, J. G.: *see* CHON, H.

AUB, M. R., POWER, E. A. and ZIENAU, S. (1957) *Phil. Mag.* **2**, 571.

AXILROD, B. M. (1951) *J. Chem. Phys.* **19**, 719.

AXILROD, B. M. (1963) Comments on the Rosen interaction potential of two helium atoms, *J. Chem. Phys.* **38**, 275.

AXILROD, B. M. and TELLER, E. (1943) *J. Chem. Phys.* **11**, 299.

AYRES, R. V. and TREDGOLD, R. H. (1956) *Proc. Phys. Soc. (London)* **B69**, 840.

BADE, W. L. (1957) Drude-model calculation of dispersion forces, I: general theory, *J. Chem. Phys.* **27**, 1280.

BADE, W. L. (1958) Drude model calculation of dispersion forces, III: the fourth-order contribution, *J. Chem. Phys.* **28**, 282.

BADE, W. L. and KIRKWOOD, J. G. (1957) Drude-model calculation of dispersion forces, II: the linear lattice, *J. Chem. Phys.* **27**, 1284.

BAGUS, P. S., GILBERT, T. L., ROOTHAAN, C. C. J. and COHEN, H. (1968) *Phys. Rev.* (to be published)

BAILEY, T. L., MAY, C. J. and MUSCHLITZ, E. E., Jr. (1957) Scattering of low energy H^- ions in helium, neon, and argon, *J. Chem. Phys.* **26**, 1446.

BAILEY, T. L.: *see* MUSCHLITZ, E. E.

BAKER, C. E., MCGUIRE, J. M. and MUSCHLITZ, E. E., Jr. (1962) Low energy collision cross-sections of H^- and OH^- ions in oxygen, *J. Chem. Phys.* **37**, 257.

BALL, M. A.: *see* MCLACHLAN, A. D.

BALLHAUSEN, C. and GRAY, H. B. (1965) *Molecular Orbital Theory*, Benjamin, New York.

BANDRAUK, A. D.: *see* THORSON, W.

BARANGER, M. (1962) In *Atomic and Molecular Processes*, ed. D. R. Bates, Academic Press, New York.

BARDEEN, J. (1940) The image and Van der Waals forces at a metallic surface, *Phys. Rev.* **58**, 727.

BARKER, J. A. and EVERETT, D. H. (1962) *Trans. Faraday Soc.* **58**, 1608.

BARKER, J. A. and LEONARD, P. J. (1964) Long-range interaction forces between inert gas atoms, *Phys. Letters* **13**, 127.

Bibliography

BARKER, R. S. and EYRING, H. (1954) *J. Chem. Phys.* **22**, 699, 1182, 2072

BARKER, R. S. and EYRING, H. (1955) *J. Chem. Phys.* **23**, 1381.

BARNETT, M. (1962) in *Methods in Computational Physics* (Eds. Alder, Fernback, and Rotenberg), Academic Press, New York.

BARRON, T. H. K. (1966) Interatomic potentials in ideal anharmonic crystals, *Disc. Faraday Soc.* **40**, 69.

BARTELL, L. S. (1960) On the effects of intramolecular Van der Waals forces, *J. Chem. Phys.* **32**, 827.

BARUA, A. K. (1959) Force parameters for some nonpolar molecules on the Exp-6-8 model, *J. Chem. Phys.* **31**, 1619.

BARUA, A. K. (1960) Intermolecular potential of helium, *Indian J. Phys.* **34**, 76.

BARUA, A. K. and CHATTERJEE, S. (1964) Repulsive energy between hydrogen and helium atoms, *Mol. Phys.* **7**, 444.

BATES, D. R. (Ed.) (1961) *Quantum Theory*, Academic Press, New York, Vol. 1.

BATES, D. R. and LEWIS, J. T. (1955) *Proc. Phys. Soc. (London)* **A68**, 173.

BATES, D. R. and ESTERMAN, I. (Eds.) (1965) *Advances in Atomic and Molecular Physics*, Vol. 1, Academic Press, New York.

BEAN, D. T. W.: *see* BRITTON, F. R.

BEENAKKER, J. J. M.: *see* KNAAP, H. F. P.

BELINFANTE, F. J. (1946) *Physica* **12**, 1.

BELL, R. J. (1965a) Calculation of the Van der Waals force constant from oscillator strength sums, *Proc. Phys. Soc.* **86**, 17.

BELL, R. J. (1965b) The long-range interaction between hydrogen atoms, *Proc. Phys. Soc.* **86**, 239.

BELL, R. J. (1965c) Long-range interaction of three atoms, *Proc. Phys. Soc.* **86**, 519.

BELL, R. J. (1966) The long-range interaction between hydrogen atoms, *Proc. Phys. Soc.* **87**, 594.

BELL, R. J. and KINGSTON, A. E. (1966) The Van der Waals interaction of two or three atoms, *Proc. Phys. Soc.* **87**, 901.

BELLEMANS, A.: "Forces in Metals", *Modern Quantum Chemistry*, Istanbul Lectures, Vol. 2, Academic Press, O. Sinanoglu, ed., p. 279.

BENNETT, A. J. and FALICOV, L. M. (1966) *Phys. Rev.* **151**, 512.

BERENCZ, F. (1960) The role of multipole interaction effects in Van der Waals interactions, *Acta Phys. Hungar.* **12**, 1.

BERKLING, K., SCHLIER, C. and TOSCHEK, P. (1962) Messung der Anisotropie der Van der Waals–Kraft, *Z. f. Physik.* **168**, 81.

BERNARDES, N. and PRIMAKOFF, H. (1959) Molecule formation in the inert gases, *J. Chem. Phys.* **30**, 691.

BERNSTEIN, R. B. (1966) Long-range interatomic forces from predissociation data and resonances, *Phys. Rev. Letters* **16**, 385.

BERNSTEIN, R. B. and O'BRIEN, T. J. P. (1966) Extrema-effect in total elastic molecular beam scattering cross-sections for characterization of the potential well, *Disc. Faraday Soc.* **40**, 35.

BERNSTEIN, R. B.: *see* FONTANA, P. R., GROBLICKI, P. J., HOSTETTLER, H. U., MORSE, F. A. and ROTHE, E. W.

BERNSTEIN, R. B. and MUCKERMAN, J. T. (1967) in *Intermolecular Forces* (Ed. Hirschfelder), Interscience, New York.

BERRY, R. S., KESTNER, N. R. and MCKOY, V. (1963) (Unpublished).

BERSOHN, R. (1962) Intermolecular bonding in the solid halogens, *J. Chem. Phys.* **36,** 3345.

BERTHIER, G. (1964) In *Molecular Orbitals in Chemistry, Physics and Biology* (eds. P. O. LÖWDIN and B. PULLMAN), Academic Press, New York.

BERTRAND, R. R.: *see* AMDUR, I.

BETHE, H. and SALPETER, E. E. (1957) *Quantum Mechanics of One and Two-Electron Atoms*, Academic Press, New York.

BIEDENHARN, L. C. (1952) *Tables of Racah Coefficients*, Oak Ridge National Laboratory, Oak Ridge, Tennessee.

BINGEL, W. (1959a) United atom treatment of the behavior of potential energy curves of diatomic molecules for small *R*, *J. Chem. Phys.* **30,** 1250.

BINGEL, W. (1959b) United atom treatment of the behavior of potential energy surfaces of polyatomic molecules at small internuclear distances, *J. Chem. Phys.* **30,** 1254.

BLACK, W., DE JONGH, J. G. V., OVERBEEK, J. T. G. and SPARNAAY, M. J. S. (1960) Measurements of retarded Van der Waals' forces, *Trans. Faraday Soc.* **56,** 1597.

BLANKEN, G.: *see* ROSENSTOCK, H. B.

BLEICK, W. E. and MAYER, J. D. (1934) *J. Chem. Phys.* **2,** 252.

DE BOER, J. H. (1936) *Trans. Faraday Soc.* **32,** 10.

DE BOER, J. H. (1942) *Physica* **9,** 363.

DE BOER, J. H. and HELLER, G. (1937) *Physica* **4,** 1045.

DE BOER, J. H. and MICHAELS, A. (1938) *Physica* **5,** 945.

DE BOER, J. H. and LUNBECK, R. I. (1948) *Physica* **14,** 510.

BORN, M. and OPPENHEIMER, J. R. (1927) *Ann. Phys.* **84,** 457.

BOUWKAMP, C. J. (1947) *Proc. Koninkl. Ned. Acad. Wetens.* **50,** 1071.

BOYS, S. F.: *see* SHAVITT, I.

BOYS, S. F. and SHAVITT, I. (1959) Univ. of Wisc. Naval Res. Lab. Tech. Report, WIS-AF-13.

BRANDT, W. (1956) Calculation of intermolecular force constants from polarizabilities, *J. Chem. Phys.* **24,** 501.

BRANT, D. A. and FLORY, P. J. (1965) *J. Am. Chem. Soc.* **87,** 2791.

BREENE, R. G., Jr. (1957) *Rev. Mod. Phys.* **29,** 94.

BRIENT, S. J.: *see* BRIGMAN, G. H.

BRIGMAN, G. H., BRIENT, S. J. and MATSEN, F. A. (1961) Interaction of a triplet and a normal helium atom, *J. Chem. Phys.* **34,** 958.

BRITTON, F. R. and BEAN, D. T. W. (1955) Long-range forces between hydrogen molecules, *Canad. J. Phys.* **33,** 668.

BROOKS, F. C. (1952) *Phys. Rev.* **86,** 92.

BROUT, R. (1954) *J. Chem. Phys.* **22,** 934.

BROWN, W. F. (1956) In *Encyclopedia of Physics* (ed. S. Flügge), Springer-Verlag, Berlin, Vol. 17, *Dielectrics*.

BROWNE, J. C.: *see* REAGAN, P. N. and SCOTT, D. R.

BROYLES, A. A.: *see* KHAN, A. A.

BRUCH, L. W. and MCGEE, I. I. (1967) *J. Chem. Phys.* **46,** 2959.

BRUNAUER, S. (1943) *The Adsorption of Gases and Vapors*, Princeton University.

BUCKINGHAM, A. D. (1966) Theory of long-range dispersion forces, *Disc. Faraday, Soc.* **40**, 232.

BUCKINGHAM, R. A. (1958) The repulsive interaction of atoms in *s* states, *Trans. Faraday Soc.* **54**, 453.

BUCKINGHAM, R. A., FOX, J. W. and GAL, E. (1965) *Proc. Roy. Soc. (London)* **A284**, 237.

BUCKINGHAM, R. A.: *see* DUPARC, D. M.

BUEHLER, R. J. and HIRSCHFELDER, J. O. (1951) *Phys. Rev.* **83**, 628.

BUEHLER, R. J. and HIRSCHFELDER, J. O. (1952) *Phys. Rev.* **85**, 149.

BULLOUGH, R., GLYDE, H. R. and VENABLES, J. A. (1966) Stacking-fault energy and many-body force effects in solid argon, *Phys. Rev. Letters* **17**, 2491.

BULUGGIU, E. and FOGLIA, C. (1965) Three-body bound state in xenon, *Phys. Letters* **14**, 26.

BYERS BROWN, W. (1966) Interatomic forces at very short range, *Disc. Faraday Soc.* **40**, 140.

BYERS BROWN, W. and ROWLINSON, J. S. (1960) A thermodynamic discriminant for the Lennard Jones potential, *Mol. Phys.* **3**, 35.

BYERS BROWN, W. and STEINER, E. (1966) On the electronic energy of a one-electron diatomic molecule near the united atom, *J. Chem. Phys.* **44**, 3934.

CADE, P. E. (1961) Thesis, University of Wisconsin, issued as Theoretical Chemistry Laboratory Report WIS–AEC–30, Series 3, June 1961.

CADE, P. E., SALES, K. D. and WAHL, A. C. (1966) *J. Chem. Phys.* **44**, 1973.

CALDWELL, P.: *see* LIPPINCOTT, E. R.

CAMPBELL, E. S. (1952) Hydrogen bonding and the interactions of water molecules, *J. Chem. Phys.* **20**, 1411.

CARLSON, B. C. and RUSHBROOKE, G. S. (1950) *Proc. Camb. Phil. Soc.* **46**, 626.

CARR, W. J., Jr. and ASHKIN, M. (1965) Long-range interaction of two electron systems, *J. Chem. Phys.* **42**, 2796.

CASANOVA, G., DE PAZ, M., DONDI, M. G., KLEIN, M. L. and SCOLES, C. (1966) Isotope effects in physical adsorption and the interaction between argon atoms and graphitized carbon blacks, *Disc. Faraday Soc.* **40**, 188.

CASHION, J. K. and HERSCHBACH, D. R. (1964) Empirical evaluation of the London potential energy surface for the H plus H_2 reaction, *J. Chem. Phys.* **40**, 2358.

CASHION, J. K.: *see* GORDON, R. G.

CASIMIR, H. B. G. and POLDER, D. (1948) *Phys. Rev.* **73**, 360.

CASTLE, B. J., JANSEN, L. and DAWSON, J. M. (1956) Second virial coefficients for assemblies of nonspherical molecules, *J. Chem. Phys.* **24**, 1078.

CHAKRABORTI, P. K. (1966) Intermolecular potential of radon, *J. Chem. Phys.* **44**, 3137.

CHAMBERLAIN, G. E. and ZORN, J. C. (1960) *Bull. Am. Phys. Soc.* **5**, 241.

CHAN, Y. M. and DALGARNO, A. (1965a) The long-range interaction of atoms and molecules, *Mol. Phys.* **9**, 349.

CHAN, Y. M. and DALGARNO, A. (1965b) Long-range interactions between three hydrogen atoms, *Mol. Phys.* **9**, 525.

CHATTERJEE, S.: *see* BARUA, A. K.

CH'EN, S. and TAKEO, M. (1957) *Rev. Mod. Phys.* **29**, 20.

CHIA–CHUNG, S. and TUNG–CHEN, K. (1961) Problem of the Van der Waals forces for a system of asymmetric top molecules, *Acta Phys. Sinica* **17**, 559.

CHON, H., FISHER, R. A., MCCAMMON, R. D. and ASTON, J. G. (1962) Interaction of helium, neon, argon, and krypton with a clean platinum surface, *J. Chem. Phys.* **36**, 1378.

CIZEK, J. (1963) *Mol. Phys.* **6**, 19.

CLAIRAULT, A. C. (1743) *Théorie de la figure de la terre*, Paris, 2nd ed., published by Courcier, Paris, 1808.

CLINTON, W. L. (1963) Forces in molecules, II: a differential equation for the potential energy function, *J. Chem. Phys.* **38**, 2339.

CLINTON, W. L. and HAMILTON, W. C. (1960) Force curves for excited electronic states, *Revs. Modern Phys.* **32**, 422.

CLONEY, R.D. and VANDERSLICE, J. T. (1962) Interaction energies from scattering cross-sections of hydrogen ions in CH_4, CF_4, C_2H_6, and C_2F_6, *J. Chem Phys.* **36**, 1866.

COLE, R. H. (1963) Dielectric constant and interactions of polar molecules, *J. Chem. Phys.* **39**, 2602.

COLE, R. H.: *see* OUDEMANS, G. J.

COLGATE, S. O.: *see* AMDUR, I.

CONROY, H. and BRUNER, B. L. (1965) *J. Chem. Phys.* **42**, 4047.

CONROY, H. and BRUNER, B. L. (1967) *J. Chem. Phys.* **47**, 921.

COOLIDGE, A. S. and JAMES, H. M. (1933) *J. Chem. Phys.* **1**, 825.

COOLIDGE, A. S. and JAMES, H. M. (1938) *J. Chem. Phys.* **6**, 730.

CORNER, J. (1948) *Trans. Faraday Soc.* **44**, 914.

COTTRELL, T. L. (1956) Intermolecular repulsive forces, *Disc. Faraday Soc.* **22**, 10.

COULSON, C. A. (1937) *Proc. Camb. Phil. Soc.* **33**, 104.

COULSON, C. A. and DAVIES, P. L. (1952) *Trans. FaradaySoc.* **48**, 777.

COULSON, C. A. and EISENBERG, D. (1965) Interactions of H_2O molecules in ice, I: the dipole moment of an H_2O molecule in ice; II: Interaction energies of H_2O molecules in ice, *Proc. Roy Soc.* **A291**, 445, 454.

COULSON· C. A. and FSHER, I. (1949) *Phil. Mag.* **40**, 386.

CRAIG, D. P., DOBUSH, P. A., MASON, R. and SANTRY, D. P. (1966) Intermolecular forces in crystals of the aromatic hydrocarbons, *Disc. Faraday Soc.* **40**, 110.

CRAMER, W. H. (1958) Elastic and inelastic scattering of low-velocity ions: He^+ in Ne, Ne^+ in He, and Ne^+ in Ne, *J. Chem. Phys.* **28**, 688.

CRAMER, W. H. (1961) Elastic and inelastic scattering of low-velocity H^+ and H_2^+ in hydrogen, *J. Chem.* **35**, 836.

CRAMER, W. H. and MARCUS, A. B. (1960) Elastic and inelastic scattering of low-velocity D and D_2 in deuterium, *J. Chem. Phys.* **32**, 186.

CRAMER, W. H. and SIMONS, J. H. (1957) Elastic end inelastic scattering of low-velocity He^+ ions in helium, *J. Chem. Phys.* **26**, 1272.

CROWELL, A. D. and STEELE, R. B. (1961) Interaction potentials of simple nonpolar molecules with graphite, *J. Chem. Phys.* **34,** 1347.

CROWELL, A. D.: *see* KRIZAN, J. E.

CUSACHS, L. C. (1962) Dependence of induction and dispersion energies at finite internuclear distances, *Phys. Rev.* **125,** 561.

CUSACHS, L. C. (1963) Induction energies: the spherical term, *J. Chem. Phys.* **38,** 2038.

CUTLER, P. H. and GIBBONS, J. J. (1958) *Phys. Rev.* **111,** 394.

DAHLER, J. S. and HIRSCHFELDER, J. O. (1956) Long-range intermolecular forces, *J. Chem. Phys.* **25,** 986.

DALGARNO, A. (1962) *Advance Phys.* **11,** 281.

DALGARNO, A. (1961) In *Quantum Theory* (ed. D. R. BATES), Vol. 1, Academic Press, New York.

DALGARNO, A. (1967) in *Intermolecular Forces* (Ed. Hirschfelder), Interscience, New York.

DALGARNO, A. and DAVISON, W. D. (1966) The calculation of Van der Waals interactions, *Advanc. Atomic and Mol. Phys.* **2,** 1.

DALGARNO, A. and KINGSTON, A. E. (1961) Van der Waals forces for hydrogen and the inert gases, *Proc. Phys. Soc.* **78,** 607.

DALGARNO, A. and LEWIS, J. T. (1956) The representation of long-range forces by series expansions, I: the divergence of series; and II, the complete perturbation calculation of long-range forces, *Proc. Phys. Soc.* **A69,** 57, 59.

DALGARNO, A. and LYNN, N. (1956) Resonance forces at large separations, *Proc. Phys. Soc.* **A69,** 821.

DALGARNO, A. and LYNN, N. (1957) *Proc. Phys. Soc.* **A70,** 802.

DALGARNO, A. and LYNN, N. (1957a) Curve crossing and orthogonality of molecular orbitals, *Proc. Phys. Soc.* **A70,** 176.

DALGARNO, A. and LYNN, N. (1957b) An exact calculation of second-order long-range forces, *Proc. Phys. Soc.* **A70,** 223.

DALGARNO, A. and MCCARROLL, R. (1956) Adiabatic coupling between electronic and nuclear motion in molecules, I and II, *Proc. Roy. Soc. (London)* **A237,** 383.

DALGARNO, A. and MCCARROLL, R. (1957) *Proc. Roy. Soc. (London)* **A239,** 413.

DANON, F. and PITZER, K. S. (1962) Volumetric and thermodynamic properties of fluids, VI: relationship of molecular properties to the acentric factor, *J. Chem. Phys.* **36,** 425.

DANON, F.: *see* ROSSI, J. C.

DAS, G. and WAHL, A. C. (1966) *J. Chem. Phys.* **44,** 87.

DASS, L., SAXENA, S. C. and KACHHAVA, C. M. (1965) Interatomic forces and binding energy of diatomic ionic molecules, *Indian J. Pure Appl. Phys.* **3,** 178.

DAVIDSON, E. R. (1960) First excited $^1\Sigma_g^+$ state of H_2: a double minimum problem, *J. Chem. Phys.* **33,** 1577.

DAVIDSON, E. R. (1961) First excited $^1\Sigma_g^+$ state of the hydrogen molecule, *J. Chem. Phys.* **35,** 1189.

DAVIDSON, E. R.: *see* ROTHENBERG, S., WRIGHT, W. M. and WAKEFIELD, C. B.

DAVISON, W. D. (1962) *Disc. Faraday Soc.* **33,** 71.

DAVISON, W. D. (1966) Variational calculations of long-range forces, *Proc. Phys. Soc.* **87,** 133.

DAVISON, W. D. and DALGARNO, A. (1966) *Advanc. Atomic Mol. Phys.* **2,** 1.

DAVISON, W. D. and DALGARNO, A. (1968) *Mol. Phys.* in press.

DAVISON, W. D.: *see* DALGARNO, A.

DAWSON, J. M.: *see* CASTLE, B. J.

DEAL, W. J. and KESTNER, N. R. (1 966) *J. Chem. Phys.* **45,** 4014.

DEBRUYN, P. L.: *see* DEVEREUX, O. F.

DEBYE, P. (1912) *Phys. Z.* **13,** 97, 295.

DEBYE, P. (1920) *Phys. Z.* **21,** 178.

DEGRAAFF, W.: *see* MICHAELS, A.

DEJONGH, J. G. V.: *see* BLACK, W.

DENBIGH, K. G. (1940) *Trans. Faraday Soc.* **36,** 936.

DEPAZ, M.: *see* CASANOVA, G.

DERJAGUIN, B. V. (1960) The forces between molecules, *Scientific American* **203,** 47.

DERJAGUIN, B. V. and ABRIKOSOVA, I. I. (1956) Direct measurements of molecular attractions between solids *in vacuo*, *Dokl. Akad. Nauk SSSR* **108,** 214.

DERJAGUIN, B. V., DZYALOSHINSKII, I. E., KOPTELOVA, M. M. and PITAEVSKII, L. P. (1966) Molecular-surface forces in binary solutions, *Disc. Faraday Soc.* **40,** 246.

DE ROCCO, A. G. and HALFORD, J. O. (1958) Intermolecular potentials of argon, methane, and ethane, *J. Chem. Phys.* **28,** 1152.

DE ROCCO, A. G. and HOOVER, W. G. (1960) *Proc. Nat. Acad. Sci.* **46,** 1057.

DE ROCCO, A. G.: *see* FEINBERG, M. J. and SHERWOOD, A. E.

DEVEREUX, O. F. and DE BRUYN, P. L. (1962) Numerical evaluation of the Van der Waals forces of interaction between two macroscopic bodies, *J. Chem. Phys.* **37,** 2147.

DE VOE, H. and TINOCO, I. (1962) *J. Mol. Biol.* **4,** 500.

DOROSH, P. A.: *see* CRAIG, D. P.

DOLGUSHIN, M. D. (1965) Construction of potential curves for diatomic molecules of copper, silver, and gold, *Optics and Spectrosc.* **19,** 289.

DONATH, W. and PITZER, K. S. (1956) Electronic correlation in molecules, I: hydrogen in the triplet state, *J. Amer. Chem. Soc.* **78,** 4562.

DONATH, W. and PITZER, K. S. (1963) Atomic calculations, II: polarizabilities and London force constants for F^-, Ne, and Na, *J. Chem. Phys.* **39,** 2685.

DONDI, M. G.: *see* CASANOVA, G.

DONIACH, S. (1963) Many-electron contribution to the cohesive energy in Van der Waals crystals, *Phil. Mag.* **8,** 129.

DOOLING, J. S.: *see* HAMPON, R.

DOWS, D. A. (1966) Vibrational spectroscopy of molecular crystals, *J. Chim. Phys.* **63,** 168.

DRESSLER, K.: *see* SCHNEPP, O.

DUEREN, R., FELTGEN, R., GAIDE, W., HELBING, R. and PAULY, H. (1965) Measurement of the velocity dependence of the total scattering cross-

section for scattering of He, H_2, and D_2 molecular beams by rare gases, *Phys. Letters* **18**, 282.
DUBE, G. P. and DAS GUPTA, H. K. (1939) *Indian J. Phys.* **13**, 411.
VAN DUIJNEVELDT, F. B. and MURRELL, N. N. (1966) *J. Chem. Phys.* **46**, 1759.
DUPARC, D. M. and BUCKINGHAM, R. A. (1964) Very short-range interaction of a hydrogen atom and a helium atom, *Proc. Phys. Soc.* **83**, 731.
DYMOND, J. M., RIGBY, M. and SMITH, E. B. (1965) Intermolecular potential-energy function for simple molecules, *J. Chem. Phys.* **42**, 2801.
DZYALOSHINSKII, I. E. (1957) Account of retardation in the interaction of neutral atoms, *Soviet Physics-JETP* **3**, 977.
DZYALOSHINSKII, I. E., LIFSHITZ, E. M. and PITAEVSKII, L. P. (1960) Van der Waals forces in liquid films, *Soviet Physics-JETP* **10**, 161.
DZYALOSHINSKII, I. E., LIFSHITZ, E. M. and PITAEVSKII, L. P. (1961) General theory of Van der Waals forces, *Soviet Physics-Uspekhi* **4**, 153; *Advanc. Phys.* **10**, 165.
DZYALOSHINSKII, I. E. and PITAEVSKII, L. P. (1959) Van der Waals forces in an inhomogeneous dielectric, *Soviet Physics–JETP* **36**, 1282.
DZYALOSHINSKII, I. E.: *see* DERJAGUIN, B. V.
EISENBERG, D.: *see* COULSON, C. A.
EISENSCHITZ, R. and LONDOŃ, F. (1930) *Z. Phys.* **60**, 491.
EISNER, M.: *see* KATTAWAR, G. W.
ELLISON, F. O. (1961) Role of exchange energy in intermediate-range interactions, *J. Chem. Phys.* **34**, 2100.
EPSTEIN, P. S. (1926) *Proc. Nat. Acad.* **12**, 629.
EPSTEIN, P. S. (1927) *Proc. Nat. Acad.* **13**, 432.
EPSTEIN, S. T. (1965) Remarks on the calculation of the Van der Waals energy, *J. Chem. Phys.* **43**, 4398.
EPSTEIN, S. T. and KARL, J. H. (1966) Generalized Sternheimer potential, *J. Chem. Phys.* **44**, 4347.
ERGINSOY, C. (1965) Repulsive atom–atom interaction and the adiabatic approximation, in *Modern Quantum Chemistry*, Istanbul Lectures, Vol. 2. Academic Press, New York, O. Sinanoglu, (ed.), p. 199.
ESSEN, L. (1953) *Proc. Phys. Soc. (London)* **B66**, 189.
EVERETT, D. H. (1966) Interactions between adsorbed molecules, *Disc. Faraday Soc.* **40**, 177.
EVETT, A. E. and MARGENAU, H. (1953) *Phys. Rev.* **90**, 1021.
EYRING, H., WALTER, J. and KIMBALL, G. E. (1944) *Quantum Chemistry*, Wiley, New York.
FALLON, R. J., MASON, E. A. and VANDERSLICE, J. T. (1960) Energies of various interactions between hydrogen and helium atoms and ions, *Astrophys. J.* **131**, 12.
FANO, U. and RACAH, G. (1959) *Irreducible Tensorial Sets*, Academic Press, New York.
FEINBERG, G. and WEINBERG, S. (1961) *Phys. Rev.* **123**, 1439.
FEINBERG, M. J. and DE ROCCO, A. G. (1964) Intermolecular forces: the triangle well and some comparisons with the square well and Lennard-Jones, *J. Chem. Phys.* **41**, 3439.
FELTGEN, R.: *see* DUEREN, R.

FENDER, B. E. (1961) Potential parameters for krypton, *J. Chem. Phys.* **35**, 2243.
FEYNMAN, R. P. (1939) *Phys. Rev.* **56**, 340.
FIRSOV, O. B. (1957a) Interaction energy of atoms at small internuclear separations, *Zhur. Eksptl. i. Teoret. Fiz.* **32**, 1464.
FIRSOV, O. B. (1957b) Calculation of the interaction potential of atoms, *Zhur. Eksptl. i. Teoret. Fiz.* **33**, 696.
FISHER, R. A.: *see* CHON, H.
FLORIN, H. (1964) Dissertation, Bonn.
FOGLIA, C.: *see* BULUGGIU, E.
FONTANA, P. R. (1961a) Theory of long-range interatomic forces, I: dispersion energies between unexcited atoms, *Phys. Rev.* **123**, 1865.
FONTANA, P. R. (1961b) Theory of long-range interatomic forces, II: first-order interaction energies in the uncoupled representation, *Phys. Rev.* **123**, 1871.
FONTANA, F. R. (1962) Theory of long-range interatomic forces, III: first-order interaction energies in the coupled representation, *Phys. Rev.* **125**, 1597.
FONTANA, P. R. and BERNSTEIN, R. B. (1964) Dipole–quadrupole and retardation effects in low-energy atom–atom scattering, *J. Chem. Phys.* **41**, 1431.
FRANCK, H. S. (1958) Covalency in the hydrogen bond and the properties of water and ice, *Proc. Roy. Soc. (London)* **A247**, 481.
FRIEDMANN, H. and KIMEL, S. (1965) Theory of shifts of vibration–rotation lines of diatomic molecules in noble gas matrices: intermolecular forces in crystals, *J. Chem. Phys.* **43**, 3925.
FRISCH, H. L. and HELFAND, E. (1960) Conditions imposed by gross properties on the intermolecular potential, *J. Chem. Phys.* **32**, 269.
FROST, A. A. and BRAUNSTEIN, J. (1951) *J. Chem. Phys.* **19**, 1133.
FUKUI, K. (1966) *Bull. Chem. Soc.* (Japan) **39**, 498.
FUKUI, K. and FUJIMOTO, H. (1966) *Bull. Chem. Soc.* (Japan) **39**, 2116.
FUNKE, P. T.: *see* POLLARA, L. Z.
FUTRELLE, R. P. (1967) *Phys. Rev. Letters*, **19**, 479.
GAIDE, W.: *see* DUEREN, R.
GASPER, R. (1960) A theoretical estimation of the interaction energy of inert gas atoms, *Acta. Phys. Acad. Sci. Hung.* **11**, 71.
GAUSS, C. F. (1830) *Principia generalia theoriae figurae fluidorum in statu aequilibru*, Mathematical Tracts, Göttingen.
GAUTSCHI, W. and CAHILL, W. F. (1964) *Handbook of Mathematical Functions* (eds. Abramowitz, M. and Stegum, I. A.), National Bureau of Standards, Washington.
GIANINETTI, E. (1964) In *Molecular Orbitals in Chemistry, Physics and Biology* (eds. P. O. LÖWDIN and B. PULLMAN), p. 57, Academic Press, New York.
GILBERT, T. L. (1964) In *Molecular Orbitals in Chemistry, Physics and Biology* (eds. P. LÖWDIN and B. PULLMAN), Academic Press, New York.
GILBERT, T. L. and WAHL, A. C. (1967) *J. Chem. Phys.* **47**, 3425.
GIRIFALCO, L. A. (1956) Potential energy of a molecule in a liquid: calculations for argon, *J. Chem. Phys.* **24**, 617.
GLYDE, H. R.: *see* BULLOUGH, R.

GOLDSTEIN, H. (1959) *Classical Mechanics*, Addison-Wesley, Reading, Mass.

GOMBEROFF, L., MCLONE, R. R. and POWER, E. A. (1966) Long-range retarded potentials between molecules, *J. Chem. Phys.* **44**, 4148.

GÖPPERT-MAYER, M. (1931) *Ann. Phys. Lpz.* **9**, 273.

GORDADSE, G. S. (1935) *Z. Physik* **96**, 542.

GORSON, R. G. and CASHION, I. K. (1966) Intermolecular potentials and the infrared spectrum of the molecular complex $(H_2)_2$, *J. Chem. Phys.* **44**, 1190.

GRABEN, H. W. and PRESENT, R. D. (1962) Evidence for three body forces from third virial coefficients, *Phys. Rev. Letters* **9**, 247.

GRABEN, H. W. and PRESENT, R. D. (1966) *Bull. Amer. Phys. Soc.* **11**, 836.

GRABEN, H. W., PRESENT, R. D. and MCCULLOCH, R. D. (1966) Intermolecular three-body forces and third virial coefficients, *Phys. Rev.* **144**, 140.

GRAHN, R. (1959) A theoretical investigation of the role of polarization in the formation of hydrogen bonds, *Arkiv. Fysik* **15**, 257.

GRAY, C. G. (1968), *Canadian Journal of Phys.*, **46**, 135.

GRAY, H. B. (1964) *Electrons and Chemical Bonding*, Benjamin, New York.

GREEN, L. C., MULDER, M. M., LEWIS, M. N. and WALL, J. W. (1954) *Phys. Rev.* **93**, 757.

GREENWALT, E. M.: *see* SCOTT, D. R.

GREGORY, R. D.: *see* MCLACHLAN, A. D.

GRIFFING, V. (1955) Interaction energies by MO theory, *Svensk. Kem. Tidskr.* **67**, 392.

GRIFFING, V., HOARE, J. P. and VANDERSLICE, J. T. (1956) Studies of the interaction between atoms and stable molecules, VI: the interaction of a Be atom with an H_2 molecule, *J. Chem. Phys.* **24**, 71.

GRIFFING, V. and VANDERSLICE, J. T. (1955) Studies of the interaction between stable molecules and atoms, V: molecular orbital approach to the H plus H_2 reaction, *J. Chem. Phys.* **23**, 1039.

GRIFFING, V. and WEHNER, J. F. (1955) Studies of the interaction between stable molecules and atoms, II: interaction between two He atoms, *J. Chem. Phys.* **23**, 1024.

GRIFFING, V.: *see* SCHEEL, N.

GROBLICKI, P. J. and BERNSTEIN, R. B. (1965) Atomic beam scattering studies on the Li–Hg system: quantum effects and velocity dependence of the cross-section, *J. Chem. Phys.* **42**, 2295.

GUGGENHEIM, E. A. and MCGLASHAN, M. L. (1966) Repulsive energy in sodium chloride and potassium chloride crystals, *Disc. Faraday Soc.* **40**, 76.

GUNDRY, P. M. and TOMPKINS, F. C. (1960a) Application of the charge transfer no-bond theory to adsorption problems, *Trans. Faraday Soc.* **56**, 846.

GUNDRY, P. M. and TOMPKINS, F. C. (1960b) Chemisorption of gases on metals, *Quarterly Reviews (London)* **14**, 257.

GURNEE, E. F. and MAGEE, J. L. (1957) Interchange of charge between gaseous, molecules in resonant and near-resonant processes, *J. Chem. Phys.* **26**, 1237.

TER HAAR, D. (1954) *Elements of Statistical Mechanics*, Rinehart, New York.

HAJJ, F. (1966) Van der Waals coefficients for the alkali halides from optical data, *J. Chem. Phys.* **44**, 4618.
HALFORD, J. O.: *see* DE ROCCO, A. G.
HALL, G. G. (1951) *Proc. Roy. Soc. (London)* **A205**, 541.
HAMAKER, H. C. (1937) *Physica* **4**, 1058.
HAMEKA, H. F. (1965) *Advanced Quantum Chemistry*, Addison-Wesley, Reading, Mass.
HAMEKA, H. F.: *see* HUTCHINSON, D. A.
HAMILTON, J. (1949) *Proc. Phys. Soc. (London)* **A62**, 12.
HAMILTON, W. C.: *see* CLINTON, W. L.
HAMPON, R. and DOOLING, J. S. (1960) Interaction of two beryllium atoms, *Bull. Amer. Phys. Soc.* **5**, 339.
HANLEY, H. J. M. (1966) Comparison of the Lennard-Jones, Exp-6, and. Kihara potential functions from viscosity data of dilute argon, *J. Chem Phys.* **44**, 4219.
HARADA, I. and SHIMANOUCHI, T. (1966) Normal vibrations and intermolecular forces of crystalline benzene and naphthalene, *J. Chem. Phys.* **44**, 2016.
HARRIS, F. E., MICHA, D. A. and POHL, H. A. (1965) The interaction potential surface for H_3, *Ark. Fys. (Sweden)* **30**, 259.
HARRIS, F. E.: *see* TAYLOR, H. S.
HARRIS, R. A. (1964) On the "average energy" in intermolecular force calculations, *J. Chem. Phys.* **41**, 901.
HARRISON, H. (1962) Total collision cross-section for scattering of thermal beams of hydrogen, hydrogen atoms, and helium by hydrogen and helium, *J. Chem. Phys.* **37**, 1164.
HASHINO, T. and HUZINAGA, S. (1958) The interaction between two normal helium atoms: a consistent treatment, *Prog. Theor. Phys.* **20**, 631.
HAUGH, E. F. and HIRSCHFELDER, J. O. (1955) Pi-electron forces between conjugated double bond molecules, *J. Chem. Phys.* **23**, 1778.
HEATH, D. F.: *see* LINNETT, J. W.
HEITLER, W. (1954) *The Quantum Theory of Radiation*, Oxford University Press.
HEITLER, W. and LONDON, F. (1927) *Z. Phys.* **44**, 455.
HELBING, R.: *see* DUEREN, R.
HELFAND, E.: *see* FRISCH, H. L.
HELLMANN, G. (1937) *Einführung in die Quantenchemie*, Leipzig.
HERMAN, R. M. (1962a) Dissertation, Yale University.
HERMAN, R. M. (1962b) *Phys. Rev.* **132**, 262.
HERMAN, R. M. (1963) *J.Q.S.R.T.* **3**, 449.
HERMAN, R. M. (1966) Center of dispersion force in HCl interacting with rare gas atoms, *J. Chem. Phys.* **44**, 1346.
HERMANS, L. J. F.: *see* KNAAP, H. F. P.
HERRING, C. (1962) *Rev. Mod. Phys.*, **34**, 631.
HERRING, C. (1966) in *Magnetism* (Eds. Rado and Suhl) Academic Press, New York, vol. IIB.
HERRING, C. and FLICKER, M. (1964) *Phys. Rev.*, **134**, A362.
HERSCHBACH, D. R.: *see* CASHION, J. K.
HERZBERG, G. (1950) *Molecular Spectra and Molecular Structure. I. Spectra of Diatomic Molecules*, Van Nostrand, Princeton, N. J.

HESSEL, M. M. and KUSCH, P. (1965) Deviation from the R^{-6} potential in the scattering of a polar molecule by nonpolar gases, *J. Chem. Phys.* **43,** 305.

HILL, T. L. (1956) *Statistical Mechanics*, McGraw-Hill, New York.

HILL, T. L. (1960) *Introduction to Statistical Thermodynamics*, Addison-Wesley, Reading, Mass.

HIRSCHFELDER, J. E., CURTIS, C. F. and BIRD, R. B. (1954) *Molecular Theory of Gases and Liquids*, Wiley, New York.

HIRSCHFELDER, J. E., EYRING, H. and ROSEN, N. (1936) *J. Chem. Phys.* **4,** 121.

HIRSCHFELDER, J. E., EYRING, H. and ROSEN, N. (1938) *J. Chem. Phys.* **6,** 795.

HIRSCHFELDER, J. O. (1965) Determination of intermolecular forces, *J. Chem. Phys.* **43,** S199.

HIRSCHFELDER, J. O. (1967) *Chem. Phys. Letters*, **1,** 325, 363.

HIRSCHFELDER, J. O. (1967) *Intermolecular Forces*, Interscience, New York.

HIRSCHFELDER, J. O., BROWN, W. B. and EPSTEIN, S. T. (1964) *Advanc. Quantum Chem.* **1,** 255.

HIRSCHFELDER, J. O., EYRING, H. and TOPLEY, B. (1936) *J. Chem. Phys.* **4,** 17.

HIRSCHFELDER, J. O. and LINNETT, J. W. (1950) *J. Chem. Phys.* **18,** 130.

HIRSCHFELDER, J. O. and LÖWDIN, P. O. (1959) Long-range interaction of two 1*s* hydrogen atoms expressed in terms of natural spin-orbitals, *Mol. Phys.* **2,** 229.

HIRSCHFELDER, J. O. and LÖWDIN, P. O. (1965) *Mol. Phys.* **9,** 491.

HIRSCHFELDER, J. O.: *see* DAHLER, J. S., HAUGH, E. F., HORNIG, J. F., LINDER, B., MASON, E. A. and MEATH, W. J.

HOARE, J. P.: *see* GRIFFING, V.

HOERNSCHEMEYER, D.: *see* LINDER, B.

HOFFMANN, R. and WOODWARD, R. B. (1968) *Accounts of Chemical Research*, **1,** 17.

HORNIG, J. F. and HIRSCHFELDER, J. O. (1952) *J. Chem. Phys.* **20,** 1812.

HORNIG, J. F. and HIRSCHFELDER, J. O. (1956) Concept of intermolecular forces in collisions, *Phys. Rev.* **103,** 908.

HOSTETTLER, H. U. and BERNSTEIN, R. B. (1959) Comparison of exp-6, L. J. (12–6), and Sutherland potential functions applied to the calculation of differential scattering cross-sections, *J. Chem. Phys.* **31,** 1422.

HOSTETTLER, H. U.: *see* MORSE, F. A.

HUGHES, V. W., MCCOLM, D. W., ZIOCK, K. and PREPOST, R. (1960) *Phys. Rev. Letters* **5,** 63.

HUO, W. M. (1965) Electronic structure of CO and BF, *J. Chem. Phys.* **43,** 624.

HURLEY, A. C. (1954) *Proc. Roy. Soc . (London)* **A226,** 170, 179, 193.

HURLEY, A. C. (1956) *Proc. Roy. Soc. (London)* **A235,** 224.

HUTCHINSON, D. A. and HAMEKA, H. F. (1964) Interaction effects on lifetimes of atomic excitations, *J. Chem. Phys.* **41,** 2006.

HUZINAGA, S. (1956) One-center expansion of molecular wave functions, I, II: lower excited states of H_2, *Prog. Theor. Phys. (Kyoto)* **15,** 501; also **17,** 162.

HUZINAGA, S. (1957a) One-center expansion of molecular wave functions, III: He–He repulsive potential, *Prog. Theor. Phys. (Kyoto)* **17,** 169.

HUZINAGA, S. (1957b) He–He repulsive potential, *Prog. Theor. Phys.* **17,** 512.

HUZINAGA, S. (1957c) The repulsive potential between two normal helium atoms, *Prog. Theor. Phys.* **18,** 139.

HUZINAGA, S.: *see* HASHINO, T.

HYLLERAAS, E. A. (1929) *Z. Physik.* **54,** 347.

ISHIGURO, E., ARAI, T., MIZUSHIMA, M. and KOTANI, M. (1952) *Proc. Phys. Soc. (London)* **A65,** 178.

ISHIGURO, E.: *see* SAKAMOTO, M.

ISIHARA, I. and KOYAMA, R. (1957) Intermolecular potential between large molecules, *J. Phys. Soc. (Japan)* **12,** 32.

JACKSON, J. D. (1962) *Classical Electrodynamics*, Wiley, New York.

JAMES, H. M. and COOLIDGE, A. S. (1938) *Astrophys. J.* **87,** 438.

JAMES, H. M., COOLIDGE, A. S. and PRESENT, R. D. (1936) *J. Chem. Phys.* **4,** 187, 193.

JAMESON, W. J. and AROESTE, H. (1960) Short-range interaction between a hydrogen molecule and a hydrogen atom, II, *J. Chem. Phys.* **32,** 374.

JAMESON, W. J.: *see* AROESTE, H.

JANSEN, L. (1958) Tensor formalism for Coulomb interactions and asymptotic properties of multipole interactions, *Phys. Rev.* **110,** 661.

JANSEN, L. (1962) Systematic analysis of many-body interactions in molecular solids, *Phys. Rev.* **125,** 1798.

JANSEN, L. (1963) *Phil. Mag.* **8,** 1305.

JANSEN, L. (1964) Stability of crystals of rare gas atoms and alkali halides in terms of three-body interactions, I: rare gas crystals, *Phys. Rev.* **135,** A1292.

JANSEN, L. (1965a) "Many-atom forces and crystal stability" in *Modern Quantum Chemistry*, Istanbul Lectures, Academic Press, Vol. 2, (O. Sinanoglu, ed.), p. 239.

JANSEN, L. (1965b) *Advances in Quantum Chemistry*, Vol. 2, (P. LÖWDIN, ed.), Academic Press, New York.

JANSEN, L. (1965c) *Disc. Faraday Soc.*, No. 40.

JANSEN, L. (1967) *Phys. Rev.*, **162,** 63.

JANSEN, L. and LOMBARDI, E. (1964) *Phys. Rev.* **136,** A1011.

JANSEN, L. and LOMBARDI, E. (1966) Three-atom and three-ion interactions and crystal stability, *Disc. Faraday Soc.* **40,** 78.

JANSEN, L. and LOMBARDI, E. (1967) *Chem. Phys. Letters*, **1,** 33.

JANSEN, L. and LOMBARDI, E. (1967) *Chem. Phys. Letters*, **1,** 417.

JANSEN, L. and MCGINNIES, R. T. (1956) Validity of the assumption of two-body interactions in molecular physics, *Phys. Rev.* **104,** 961.

JANSEN, L. and ZIMERING, S. (1963) *Phys. Letters* **4,** 95.

JANSEN, L.: *see* CASTLE, B. J. and MCGINNIES, R. T.

JEFIMENKO, O. (1962) Semiclassical model of atomic interactions, *J. Chem. Phys.* **37,** 2125.

JEFIMENKO, O. (1965) Potential curves for neutral atoms and molecules in the state of collision, *J. Chem. Phys.* **42,** 205.

JEHLE, H., PARKE, W. C. and SALYERS, A. (1964) Intermolecular charge fluctuation interactions in molecular biology, in *Electronic Aspects of Biochemistry*, Academic Press (B. PULLMAN, ed.), p. 313.

JOHNSON, J. D. and KLEIN, M. L. (1964) *Trans. Faraday Soc.* **60,** 1964.

JONKMAN, R. M.: *see* KNAAP, H. F. P.

JORDAN, J. E. and AMDUR, I. (1967) Scattering of high velocity neutral particles. XIV. He–He interaction below 1.1 Å. *J. Chem. Phys.* **46,** 165.

JORDAN, J. E.: *see* AMDUR, I.

KACHHAUA, C. M.: *see* DASS, L.

KAHALAS, S. L. and NESBET, R. K. (1963) *J. Chem. Phys.* **39,** 529.

KAMP, B. (1965) Overlap interaction of water molecules, *J. Chem. Phys.* **43,** 3917.

KANEKO, S.: *see* KIHARA, T. and KOBA, S.

KARL, J. H.: *see* EPSTEIN, S. T.

KARPLUS, M. (1960) Weak interactions in molecular quantum mechanics, *Revs. Modern. Phys.* **32,** 455.

KARPLUS, M. and KOLKER, H. J. (1964) Van der Waals forces in atoms and molecules, *J. Chem. Phys.* **41,** 3955.

KARPLUS, M., PORTER, R. N. and SHARMA, R. D. (1965) Exchange reactions with activation energy, I: simple barrier potential for (H, H_2), *J. Chem. Phys.* **43,** 3259.

KATTAWAR, G. W. and EISNER, M. (1965) Van der Waals forces between conducting bodies, *J. Chem. Phys.* **43,** 863.

KAUZMAN, W. (1957) *Quantum Chemistry*, Academic Press, New York.

KEBBEKUS, E. R.: *see* STEELE, W. A.

KEESOM, W. H. (1921) *Phys. Z.* **22,** 129.

KELLER, J. B. and ZUMINO, B. (1959) Determination of intermolecular potentials from thermodynamic data and the law of corresponding states, *J. Chem. Phys.* **30,** 1351.

KESTNER, N. R. (1966a) Long-range interaction of helium atoms, A: calculations, *J. Chem. Phys.* **45,** 208.

KESTNER, N. R. (1966b) Long-range interaction of helium atoms, B: study of approximations, *J. Chem. Phys.* **45,** 213.

KESTNER, N. R. (1966c) *J. Chem. Phys.* **45,** 3121.

KESTNER, N. R. (1968) *J. Chem. Phys.* in press.

KESTNER, N. R. and SINANOGLU, O. (1962) *Phys. Rev.* **128,** 2687.

KESTNER, N. R. and SINANOGLU, O. (1963) Effective intermolecular pair potentials in nonpolar media, *J. Chem. Phys.* **38,** 1730.

KESTNER, N. R. and SINANOGLU, O. (1966a) Intermolecular forces in dense media, *Disc. Faraday Soc.* **40,** 266.

KESTNER, N. R. and SINANOGLU, O. (1966b) Intermolecular–potential-energy curves–theory and calculations on the helium–helium potential, *J. Chem. Phys.* **45,** 194.

KESTNER, N. R., RANSIL, B. J. and ROOTHAAN, C. C. J. (to be published).

KESTNER, N. R.: *see* SINANOGLU, O. and DEAL, W. J.

KHAN, A. A. and BROYLES, A. A. (1965) Interatomic potentials and X-ray diffraction intensities for liquid xenon, *J. Chem. Phys.* **43,** 43.

KIHARA, T. (1951) The second virial coefficient of non-spherical molecules, *J. Phys. Soc. (Japan)* **6,** 289.

KIHARA, T. (1953) Virial coefficients and models of molecules in gases, *Revs. Modern Phys.* **25**, 831.

KIHARA, T. (1955) Virial coefficients and models of molecules in gases, B, *Revs. Modern. Phys.* **27**, 412.

KIHARA, T. (1957) Geometrical theory of convex molecules in non-uniform gases, *J. Phys. Soc. (Japan)* **12**, 564.

KIHARA, T. (1958a) Intermolecular forces and equation of state of gases, *Advanc. Chem. Phys.* **1**, 267.

KIHARA, T. (1958b) *Advanc. Chem. Phys.* **1**, 267.

KIHARA, T. (1960) Multipole interaction stabilizing cubic molecular crystals, *J. Phys. Soc. (Japan)* **15**, 1920.

KIHARA, T. (1963) Convex molecules in gaseous and crystalline states, *Advanc. Chem. Phys.* **5**, 147.

KIHARA, T., MIDZUNO, Y. and KANEKO, S. (1956) Virial coefficients and intermolecular potential for small nonspherical molecules, *J. Phys. Soc. (Japan)* **11**, 362.

KIHARA, T. and OUCHI, A. (1957) Convex molecules in non-uniform gases supplement, *J. Phys. Soc. (Japan)* **12**, 1052.

KIHARA, T.: *see* KOBA, S. and MIDZUNO, Y.

KILPATRICK, J. E., KELLER, W. and HAMMEL, E. F. (1955) *Phys. Rev.* **97**, 9.

KIM, D. Y. (1962) Ueber die Wechselwirkung zweier Helium atome, *Z. f. Physik*, **166**, 359.

KIMBALL, G. E. and TRULIO, J. G. (1958) Quantum mechanics of the H_3 complex, *J. Chem. Phys.* **28**, 493.

KIMEL, S.: *see* FRIEDMAN, H.

KINGSTON, A. E. (1964) Van der Waals forces for the inert gases, *Phys. Rev.* **135A**, 1018.

KINGSTON, A. E. (1965) Derivation of interatomic potentials for inert gas atoms from the second virial coefficient, *J. Chem. Phys.* **42**, 719.

KINGSTON, A. E.: *see* BELL, R. J. and DALGARNO, A.

KIRKWOOD, J. G. (1932) *Phys. Z.* **33**, 57.

KIRKWOOD, J. G. (1955) The influence of fluctuations in protein charge and charge configuration on the rates of enzymatic reactions, *Disc. Faraday Soc.* **20**, 78.

KIRKWOOD, J. G. and SHUMAKER, J. B. (1952) *Proc. Nat'l. Acad. Sci. U.S.* **38**, 863.

KIRKWOOD, J. G.: *see* BADE, W. L.

KISELEV, A. V. (1966) Non-specific and specific interactions of molecules of different electronic structures with solid surfaces, *Disc. Faraday Soc.* **40**, 205.

KITAIGORODSKY, A. (1966) Stacking of molecules in a crystal, interaction potential of atoms not bonded by valence bonds, and calculation of the motion of the molecules, *J. Chim. Phys. (France)* **63**, 9.

KITCHNER, J. A. and PROSSER, A. P. (1957) Direct measurement of the long-range Van der Waals forces, *Proc. Roy. Soc. (London)* **A242**, 403.

KLEIN, L. and MARGENAU, H. (1959) *J. Chem. Phys.* **30**, 1556.

KLEIN, M. L.: *see* CASANOVA, G.

KNAAP, H. F. P., HERMANS, L. J. F., JONKMAN, R. M. and BEENAKKER, J. J. M.

(1966) Interactions between the hydrogen isotopes and their ortho and para modifications, *Disc. Faraday Soc.* **40,** 135.

KNOF, H., MASON, E. A. and VANDERSLICE, J. T. (1964) Interaction energies, charge exchange cross-sections, and diffusion cross-sections for N^+–N and O^+–O collisions, *J. Chem. Phys.* **40,** 3547.

KNOX, R. S. and REILLY, M. H. (1964) Atomic multipole interactions in rare-gas crystals, *Phys. Rev.* **135,** A166.

KOBA, S., KANEKO, S. and KIHARA, T. (1956) Non-additive intermolecular potential in gases, II: cluster integrals, *J. Phys. Soc. (Japan)* **11,** 1050.

KOLKER, H. J.: *see* KARPLUS, M.

KOLOS, W. and ROOTHAAN, C. C. J. (1960) *Rev. Mod. Phys.* **32,** 219.

KOLOS, W. and WOLNIEWICZ, L. (1963) *Rev. Mod. Phys.* **35,** 473.

KOLOS, W. and WOLNIEWICZ, L. (1964) *J. Chem. Phys.* **41,** 3663, 3674.

KOLOS, W. and WOLNIEWICZ, L. (1965) Potential energy curves for the $X\Sigma_g^+$, $b\,^3\Sigma_u^+$, and $C\,'\Pi_u$ states of the hydrogen molecule, *J. Chem. Phys.* **43,** 2429.

KOLOS, W. and WOLNIEWICZ, L. (1966) Potential energy curve for the $B'\Sigma_u^+$ state of the hydrogen molecule, *J. Chem. Phys.* **45,** 509.

KONOWALOW, D. D. (1966) Central potentials for nonpolar polyatomic molecules, *Phys. Fluids* **9,** 23.

KOPTELOVA, M. M.: *see* DERJAGUIN, B. V.

KOTANI, M., OHNO, K. and KAYAMA, K. (1961) In vol. 17/2 of *Handbuch der Physik* (ed. S. FLÜGGE), Springer–Verlag, Berlin.

KOYAMA, R.: *see* ISIHARA, I.

KRAUSS, M. and MIES, F. H. (1965) Interaction potential between He and H_2, *J. Chem. Phys.* **42,** 2703.

KRISHNAJI, and SRIVASTAVA, S. L. (1964) First-order London dispersion forces and microwave spectral line width, *J. Chem. Phys.* **41,** 2266.

KRIZAN, J. E. and CROWELL, A. D. (1964) Two-dimensional second virial coefficient in physical adsorption, *J. Chem. Phys.* **41,** 1322.

KRUPENIE, P. H., MASON, E. A. and VANDERSLICE, J. T. (1963) Interaction energies and transport coefficients for Li plus H and O plus H gas mixtures at high temperatures, *J. Chem. Phys.* **39,** 2399.

KUHN, H. (1948a) *Helv. Chim. Acta* **31,** 1441 and 1780.

KUHN, H. (1948b) *J. Chem. Phys.* **16,** 1840.

KUHN, H. (1949) *J. Chem. Phys.* **17,** 1198.

KUNIMUNE, M. (1950a) *J. Chem. Phys.* **18,** 754.

KUNIMUNE, M. (1950b) *Prog. Theo. Phys. (Japan)* **5,** 412.

KUSCH, P.: *see* HESSEL, M. M.

LANDAU, L. (1932) *Physik. Z. Sowjetunion* **2,** 461.

LANDAU, L. (1965) *Collected Papers*, Pergamon, Oxford.

LANDAU, L. and LIFSHITZ, E. M. (1958) *Quantum Mechanics*, Addison-Wesley, Reading, Mass.

LANDAU, L. D. and LIFSHITZ, E. M. (1960) *Electrodynamics of Continuous Media*, Pergamon, Oxford.

LANGEVIN, P. (1905) *Ann. d. Chim. et. Phys.* (VIII) **5,** 245.

LAPLACE, M. (1805) *Traité de Mécanique Céleste*, Courcier, Paris.

LAUE, H. (1967) *J. Chem. Phys.* **46,** 3034.

LEACH, S. J.: *see* SCHERAGA, H. A.

LEBEDEFF, S. A. (1964) Determination of the Lennard-Jones parameters from total scattering cross-section measurements, *J. Chem. Phys.* **40**, 2716.

LECKENBY, R. E. and ROBBINS, E. J. (1966) The observation of double molecules in gases, *Proc. Roy. Soc. (London)* **A291**, 389.

LEGVOLD, S.: *see* AMME, R. C.

LENNARD-JONES, J. E. (1932) *Trans. Faraday Soc.* **28**, 334.

LENNARD-JONES, J. and POPLE, J. (1950) *Proc. Roy. Soc.* **A202**, 155, 166.

LEONARD, P. J.: *see* BARKER, J. A.

LE SAGE, G. (1862) *Poggendorf's Annalen.*

LEVINE, I. N. (1964) Comment on united atom expansions, *J. Chem. Phys.* **40**, 3444.

LEWIS, J. T.: *see* DALGARNO, A.

LIFSHITZ, E. M. (1955) *Zh. Eksp. Teor. Fiz.* **29**, 94.

LIFSHITZ, E. M. (1956) The theory of molecular attractive forces between solids, *Soviet Physics-JETP* **2**, 73.

LIFSHITZ, E. M.: *see* DZYALOSHINSKII, I. E.

LINDER, B. (1960) Continuum model treatment of long-range intermolecular forces, I: pure substances, *J. Chem. Phys.* **33**, 668.

LINDER, B. (1962) Generalized form for dispersion interactions, *J. Chem. Phys.* **37**, 963.

LINDER, B. (1964) Van der Waals interaction potential between polar molecules, pair potential, and (Nonadditive) Triple Potential, *J. Chem. Phys.* **40**, 2003.

LINDER, B. (1966a) Temperature-dependent potentials, *Disc. Faraday Soc.* **40**, 164.

LINDER, B. (1966b) Interaction between two rotating dipolar systems and a generalization to the rotational double-temperature potential, *J. Chem. Phys.* **44**, 265.

LINDER, B. (1967) in *Intermolecular Forces* (Ed. Hirschfelder), Interscience, New York.

LINDER, B. and HIRSCHFELDER, J. O. (1958) Energy of interaction between two excited hydrogen atoms in either 2*s* or 2*p* states, *J. Chem. Phys.* **28**, 197.

LINDER, B. and HOERNSCHEMEYER, D. (1964a) Many-body aspects of intermolecular forces, *J. Chem. Phys.* **40**, 622.

LINDER, B. and HOERNSCHEMEYER, D. (1964b) *J. Chem. Phys.* **40**, 622.

LINDERBERG, J. (1964) *Ark. Fys.* **26**, 323.

LINDERBERG, J. and BYSTRAND, F. W. (1964) *Ark. Fys.* **26**, 383.

LING, R. C. (1956) Interatomic potential functions of sodium and potassium, *J. Chem. Phys.* **25**, 609.

LINNETT, J. W. and HEATH, D. F. (1952) Molecular force fields, Part XIV: intermolecular forces between non-bonded atoms, *Trans. Faraday Soc.* **48**, 592.

LIPPINCOTT, E. R. (1957) Derivation of an internuclear potential function from a quantum mechanical model, *J. Chem. Phys.* **26**, 1678.

LIPPINCOTT, E. R., STEELE, D. and CALDWELL, P. (1961) General relation between potential energy and internuclear distance for diatomic molecules, III: excited states, *J. Chem. Phys.* **35**, 123.

LIPPINCOTT, E. R.: *see* VANDERSLICE, J. T.

LOMBARDI, E.: *see* JANSEN, L.

LONDON, F. (1930a) *Z. Phys.* **63,** 245.
LONDON, F. (1930b) *Z. Phys. Chem.* (B) **11,** 222.
LONDON, F. (1937) *Trans. Faraday Soc.* **33,** 8.
LONDON, F. (1942) *J. Phys. Chem.* **46,** 305.
LONGMIRE, M. S.: *see* AMDUR, I.
LONQUET-HIGGINS, H. C. (1966) Intermolecular forces, *Disc. Faraday Soc.* **40,** 7.
LONQUET-HIGGINS, H. C. and SALEM, L. (1961) The forces between polyatomic molecules, I: long-range forces, *Proc. Roy. Soc. (London)* **A259,** 433.
LOUISELL, W. (1964) *Radiation and Noise in Quantum Electronics*, McGraw-Hill, New York.
LÖWDIN, P. O. (1948) A *Theoretical Investigation into Some Properties of Ionic Crystals*, Almquist and Wiksells Boktryckeri A. B., Uppsala.
LÖWDIN, P. O. (1950) *J. Chem. Phys.* **18,** 364.
LÖWDIN, P. O. (1956) *Advanc. Phys.* **5,** 1.
LÖWDIN, P. O. (1959) *Advanc. Chem. Phys.* **2,** 207.
LÖWDIN, P. O. (1966) *Advanc. Quant. Chem.* **2,** 213.
LÖWDIN, P. O.: *see* HIRSCHFELDER, J. O.
LUCAS, A. (1967), *Physica*, **35,** 353.
LUOMA, J. and MUELLER, C. R. (1966) Inversion problem in molecular beam determinations of the intermolecular potential, *Disc. Faraday Soc.* **40,** 45.
LYNN, N. (1958) A second-order calculation of the energy of interaction between two normal helium atoms, *Proc. Phys. Soc.* **A72,** 201.
LYNN, N.: *see* DALGARNO, A.
MCCAMMON, R. D.: *see* CHON, H.
MCCARROLL, R.: *see* DALGARNO, A.
MCCULLOCH, R. D.: *see* GRABEN, H. W.
MCGINNIES, R. T. and JANSEN, L. (1956a) Validity of the assumption of two-body interactions in molecular physics, I, *Phys. Rev.* **101,** 1301.
MCGINNIES, R. T. and JANSEN, L. (1956b) *Phys. Rev.* **104,** 961.
MCGINNIES, R. T.: *see* JANSEN, L.
MCGLASHAN, M. L. (1966) Effective pair interaction energy in crystalline argon, *Disc. Faraday Soc.* **40,** 59.
MCGLASHAN, M. L.: *see* GUGGENHEIM, E. A.
MCGUIRE, J. M.: *see* BAKER, C. E.
MCKINLEY, M. D. and REED, T. M. III (1965) Intermolecular potential energy functions for pairs of simple polyatomic molecules, *J. Chem. Phys.* **42,** 3891.
MCKOY, V. (1965) *J. Chem. Phys.* **43,** 1605.
MCLACHLAN, A. D. (1963a) Retarded dispersion forces between molecules, *Proc. Roy. Soc. (London)* **A271,** 387.
MCLACHLAN, A. D. (1963b) Three-body dispersion forces, *Mol. Phys.* **6,** 423.
MCLACHLAN, A. D. (1963c) *Proc. Roy. Soc.* **A274,** 80.
MCLACHLAN, A. D. (1964a) Van der Waals forces between an atom and a surface, *Mol. Phys.* **7,** 381.
MCLACHLAN, A. D. (1964b) *Mol. Phys.* **8,** 409.

McLachlan, A. D. (1966) Effect of medium on dispersion forces in liquids, *Disc. Faraday Soc.* **40,** 239.

McLachlan, A. D., Gregory, R. D. and Ball, M. A. (1964) Molecular interactions by the time-dependent Hartree method, *Mol. Phys.* **7,** 119.

McLone, R. R. and Power, E. A. (1964) On the interaction between two identical neutral dipole systems, one in an excited state and the other in the ground state, *Mathematika* **11,** 91.

McLone, R. R. and Power, E. A. (1965) The long-range Van der Waals forces between non-identical systems, *Proc. Soc. (London)* **A286,** 573.

McLone, R. R.: *see* Gomberoff, L.

Madan, M. P. (1957) Potential parameters for krypton, *J. Chem. Phys.* **27,** 113.

Magee, J. L.: *see* Gurnee, E. F.

Magnasco, V. and Musso, G. F. (1967) *J. Chem. Phys.*, **46,** 4015.

Mahan, G. D. (1965) Van der Waals forces in solids, *J. Chem. Phys.* **43,** 1569.

Maisch, W. G.: *see* Mason, E. A. and Vanderslice, J. T.

Manev, E.: *see* Scheldko, A.

Marcus, A. B.: *see* Cramer, W. H.

Marcus, R. A. (1963a) Interactions in polar media, I: interparticle interaction energy, *J. Chem. Phys.* **38,** 1335.

Marcus, R. A. (1963b) Interactions in polar media, II: continua, *J. Chem. Phys.* **39,** 460.

Margenau, H. (1931) *Phys. Rev.* **38,** 747.

Margenau, H. (1938) *J. Chem. Phys.* **6,** 896.

Margenau, H. (1939a) *Rev. Mod. Phys.* **11,** 1.

Margenau, H. (1939b) *Rev. Mod. Phys.* **11,** 28.

Margenau, H. (1939c) *Phys. Rev.* **56,** 1000.

Margenau, H. (1943) *Phys. Rev.* **64,** 131.

Margenau, H. (1944) *Phys. Rev.* **66,** 303.

Margenau, H. (1966) Exclusion principle and measurement theory, in *Quantum Theory of Atoms, Molecules, and the Solid state*, Academic Press (P. O. Löwdin, ed.) p. 81.

Margenau, H. and Lewis, M. (1959) *Rev. Mod. Phys.* **31,** 569.

Margenau, H. and Meyers, V. W. (1944) *Phys. Rev.* **66,** 307.

Margenau, H. and Murphy, G. (1964) *Mathematics of Physics and Chemistry*, Vol. 2, Van Nostrand, New York.

Margenau, H. and Pollard, W. G. (1941) *Phys. Rev.* **60,** 128.

Margenau, H. and Rosen, P. (1953) *J. Chem. Phys.* **21,** 394.

Margenau, H. and Watson, W. W. (1936) *Rev. Mod. Phys.* **8,** 22.

Marino, L. L.: *see* Rothe, E. W.

Mason, E. A. (1957) Scattering of low velocity molecular beams in gases, *J. Chem. Phys.* **26,** 667.

Mason, E. A. (1960) Redetermination of the intermolecular potential for krypton, *J. Chem. Phys.* **32,** 1832.

Mason, E. A. (1961) Intermolecular potential for krypton, *J. Chem. Phys.* **35,** 2245.

Mason, E. A. (1964) Determination of intermolecular forces, in *Proc. Internat. Seminar on the Transport Properties of gases*, Brown University, p. 249.

MASON, E. A. and AMDUR, I. (1964) Scattering of high velocity neutral particles, XIII: Ar–CH_4: a test of the peripheral-force approximation, *J. Chem. Phys.* **41,** 2695.

MASON, E. A., AMDUR, I. and OPPENHEIM, I. (1965) Differences in the spherical intermolecular potentials of hydrogen and deuterium, *J. Chem. Phys.* **43,** 4458.

MASON, E. A. and HIRSCHFELDER, J. O. (1957a) Short-range intermolecular forces, I, *J. Chem. Phys.* **26,** 173.

MASON, E. A. and HIRSCHFELDER, J. O. (1957b) Short-range intermolecular forces, II: H_2–H_2 and H_2–H, *J. Chem. Phys.* **26,** 756.

MASON, E. A. and MAISCH, W. G. (1959) Interactions between oxygen and nitrogen: O–N, O–N_2, and O_2–N_2, *J. Chem. Phys.* **31,** 738.

MASON, E. A., MUNN, R. J. and SMITH, F. J. (1966) Recent work on the determination of the intermolecular potential function, *Disc. Faraday Soc.* **40,** 27.

MASON, E. A. and RICE, E. W. (1954) *J. Chem. Phys.* **22,** 522.

MASON, E. A., ROSS, J. and SCHATZ, P. N. (1956) Energy of interaction between a hydrogen atom and a helium atom, *J. Chem. Phys.* **25,** 626.

MASON, E. A. and VANDERSLICE, J. T. (1957) Interaction energies and scattering cross-sections of hydrogen ions in helium, *J. Chem. Phys.* **27,** 1284.

MASON, E. A. and VANDERSLICE, J. T. (1958a) Interaction energy and scattering cross-sections of H^- ions in helium, *J. Chem. Phys.* **28,** 253.

MASON, E. A. and VANDERSLICE, J. T. (1958b) Determination of the binding energy of He_2^+ from ion scattering data, *J. Chem. Phys.* **29,** 361.

MASON, E. A. and VANDERSLICE, J. T. (1958c) Delta function model for short-range intermolecular forces, I: rare gases, *J. Chem. Phys.* **28,** 432.

MASON, E. A. and VANDERSLICE, J. T. (1958d) Interactions of H^- ions and H atoms with Ne, Ar, and H_2, *J. Chem. Phys.* **28,** 1070.

MASON, E. A.: *see* AMDUR, I., FALLON, R. J., KNOF, H., KRUPENIE, P. H., ROSS, J., SHERWOOD, A. E. and VANDERSLICE, J. T.

MASON, R.: *see* CRAIG, D. P.

MASSEY, H. S. W. and MOHR, C. B. O. (1934) *Proc. Roy. Soc.* **A144,** 188.

MATSEN, F. A.: *see* BRIGMAN, G. H., POSHUSTA, R. D., REAGAN, P. N. and SCOTT, D. R.

MATSUMOTO, G. H., BENDER, C. F. and DAVIDSON, E. R. (1967) *J. Chem. Phys.* **46,** 402.

MAVROYANNIS, C. (1963) The interaction of neutral molecules with dielectric surfaces, *Mol. Phys.* **6,** 593.

MAVROYANNIS, C. and STEPHEN, M. J. (1962) Dispersion forces, *Mol. Phys.* **5,** 629.

MAXWELL, J. C. (1868) *Phil. Mag.* **4,** 129, 185.

MAY, C. J.: *see* BAILEY, T. L.

MAYER, J. E. (1934) *J. Chem. Phys.* **2,** 252.

MEATH, W. J. (1966) Theoretical Chemistry Institute, Report No. 174 and paper to be published in *J. Chem. Phys.*

MEATH, W. J. and HIRSCHFELDER, J. O. (1965) Technical Report of the Theoretical Chemistry Institute at the University of Wisconson, Wis–TCI–124, September 1965.

MEATH, W. J. and HIRSCHFELDER, J. O. (1966a) Relativistic intermolecular forces: moderately long range, *J. Chem. Phys.* **44**, 3197.

MEATH, W. J. and HIRSCHFELDER, J. O. (1966b) Long-range (retarded) intermolecular forces, *J. Chem. Phys.* **44**, 3210.

MERWE, A. VAN DER (1966) Dispersion energies of interaction between asymmetric molecules, *Z. f. Phys.* **196**, 212, 322.

MEYER, L., BARRETT, C. S. and HAASEN, P. (1965) *J. Chem. Phys.* **42**, 107.

MICHA, D. A.: *see* HARRIS, F. E.

MICHAELS, A., DE GRAAFF, W. and TEN SELDAM, C. A. (1960) Virial coefficients of hydrogen and deuterium at temperatures between −175°C and 150°C: conclusions from the second virial coefficient with regard to the intermolecular potential, *Physica* **26**, 393.

MICHELS, H. H.: *see* SCHNEIDERMAN, S. B.

MIDZUNO, Y. and KIHARA, T. (1956) Non-additive intermolecular potential in gases, I: Van der Waals interactions, *J. Phys. Soc. (Japan)* **11**, 1045.

MIDZUNO, Y.: *see* KIHARA, T.

MIES, F. H.: *see* KRAUSS, M.

MIKOLAJ, P. G. and PINGS, C. J. (1966) Direct determination of the intermolecular potential function for argon from X-ray scattering data, *Phys. Rev. Letters* **16**, 4.

MILLER, R. C. and SMYTH, C. P. (1956) Microwave absorption and molecular structure in liquids, XII: critical wave lengths and intermolecular forces in the tetrahalogenated molecules, *J. Chem. Phys.* **24**, 814.

MILLER, R. V. and PRESENT, R. D. (1963) United-atom configuration–interaction treatment of the He–He repulsion, *J. Chem. Phys.* **38**, 1179.

MILLER, W. H. (1966) *J. Chem. Phys.* **44**, 2198.

MOISEIWITSCH, B. I. (1956) Interaction energy and charge exchange between helium atoms and ions, *Proc. Phys. Soc.* **A69**, 653.

MOORE, N. (1960) Energy of interaction of two helium atoms, *J. Chem. Phys.* **33**, 471.

MORGAN, D. L. (1966) Dissertation, Yale University.

MORSE, F. A., BERNSTEIN, R. B. and HOSTETTLER, H. U. (1962) Evaluation of the intermolecular well depth from observations of rainbow scattering: Cs–Hg and K–Hg, *J. Chem. Phys.* **36**, 1947.

MORSE, P. M. and FESHBACH, F. (1953) *Methods of Theoretical Physics*, McGraw-Hill.

MOSKOWITZ, J. and HARRISON, M. C. (1965) *J. Chem. Phys.* **42**, 1726.

MUELLER, C. R. and BRACKETT, J. W. (1964) Quantum calculation of the sensitivity of diffusion, viscosity, and scattering experiments to the intermolecular potential, *J. Chem. Phys.* **40**, 654.

MUELLER, C. R.: *see* LUOMA, J. and SANDERS, W. A.

MULLER, A. (1936) *Proc. Roy. Soc. (London)* **A154**, 624.

MULLER, A. (1941) *Proc. Roy. Soc. (London)* **A178**, 227.

MULLIKEN, R. S. (1950), *J. Am. Chem. Soc.* **72**, 600.

MULLIKEN, R. S. (1952), *J. Am. Chem. Soc.* **74**, 811.

MULLIKEN, R. S. (1960) The interaction of differently excited like atoms at large distances, *Phys. Rev.* **120**, 1674.

MUNN, R. J. (1964) Interaction potential of the inert gases, I, *J. Chem. Phys.* **40**, 1439.

MUNN, R. J. (1965) On the calculation of the dispersion forces coefficient directly from experimental transport data, *J. Chem. Phys.* **42,** 3032.

MUNN, R. J. and SMITH, F. J. (1965) Interaction potential of the inert gases, II, *J. Chem. Phys.* **43,** 3998.

MUNN, R. J.: *see* MASON, E. A.

MURRELL, J. N., RANDIC, M. and WILLIAMS, D. R. (1965) The theory of intermolecular forces in the region of small orbital overlap, *Proc. Roy. Soc. (London)* **A284,** 566.

MURRELL, J. N. and SHAW, G. (1967a) *J. Chem. Phys.* **46,** 1768.

MURRELL, J. N. and SHAW, G. (1967b) *Mol. Phys.* **7,** 475.

MUSCHLITZ, E. E. Jr. (1966) Collisions of electronically excited atoms and molecules, *Advanc. Chem. Phys.* **10,** 171.

MUSCHLITZ, E. E. JR., BAILEY, T. L. and SIMONS, J. H. (1957) Elastic and inelastic scattering of low-velocity H^- ions in hydrogen, II, *J. Chem. Phys.* **26,** 711.

MUSCHLITZ, E. E. JR.: *see* BAILEY, T. L., BAKER, C. E. and SMITH, G. M.

MUSHER, J. I. (1963) Calculation of London-Van der Waals energies, *J. Chem. Phys.* **39,** 2409.

MUSHER, J. I. (1964) Calculation of London energies, II: an iteration procedure, *J. Chem. Phys.* **41,** 2671.

MUSHER, J. I. (1965), *J. Chem. Phys.* **42,** 2633.

MUSHER, J. I. (1967), *Rev. Mod. Phys.* **39,** 203.

MUSHER, J. I. and SALEM, L. (1966) Energy of interaction between two molecules, *J. Chem. Phys.* **44,** 2943.

MUTO, Y. (1943) *Proc. Phys. Math. Soc. Japan* **17,** 629.

MYERS, V. (1950) Intermolecular forces in benzene and ammonia, *J. Chem. Phys.* **18,** 1442.

MYERS, V. (1956) Intermolecular forces in ethane, *J. Chem. Phys.* **24,** 924.

NAGAMIYA, T. and KISHI, H. (1951) *Busseiron–Kenkyu* **39,** 64.

NEMETHY, G.: *see* SCHERAGA, H. A.

NESBET, R. K. (1962) Interaction of two ethylene molecules, *Mol. Phys.* **5,** 63.

NEUMANN, J. V. and WIGNER, E. (1929) *Physik Z.* **30,** 467.

NEYNABER, R. H.: *see* ROTHE, E. W.

O'BRIEN, T. J. P.: *see* BERNSTEIN, R. B.

OPPENHEIM, I.: *see* MASON, E. A.

OUCHI, A.: *see* KIHARA, T.

OUDEMANS, G. J. and COLE, R. H. (1959) Dielectric constant and pair interactions in gaseous helium and argon, *J. Chem. Phys.* **31,** 843.

OVERBEEK, J. T. G.: *see* BLACK, W.

PAGE, C. H. (1938) *Phys. Rev.* **53,** 426.

PANOFSKY, W. K. H. and PHILLIPS, M. (1955) *Classical Electricity and Magnetism*, Addison-Wesley, Reading, Mass.

PARKE, W. C.: *see* JEHLE, H.

PAULING, L. and BEACH, J. Y. (1935) *Phys. Rev.* **47,** 686.

PAULING, L. and WILSON, E. B. (1935) *Introduction to Quantum Mechanics*, McGraw-Hill, New York.

PAULSON, R. H.: *see* ALDER, B. J.

PAULY, H. (1960) *Z. Naturforsch* **15A**, 277.

PAULY, H. and TOENNIES, J. P. (1965) The study of intermolecular potentials with molecular beams at thermal energies, *Advanc. Atomic Molecular Phys.* **1**, 195.

PAULY, H.: *see* DUEREN, R.

PEEK, J. M. (1966) Proton–hydrogen-atom system at large distances: resonant charge transfer and the $1s\sigma_g - 2p\sigma_u$ eigenenergies of H_2^+, *Phys. Rev.* **143**, 33.

PEKERIS, C. L. (1959) *Phys. Rev.* **115**, 1216.

PENFIELD, R. H. and ZATZKIS, H. (1957) Quantization of the relativistic harmonic oscillator by perturbative methods with application to Van der Waals forces, *J. Franklin Inst.* **263**, 331.

PETERSON, D. L.: *see* SIMPSON, W. T.

PEYERIMLOFF, S. (1965) Hartree-Fock Roothaan wave functions, potential curves, and charge density contours for the He–H ($X^1\Sigma_g$) and Ne–H ($X\,^1\Sigma_g$) molecular ions, *J. Chem. Phys.* **43**, 998.

PHILLIPSON, P. E. (1962) Repulsive interaction between two ground state helium atoms, *Phys. Rev.* **125**, 1981.

PHILPOTT, M. R. (1966) The retarded interaction between electronically excited molecules, *Proc. Phys. Soc.* **87**, 619.

PINES, D. (1963) *Elementary Excitations in Solids*, Benjamin, New York.

PINGS, C. J. (1967) *Mol. Phys.* **12**, 501.

PINGS, C. J.: *see* MIKOLAJ, P. G.

PITAEVSKII, L. P. (1960) Attraction of small particles suspended in a liquid at large distances, *Soviet Physics–JETP* **10**, 408.

PITAEVSKII, L. P.: *see* DERJAGUIN, B. V. and DZYALOSHINSKII, I. E.

PITZER, K. S. (1956) Electronic correlation in molecules, II: the rare gases, *J. Amer. Chem. Soc.* **78**, 4565.

PITZER, K. S. (1959) Inter- and intramolecular forces and molecular polarizability, *Advanc. Chem. Phys.* **2**, 59.

PITZER, K. S. and CATALANO, E. (1956) *J. Am. Chem. Soc.* **78**, 4844.

PITZER, K. S.: *see* DONATH, W., DANON, F. and SINANOGLU, O.

PITZER, R. S. and LIPSCOMB, W. N. (1963) *J. Chem. Phys.* **39**, 1995.

PLATIKANOV, D.: *see* SCHELUDKO, A.

PODLUBNYA, L. I. (1960) The non-additivity of London-Van der Waals forces, *Soviet Physics–JETP* **10**, 633.

POHL, H. A.: *see* HARRIS, F. E.

POLLARA, L. Z. and FUNKE, P. T. (1959) Note on a new potential function, *J. Chem. Phys.* **31**, 855.

POLLARD, W. G. (1941) Exchange forces between neutral molecules and a metal surface, *Phys. Rev.* **60**, 578.

POLLARD, W. G.: *see* MARGENAU, H.

POPLE, J. R.: *see* ALDER, B. J.

PORTER, R. N. and KARPLUS, M. (1964) *J. Chem. Phys.* **40**, 1105.

PORTER, R. N.: *see* KARPLUS, M.

POSHKUS, D. P. (1966) Interaction energy of non-polar molecules with graphite surface and in the gas phase, *Disc. Faraday Soc.* **40**, 195.

POSHUSTA, R. D. and MATSEN, F. A. (1963) Potential curve of the metastable helium molecule, *Phys. Rev.* **132**, 307.

POWER, E. A. (1965) *Introductory Quantum Electrodynamics*, American Elsevier, New York.

POWER, E. A. and SHAIL, R. (1959) *Proc. Camb. Phil. Soc.* **55**, 87.

POWER, E. A. and ZIENAU, S. (1957a) On the radiative contributions to the Van der Waals force, *Nuovo Cimento* **6**, 7.

POWER, E. A. and ZIENAU, S. (1957b) On the physical interpretation of the relativistic corrections to the Van der Waals force found by Penfield and Zatskis, *J. Franklin Inst.* **263**, 403.

POWER, E. A. and ZIENAU, S. (1959) *Trans. Roy. Soc.* **A251**, 427.

POWER, E. A.: *see* GOMBEROFF, L. and MCLONE, R. R.

PRAUSNITZ, J. M.: *see* SHERWOOD, A. E.

PRESENT, R. D. (1935) *J. Chem. Phys.* **3**, 122.

PRESENT, R. D. (1958) *Kinetic Theory of Gases*, McGraw-Hill, New York.

PRESENT, R. D. (1967) *J. Chem. Phys.* **47**, 1793.

PRESENT, R. D.: *see* GRABEN, H. W. and MILLER, R. V.

PRIMAKOFF, H.: *see* BERNARDES, N.

PROSEN, E. J. R. and SACHS, R. G. (1942) The interaction between a molecule and a metal surface, *Phys. Rev.* **61**, 65.

PROSSER, A. P.: *see* KITCHNER, J. A.

RACAH, G. (1942) *Phys. Rev.* **62**, 438.

RACAH, G. (1943) *Phys. Rev.* **63**, 367.

RAI, D. K.: *see* SINGH, R. B.

RANDIC, M.: *see* MURRELL, J. N.

RANSIL, B. J. (1957) Application of configuration-interaction to the H_3 complex, *J. Chem. Phys.* **26**, 971.

RANSIL, B. (1960) *Rev. Mod. Phys.* **32**, 239, 245.

RANSIL, B. J. (1961) Studies in molecular structure, IV: potential curve for the interaction of two helium atoms in the single configuration LCAO–MO–SCF approximation, *J. Chem. Phys.* **34**, 2109.

REAGAN, P. N., BROWNE, J. C. and MATSEN, F. A. (1963) Dissociation energy of He_2^+ $(^2\Sigma_u^+)$, *Phys. Rev.* **132**, 304.

REED, T. M. III.: *see* MCKINLEY, M. D.

REILLY, M. H.: *see* KNOX, R. S.

REINGANUM, M. (1903) *Ann. d. Phys.* **10**, 334.

REINGANUM, M. (1912) *Ann. d. Phys.* **38**, 649.

RICE, W. E. and HIRSCHFELDER, J. O. (1954) *J. Chem. Phys.* **22**, 187.

RIGBY, M.: *see* DYMOND, J. H.

ROBBINS, E. J.: *see* LECKENBY, R. E.

ROBERTS, C. S. (1963) Interaction energy between a helium atom and a hydrogen molecule, *Phys. Rev.* **131**, 203.

ROE, G. M. (1952) *Phys. Rev.* **88**, 659.

ROL, P. K.: *see* ROTHE, E. W.

ROOTHAAN, C. C. J (1951a) *J. Chem. Phys.* **19**, 1445.

ROOTHAAN, C. C. J. (1951b) *Rev. Mod. Phys.* **23**, 69.

ROOTHAAN, C. C. J. (1960) *Rev. Mod. Phys.* **32**, 179.

ROOTHAAN, C. C. J. and BAGUS, P. S. (1963) In *Methods of Computational Physics*, Vol. 2, Academic Press, New York.

ROOTHAAN, C. C. J. and WEISS, A. W. (1960) *Rev. Mod. Phys.* **32**, 194.

ROSE, M. E. (1957) *Elementary Theory of Angular Momentum*, Wiley, New York.

ROSE, M. E. (1958) *J. Math. Phys.* **37**, 215.

ROSEN, N. (1931) *Phys. Rev.* **38**, 255.

ROSEN, P. (1950) *J. Chem. Phys.* **18**, 1182.

ROSEN, P. (1953) The nonadditivity of the repulsive potential of helium, *J. Chem. Phys.* **21**, 1007.

ROSENSTOCK, H. B. and BLANKEN, G. (1966) Interatomic forces in various solids, *Phys. Rev.* **145**, 546.

ROSS, J. and MASON, E. A. (1956) The energy of interaction of He^+ and H, *Astrophys. J.* **124**, 485.

ROSS, J. and MASON, E. A.: *see* MASON, E. A.

ROSS, M. and ALDER, B. (1967) *J. Chem. Phys.* **46**, 4203.

ROSSI, J. C. and DANON, F. (1965) Intermolecular forces of the heavy rare gases, *J. Chem. Phys.* **43**, 762.

ROSSI, J. C. and DANON, F. (1966) Molecular interactions in the heavy rare gases, *Disc. Faraday Soc.* **40**, 97.

ROTHE, E. W. and BERNSTEIN, R. B. (1959) Total collision cross-sections for the interaction of atomic beams of alkali metals with gases, *J. Chem. Phys.* **31**, 1619.

ROTHE, E. W., MARINO, L. L., NEYNABER, R. H., ROL, P. K. and TRUJILLO, S. M. (1962) Scattering of thermal rare gas beams by argon: influence of the long-range dispersion forces, *Phys. Rev.* **126**, 598.

ROTHE, E. W. and NEYNABER, R. H. (1965a) Atomic beam measurements of Van der Waals forces, *J. Chem. Phys.* **42**, 3306.

ROTHE, E. W. and NEYNABER, R. H. (1965b) Measurements of absolute total cross-sections for rare gas scattering, *J. Chem. Phys.* **43**, 4177.

ROTHE, E. W., NEYNABER, R. H., SCOTT, B. W., TRUJILLO, S. M. and ROL, P. K. (1961) Substantiation of a method for obtaining interatomic potential energy parameters, *J. Chem. Phys.* **39**, 493.

ROTHE, E. W., NEYNABER, R. H. and TRUJILLO, S. M. (1965) Velocity dependence of the total cross-section for the scattering of metastable He ($^3S_1$) by helium, argon, and krypton, *J. Chem. Phys.* **42**, 3310.

ROTHE, E. W., ROL, P. K., TRUJILLO, S. M. and NEYNABER, R. H. (1962) *Phys. Rev.* **128**, 659.

ROTHENBERG, S. and DAVIDSON, E. R. (1966) Hydrogen molecule excited states, *J. Chem. Phys.* **44**, 730.

ROWLINSON, J. S. (1960) A test of the Lennard–Jones potential for nitrogen and methane, *Mol. Phys.* **3**, 265.

ROWLINSON, J. S. (1963a) Limits of the third virial coefficient at low and high temperatures for a Lennard-Jones potential, *Mol. Phys.* **6**, 75.

ROWLINSON, J. S. (1963b) Limits of the fourth virial coefficient at low and high temperatures for a Lennard-Jones potential, *Mol. Phys.* **6**, 429.

ROWLINSON, J. S (1965) A test of Kihara's intermolecular potential, *Mol. Phys.* **9**, 197.

ROWLINSON, J. S. (1966) Determination of intermolecular forces from macroscopic properties, *Disc. Faraday Soc.* **40**, 19.

ROWLINSON, J. S. (1967) *Mol. Phys.* **12**, 513.

ROWLINSON, J. S.: *see* BYERS BROWN, W.

RUEDENBERG, K. and EDMISTON, C. (1963) *Rev. Mod. Phys.* **35,** 457.
RUSHBROOKE, G. S. and SILBERT, M. (1967) *Mol. Phys.* **12,** 505.
RUTHERFORD, E. and MCKLING, J. (1900) *Phys. Z.* **2,** 53.
SACHS, R. G.: *see* PROSEN, E. J. R.
SAKAMOTO, M. and ISHIGURO, E. (1956) He–He repulsive potential, I, *Prog. Theor. Phys.* **15,** 37.
SALEM, L. (1950) *Mol. Phys.* **3,** 441.
SALEM, L. (1960) The calculation of dispersion forces, *Mol. Phys.* **3,** 441.
SALEM, L. (1961) The forces between polyatomic molecules, II: short range repulsive forces, *Proc. Roy. Soc. (London)* **A264,** 379.
SALEM, L. (1962) Attractive forces between long chains at short distances, *J. Chem. Phys.* **37,** 2100.
SALEM, L. (1964) Intermolecular forces in biological systems in *Electronic Aspects of Biochemistry*, Academic Press (B. Pullman, ed.), p. 293.
SALEM, L. (1966) Second-order exchange forces, *Disc. Faraday Soc.* **40,** 150.
SALEM, L. (1968) *J. Am. Chem. Soc.* **90,** 543, 553.
SALEM, L.: *see* LONQUET-HIGGINS, H. C. and MUSHER, J. I.
SALYERS, A.: *see* JEHLE, H.
SAMS, J. R. (1964) *Trans. Faraday Soc.* **60,** 149.
SAMS, J. R. JR., CONSTABARIS, G. and HALSEY, G. D. JR. (1962) *J. Chem. Phys.* **36,** 1334.
SAMS, J. R.: *see* WOLFE, R.
SANDERS, W. A. and MUELLER, C. R. (1963) Direct determination of intermolecular potential parameters from scattering phase shifts, *J. Chem. Phys.* **39,** 2572.
SANDERS, W. A.: *see* BRACKETT, J. W.
SANKER, R. (1960) The Van der Waals interaction of particles, *J. Indiana Inst. Sci.* **42,** 17.
SANTRY, D. P.: *see* CRAIG, D. P.
SARAN, A. (1965) A hybrid potential for inert gas atoms, *Indian J. Phys.* **39,** 72.
SAXENA, S. C.: *see* DASS, L.
SCHATZ, P. N.: *see* MASON, E. A.
SCHEEL, N. and GRIFFING, V. (1962) Interaction energy of an alkali metal with a rare gas, *J. Chem. Phys.* **36,** 1453.
SCHELUDKO, A., PLATIKANOV, D. and MANEV, E. (1966) Disjoining pressure in the thin liquid films and the electro-magnetic retardation effect of the molecule dispersion interactions, *Disc. Faraday Soc.* **40,** 253.
SCHERAGA, H. A., LEACH, S. J., SCOTT, R. A. and NEMETHY, G. (1966) Intramolecular forces and protein conformation, *Disc. Faraday Soc.* **40,** 268.
SCHERR, C. W. (1952) *J. Chem. Phys.* **21,** 1582.
SCHLIER, C.: *see* BERKLING, K.
SCHNEIDERMAN, S. B. and MICHELS, H. H. (1965) Quantum-mechanical studies of several helium–lithium interaction potentials, *J. Chem. Phys.* **42,** 3706.
SCHNEPP, O. and DRESSLER, K. (1965) Schumann–Runge bands of O_2 in solid phases: spectroscopic measurement of intermolecular potentials, *J. Chem. Phys.* **42,** 2482.

SCHOOMAKER, R. C. (1961) *J. Phys. Chem.* **65,** 892.

SCOLES, C.: *see* CASANOVA, G.

SCOTT, B. W.: *see* ROTHE, E. W.

SCOTT, D. R., GREENWALT, E. M., BROWNE, J. C. and MATSEN, F. A. (1966) Quantum-mechanical potential energy curve for the lowest $^1\Sigma_u^+$ state of He_2, *J. Chem. Phys.* **44,** 2981.

SCOTT, R. A.: *see* SCHERAGA, H. A.

SHARMA, R. D.: *see* KARPLUS, M.

SHAVITT, I. (1962) in *Methods in Computational Physics* (Eds. Alder, Fernbach and Rotenberg), Academic Press, New York.

SHAVITT, I. and BOYS, S. F. (1956) A general expression for intermolecular potentials, *Nature (London)* **178,** 1340.

SHERWOOD, A. E. and PRAUSNITZ, J. M. (1964a) Third virial coefficients for the Kihara, exp-6, and square well potentials, *J. Chem. Phys.* **41,** 413.

SHERWOOD, A. E. and PRAUSNITZ, J. M. (1964b) Intermolecular potential functions and second and third virial coefficients, *J. Chem. Phys.* **41,** 429.

SHERWOOD, A. E., DE ROCCO, A. G. and MASON, E. A. (1966) Nonadditivity of intermolecular forces: effects on the third virial coefficient, *J. Chem. Phys.* **44,** 2984.

SHIMANOUCHI, T.: *see* HARADA, I.

SHOSTAK, A. (1955) Interaction energy among three helium atoms, *J. Chem. Phys.* **23,** 1808.

SIMONETTA, M. (1964) In *Molecular Orbitals in Chemistry, Physics and Biology* (eds. P. O. LÖWDIN and B. PULLMAN), Academic Press, New York.

SIMONS, J. H.: *see* CRAMER, W. H. and MUSCHLITZ, E. E., JR.

SIMPSON, W. T. (1962) *Electrons in Molecules*, Prentice-Hall, Englewood Cliffs, New Jersey.

SIMPSON, W. T. and PETERSON, D. L. (1957) Coupling strength for resonance force transfer of electronic energy in Van der Waals solids, *J. Chem. Phys.* **26,** 588.

SINANOGLU, O. (1959) Scattering of high velocity neutral particles by an orientation dependent potential, *J. Chem. Phys.* **30,** 850.

SINANOGLU, O. (1960) Inter- and intra atomic correlation energies and the theory of core polarization, *J. Chem. Phys.* **33,** 1212.

SINANOGLU, O. (1961a) *Proc. Roy. Soc. (London)* **A260,** 379.

SINANOGLU, O. (1961b) *Phys. Rev.* **122,** 491.

SINANOGLU, O. (1961c) *Proc. Natl. Acad. Sci. (US)* **47,** 1217.

SINANOGLU, O. (1961d) *J. Chem. Phys.* **34,** 1237.

SINANOGLU, O. (1962) Many electron theory of atoms and molecules, I: shells, electron pairs, *vs.* many electron correlations, II, *J. Chem. Phys.* **36,** 706, 3198.

SINANOGLU, O. (1963) Bonds and intramolecular forces, *J. Chem. Phys.* **37,** 191.

SINANOGLU, O. (1964) Many electron theory of atoms, molecules, and their interactions, *Advanc. Chem. Phys.* **6,** 315.

SINANOGLU, O. (1965) Intermolecular forces in gases and dense media, in *Modern Quantum Chemistry*, Istanbul Lectures, Academic Press, Vol. 2 (Ed. O. SINANOGLU), p. 221.

SINANOGLU, O. (1967) in *Intermolecular Forces* (Ed. Hirschfelder), Interscience, New York.

SINANOGLU, O., ABDULNUR, S. and KESTNER, N. R. (1964) "Solvent effects on Van der Waals dispersion attractions particularly in DNA" in *Electronic Aspects of Biochemistry*, Academic Press (Ed. B. PULLMAN), p. 301.

SINANOGLU, O. and KESTNER, N. R. (1966) *J. Chem. Phys.* **45**, 194.

SINANOGLU, O. and PITZER, K. S. (1960) Interactions between molecules adsorbed on a surface, *J. Chem. Phys.* **32**, 1279.

SINANOGLU, O. and TUAN, D. F. (1964) *Ann. Revs. Phys. Chem.* **15**, 251.

SINANOGLU, O.: *see* KESTNER, N. R.

SINGER, K. (1958) Use of Gaussian functions for intermolecular potentials, *Nature (London)* **181**, 262.

SINGH, R. B. and RAI, D. K. (1965) Potential curves for Na_2 molecule, *Indian J. Pure Appl. Phys.* **3**, 475.

SLATER, J. C. (1928) *Phys. Rev.* **32**, 329.

SLATER, J. C. (1960) *Quantum Theory of Atomic Structure*, Vol. 2, McGraw-Hill, New York.

SLATER, J. C. (1963) *Quantum Theory of Molecules and Solids*, Vol. 1, McGraw-Hill, New York.

SLATER, J. C. (1964) In *Molecular Orbitals in Chemistry, Physics and Biology* (Eds. P. LÖWDIN and B. PULLMAN), Academic Press, New York.

SLATER, J. C. and KIRKWOOD, J. G. (1931) *Phys. Rev.* **37**, 682.

SMITH, E. B.: *see* DYMOND, J. H.

SMITH, F. J.: *see* MASON, E. A. and MUNN, R. J.

SMITH, G. M. and MUSCHLITZ, E. E., JR. (1960) Scattering of metastable helium atoms in helium, neon, and argon, *J. Chem. Phys.* **33**, 1819.

SMYTH, C. P.: *see* MILLER, R. C.

SNOW, R. and EYRING, H. (1957) *J. Chem. Phys.* **61**, 1.

SOKOLOV, N. D. (1955) The hydrogen bond, *Uspekhi. Fiz. Nauk.* **57**, 204.

SPARNAAY, M. J. (1959a) On the additivity of London–Van der Waals forces: an extension of London's oscillator model, *Physica* **25**, 217.

SPARNAAY, M. J. (1959b) Van der Waals forces and fluctuation phenomena, *Physica* **25**, 444.

SPARNAAY, M. J. (1959c) *Rev. Trav. Chim.* **78**, 680.

SPARNAAY, M. J.: *see* BLACK, W.

SRIVASTAVA, B. N. and MADON, M. P. (1953) *Proc. Phys. Soc. (London)* **A66**, 278.

SRIVASTAVA, K. R. (1958) Unlike molecular interactions and properties of gas mixtures, *J. Chem. Phys.* **28**, 543.

SRIVASTAVA, S. L.: *see* KRISHNAJI.

STAMPER, J. (1965) Dissertation, Yale University.

STANTON, R. E. (1965) *J. Chem. Phys.* **42**, 2353.

STEELE, D.: *see* LIPPINCOTT, E. R.

STEELE, R. B.: *see* CROWELL, A. D.

STEELE, W. A. and KEBBEKUS, E. R. (1965) Theoretical study of the monolayer adsorption of argon on a xenon surface, *J. Chem. Phys.* **43**, 292.

STEINER, E.: *see* BYERS BROWN, W.

STENSCHKE, H.: *see* ZICKENDRAHT, W.

STEPHEN, M. J. (1964) First-order dispersion forces, *J. Chem. Phys.* **40**, 669.

STEPHEN, M. J.: *see* MAVROYANNIS, C.

STEPHENSON, G. (1951a) *Nature* **167**, 156.

STEPHENSON, G. (1951b) *Proc. Phys. Soc. (London)* **A64**, 458.

STERNLICHT, H. (1964) Attraction between long polyconjugated chain molecules, *J. Chem. Phys.* **40**, 1175.

STILLINGER, F. H. JR., SALSBURG, Z. W. and KORNEGAY, R. L. (1965) *J. Chem. Phys.* **43**, 932.

SUTHERLAND, W. (1886) *Phil. Mag.* **22**, 81.

SUTHERLAND, W. (1887) *Phil. Mag.* **24**, 113.

SUTHERLAND, W. (1893a) *Phil. Mag.* **35**, 211.

SUTHERLAND, W. (1893b) *Phil. Mag.* **36**, 507.

SWENBERG, C. E. (1967) *Phys. Letters* **A24**, 163.

TAKAYANAGI, K. (1957) *Proc. Phys. Soc. (London)* **A70**, 348.

TAKAYANAGI, K. (1963) *Prog. Theo. Phys. (Kyoto)*, Suppl., No. 25, 1.

TAKAYANAGI, K. (1965) in *Atomic and Molecular Physics* (Eds. Bates and Ectermann), Academic Press, New York.

TAYLOR, H. S. (1963) Potential energy curve of the lowest lying triplet sigma state of lithium hydride, *J. Chem. Phys.* **39**, 3382.

TAYLOR, H. S. and HARRIS, F. E. (1964) A quantum mechanical study of the He, H^-, He–He, and He–H systems, *Mol. Phys.* **7**, 287.

TEN SELDAM, C. A.: *see* MICHAELS, A.

THOBURN, W. C. (1966) Dynamic consequence of intermolecular attraction in gases, *Amer. J. Phys.* **34**, 136.

THORSON, W. R. (1963) Nonadiabatic effects in the high-energy scattering of normal helium atoms, *J. Chem. Phys.* **39**, 1431.

THORSON, W. R. (1964) Nonadiabatic elastic scattering, II: the scattering process, *J. Chem. Phys.* **41**, 3881.

THORSON, W. R. and BANDRAUK, A. D. (1964) Coriolis shifts in elastic scattering potentials, *J. Chem. Phys.* **41**, 2503.

TIETZ, T. (1961) Interaction energy in the Amaldi–Fermi theory for a pair of simple negative ions of the same kind with filled electronic shells, *J. Chem. Phys.* **34**, 1848.

TOENNIES, J. P.: *see* PAULY, H.

TOMPKINS, F. C.: *see* GUNDRY, P. M.

TOSCHEK, P.: *see* BERKLING, K.

TOYA, T. (1958) *J. Res. Inst. Catalysis, Hokkaido Univer.* **6**, 308.

TRUJILLO, S. M.: *see* ROTHE, E. W.

TRULIO, J. G.: *see* KIMBALL, G. E.

TUCK, D. G. (1958) Application of charge-transfer theory to the heats of adsorption of the inert gases on charcoal, *J. Chem. Phys.* **29**, 724.

TUNG-CHEN, K.: *see* CHIA-CHUNG, S.

VAN DER AVOIRD, A. (1967) *Chem. Phys. Letters* **1**, 24, 411.

VANDERSLICE, J. T. and MASON, E. A. (1960a) Quantum mechanical calculations of short-range intermolecular forces, *Revs. Mod. Phys.* **32**, 417.

VANDERSLICE, J. T. and MASON, E. A. (1960b) Interaction energies for the $H–H_2$ and $H_2–H_2$ systems, *J. Chem. Phys.* **33**, 492.

VANDERSLICE, J. T., MASON, E. A. and LIPPINCOTT, E. R. (1959) Interactions between ground-state nitrogen atoms and molecules: the N–N, N–N_2, N_2–N_2 interactions, *J. Chem. Phys.* **30,** 129.

VANDERSLICE, J. T., MASON, E. A. and MAISCH, W. G. (1960) Interactions between ground state oxygen atoms and molecules: O–O and O_2–O_2, *J. Chem. Phys.* **32,** 515.

VANDERSLICE, J. T., MASON, E. A., MAISCH, W. G. and LIPPINCOTT, E. R. (1960) Potential curves for N_2, NO, and O_2, *J. Chem. Phys.* **33,** 614.

VANDERSLICE, J. T.: *see* CLONEY, R. D., FALLON, R. J., GRIFFING, V., KNOF, H., KRUPENIE, P. H. and MASON, E. A.

VENABLES, J. A.: *see* BULLOUGH, R.

VERWEY, E. J. W. (1947) *J. Phys. Colloid Chem.* **51,** 631.

VICKERY, B. C. and DENBIGH, K. B. (1949) *Trans. Faraday Soc.* **45,** 61.

VINEYARD, G. H. (1956) Potential energy of a molecule in a liquid, *J. Chem. Phys.* **24,** 617.

VOLD, M. J. (1954) *J. Colloid Sci.* **9,** 451.

VAN DER WAALS, J. D. JR. (1909) *Amst. Acad. Proc.*, pp. 132, 315.

VAN DER WAALS, J. H. (1908) *Lehrbuch der Thermodynamik*, Maas and Van Suchtelen, Leipzig, Part 1.

WAHL, A. C. (1964) *J. Chem. Phys.* **41,** 2600.

WAKEFIELD, C. B. and DAVIDSON, E. R. (1965) Some triplet states of the hydrogen molecule, *J. Chem. Phys.* **43,** 834.

WALKLEY, J. (1966) Reduction parameters for an unspecified intermolecular potential: a harmonic-oscillator approximation, *J. Chem. Phys.* **44,** 2417.

WANG, S. C. (1927) *Phys. Z.* **28,** 663.

WANG, S. C. (1928) *Phys. Rev.* **31,** 579.

WEBER, G. G. (1964) Delta-function model for interactions of two-electron atoms and ions, *J. Chem. Phys.* **40,** 1762.

WEHNER, J. F.: *see* GRIFFING, V.

WHEELER, A. (1964) Third virial coefficients and intermolecular forces for the simple gases, *J. Chem. Phys.* **41,** 2219.

WILLIAMS, R. R., SCHAAD, L. J. and MURRELL, J. N. (1967) *J. Chem. Phys.* **47,** 4916.

WILLIAMS, D. R.: *see* MURRELL, J. N.

WILSON, J. N. (1965) On the London potential between pairs of rare-gas atoms, *J. Chem. Phys.* **43,** 2564.

WOJCIECHOWSKI, K. F. (1966) The quantum theory of adsorption on metals, I, *Acta Physica Polonica* **29,** 119.

WOJTALA, J. (1964) *Acta Phys. Polonica* **25,** 27.

WOLFE, R. and SAMS, J. R. (1966) Three-body effects in physical adsorption, *J. Chem. Phys.* **44,** 2181.

WOLFSBERG, M. (1963) Isotope effects on intermolecular interactions and isotopic vapor pressure differences, *J. Chim. Phys.* **60,** 15.

WOLNIEWICZ, L.: *see* KOLOS, W.

WOODWARD, R. B. and HOFFMANN, R. (1965) *J. Am. Chem. Soc.* **87,** 395, 2511, 2046, 4388, 4389.

WOOLLEY, H. W. (1960) Empirical intermolecular potential for inert gas atoms, *J. Chem. Phys.* **32,** 405.

WRIGHT, W. M. and DAVIDSON, E. R. (1965) $1s3d\ ^3\Pi_g$ state of the hydrogen molecule, *J. Chem. Phys.* **43**, 840.

WU, T. (1956) Coupling between electronic and nuclear motion and Van der Waals interaction between helium atoms, *J. Chem. Phys.* **24**, 444.

WU, T. (1958) Note on the He–He interaction potential and its determination from Amdur's scattering measurements, *J. Chem. Phys.* **28**, 986.

YOUNG, D. M. and CROWELL, A. D. (1962) *Physical Adsorption of Gases*, Butterworth, London.

YOURGRAU, W., VAN DER MERWE, A. and RAW, G. (1966). *Irreversible and Statistical Thermophysics*, MacMillan, New York.

ZATZKIS, H.: *see* PENFIELD, R. H.

ZENNER, C. (1932) *Proc. Roy. Soc. (London)* **A137**, 696.

ZICKENDRAHT, W. and TENSCHKE, H. (1965) Three-body bound state in inert gases, *Phys. Letters (Netherlands)* **17**, 243.

ZIENAU, S.: *see* POWER, E. A.

ZIMERING, S. (1965) *J. Math. Phys.* **6**, 336.

ZUCKER, I. J. (1956) Intermolecular potentials of the inert gase from solid state data, *J. Chem. Phys.* **25**, 915.

ZUMINO, B.: *see* KELLER, J. B.

ZWANZIG, R. W. (1960) Intermolecular forces from optical spectra of impurities in molecular crystals, *Mol. Phys.* **3**, 305.

ZWANZIG, R. W. (1963) Two assumptions in the theory of attractive forces between long saturated chains, *J. Chem. Phys.* **39**, 2251.

Index

Index

Index

Index